前　言

　　網際網路技術迅猛發展，前端技術更新迅速，業務場景變得越來越複雜。JavaScript 語言從早期的網頁指令碼語言逐漸發展，基於 Node.js 的服務端框架和函數庫為前端開發開啟了新的視野。不同框架和函數庫的設計思想和理念各有特色，它們針對不同場景設計，推動了 Node.js 領域的技術發展。

　　Nest 是基於 TypeScript 的開放原始碼企業級框架，建立在 Express 之上，提供了一層抽象。它憑藉模組化、強大的相依注入系統和優秀的設計思想，成為許多 Node.js 開發者的首選。

　　本書致力於普及 Nest 從基礎到實戰的核心知識，對每個核心基礎知識透過理論結合程式範例的方式進行講解。本書透過完整的專案實踐，將帶你快速上手 Nest 並將其應用於實際專案中，幫助開發者建構堅實的獨立開發基礎。

本書內容

　　本書採用循序漸進的行文模式，從前端基礎知識到後端中介軟體的使用和開發，全面覆蓋。內容分為基礎篇、進階篇、擴充篇和專案實戰篇，每篇包含若干章節。

　　基礎篇介紹了學習 Nest 所需的前置知識，包括 Node.js 的請求回應物件和 TypeScript 基礎知識，然後介紹 Nest 的核心設計理念、建立和偵錯 Nest 應用，並結合實際程式範例深入解讀 Nest 的核心概念。

　　進階篇詳細講解了後端中介軟體服務，如 MySQL 資料庫、Redis 快取等，並指導如何在 Nest 中整合和使用這些服務。此外，還介紹了企業級應用中實現身份驗證與授權的方法，以及如何透過 Docker 高效部署和管理中介軟體服務。

　　擴充篇討論了對系統穩定性至關重要的系統測試與日誌管理，包括開發階段的單元測試、整合測試、端對端測試等，以及生產應用階段的日誌統計實踐。

　　專案實戰篇提供了一個完整的數字門店管理平臺開發實戰案例。

書附資源

為了讓讀者能夠更進一步地理解和實踐所學的知識，本書提供了原始程式碼（本書中的程式範例都可以直接用於實際工作中）。如果在學習和資源下載的過程中遇到問題，可以發送郵件至 booksaga@126.com，郵件主題為「NestJS 全端開發解析：快速上手與實踐」。

面向的閱讀群眾

本書主要面向所有前端和 Node.js 開發工程師，以及有意向學習全端知識的 IT 專業人員。要求讀者具備 Node.js 和 TypeScript 語言的基礎。

如何使用這本書

希望本書能為你帶來一段愉快的學習之旅。

如果你是 Nest 初學者或想更加了解本書內容，建議按照章節順序進行閱讀和實踐。對於有實際後端開發經驗並且了解 Nest 開發的讀者，可以選擇感興趣的章節閱讀，因為每個章節的設計相對獨立。

在學習本書的過程中，建議在閱讀完每個基礎知識後，按照書中的程式案例進行實踐，將其轉化為自己的儲備知識。如果遇到問題，建議首先偵錯分析問題所在，總結經驗，然後可以與作者溝通解答疑惑。

希望本書能給你帶來一段愉快的學習之旅。

致謝

感謝北京清華大學出版社提供這次創作機會，編輯們花了大量時間與我溝通和修改。感謝這次相遇，沒有這次相遇，我也無法以圖文形式與大家交流，這本書也不會誕生。

感謝我的妻子 Minnie 在背後一如既往地支持我，她把家庭打理得井然有序，給了我足夠的時間來完成創作。同時感謝我的兒子甜筒，他四個月大時我開始創作，他的懂事和健康成長讓我深感欣慰。

感謝自己，成為時間的主人，在保證工作和休息的同時，利用每天 3 小時的通勤時間、1 小時的午休時間，以及陪伴家人的間隙完成了本次創作；不斷改變，擁抱不確定性，是為了能夠有更多時間陪伴身邊親近的人。

感謝每一位閱讀本書的讀者朋友，你們的支持與回饋是我不斷進步、持續創作的最大動力。

　最後，由於筆者能力有限，書中基礎知識可能存在疏忽和遺漏，懇請讀者指正。

編者

目錄

第 1 部分　基礎篇

第 1 章　需要提前掌握的知識

1.1　Node 中的請求與回應物件 1-1
　　1.1.1　原生 Node 處理 HTTP 請求 1-1
　　1.1.2　Express 處理 HTTP 請求 1-4
　　1.1.3　Nest 處理 HTTP 請求 1-7
1.2　TypeScript 基礎與應用 1-8
　　1.2.1　TypeScript 編譯 1-8
　　1.2.2　TypeScript 類型系統 1-10

第 2 章　Nest 初識

2.1　什麼是 Nest 2-1
　　2.1.1　Nest 概述 2-1
　　2.1.2　Nest 的主要特點 2-1
　　2.1.3　Nest 的應用場景 2-2
2.2　快速上手 Nest CLI 2-3
　　2.2.1　Nest CLI 的安裝 2-3
　　2.2.2　建立專案 2-4
　　2.2.3　生成指定的程式部分 2-6
　　2.2.4　建構應用 2-10
　　2.2.5　啟動開發偵錯 2-12
　　2.2.6　查看專案資訊 2-13
2.3　建立第一個 Nest 應用 2-14
　　2.3.1　生成後端專案 2-14

	2.3.2	生成前端專案 2-15
	2.3.3	準備工作 2-16
	2.3.4	執行結果 2-19
	2.3.5	模組化開發 2-19
2.4	Nest 的 AOP 架構理念 2-21
	2.4.1	MVC 架構概述 2-21
	2.4.2	AOP 解決的問題 2-22
	2.4.3	AOP 在 Nest 中的應用 2-23
2.5	IoC 思想解決了什麼問題 2-31
	2.5.1	IoC 核心思想概述 2-31
	2.5.2	IoC 在 Nest 中的應用 2-32
2.6	學會偵錯 Nest 應用 2-36
	2.6.1	Chrome DevTools 偵錯 2-37
	2.6.2	VS Code 偵錯 2-39
	2.6.3	擴充偵錯技巧 2-42

第 3 章　Nest 核心概念介紹

3.1	貫穿全書的裝飾器 3-1
	3.1.1	基本概念 3-1
	3.1.2	裝飾器的種類 3-2
	3.1.3	Nest 中的裝飾器 3-10
3.2	井然有序的模組化 3-11
	3.2.1	基本概念 3-11
	3.2.2	建立模組 3-12
	3.2.3	共用模組 3-13
	3.2.4	全域模組 3-16
	3.2.5	動態模組 3-16
3.3	控制器與服務的默契配合 3-18
	3.3.1	基本概念 3-18
	3.3.2	Controller 管理請求路由 3-19

	3.3.3	Controller 處理請求參數與請求本體 3-21
	3.3.4	Service 處理資料層 3-22
	3.3.5	服務與服務提供者 3-24
3.4	耳熟能詳的中介軟體 3-24
	3.4.1	類別中介軟體 3-25
	3.4.2	函數式中介軟體 3-27
	3.4.3	局部中介軟體 3-28
	3.4.4	全域中介軟體 3-28
3.5	攔截器與 RxJS 知多少 3-29
	3.5.1	基本概念 3-29
	3.5.2	建立專案 3-30
	3.5.3	攔截器的基本使用方法 3-31
3.6	資料之源守護者：管道 3-34
	3.6.1	基本概念 3-34
	3.6.2	內建管道 3-35
	3.6.3	自訂管道 3-45
3.7	Nest 實現檔案上傳 3-46
	3.7.1	初識 Multer 3-46
	3.7.2	單檔案上傳 3-49
	3.7.3	多檔案上傳 3-52
	3.7.4	上傳任意檔案 3-56
	3.7.5	檔案驗證 3-58

第 2 部分　進階篇

第 4 章　Nest 與資料庫

4.1	快速上手 MySQL 4-1
	4.1.1	安裝和執行 4-1
	4.1.2	MySQL 的常用命令 4-4
	4.1.3	視覺化操作 MySQL 4-8
4.2	MySQL 表之間的關係 4-14

	4.2.1 一對一關聯性 4-14
	4.2.2 一對多 / 多對一關係 4-22
	4.2.3 多對多關係 4-25
4.3	快速上手 TypeORM 4-31
	4.3.1 基本概念 4-31
	4.3.2 專案準備 4-31
	4.3.3 建立模型及實體 4-32
	4.3.4 定義資料列及類型 4-32
	4.3.5 連接資料庫 4-33
	4.3.6 使用 Repository 操作 CRUD 4-35
	4.3.7 使用 QueryBuilder 操作 CRUD 4-37
4.4	使用 TypeORM 處理多表關係 4-41
	4.4.1 一對一關聯性 4-41
	4.4.2 一對多 / 多對一關係 4-47
	4.4.3 多對多關係 4-49
4.5	在 Nest 中使用 TypeORM 操作 MySQL 4-52
	4.5.1 專案準備 4-52
	4.5.2 使用 EntityManager 操作實體 4-55
	4.5.3 使用 Repository 操作實體 4-57
	4.5.4 使用 QueryBuilder 操作實體 4-60

第 5 章　性能最佳化之資料快取

5.1	快速上手 Redis 5-1
	5.1.1 安裝和執行 5-1
	5.1.2 Redis 的常用命令 5-2
5.2	在 Nest 中使用 Redis 快取 5-11
	5.2.1 專案準備 5-11
	5.2.2 Redis 初始化 5-13
	5.2.3 建表並建構快取 5-14
	5.2.4 執行程式 5-17

5.2.5　設置快取有效期...5-19
　　　5.2.6　選擇合理的有效期...5-20

第 6 章　身份驗證與授權

　6.1　Cookie、Session、Token、JWT、SSO 詳解................................6-1
　　　6.1.1　什麼是身份驗證...6-2
　　　6.1.2　什麼是授權...6-2
　　　6.1.3　什麼是憑證...6-2
　　　6.1.4　什麼是 Cookie..6-2
　　　6.1.5　什麼是 Session...6-3
　　　6.1.6　Session 與 Cookie 的區別..6-4
　　　6.1.7　什麼是 Token..6-5
　　　6.1.8　什麼是 JWT..6-8
　　　6.1.9　JWT 與 Token 的區別..6-9
　　　6.1.10　什麼是 SSO..6-10
　6.2　基於 Passport 和 JWT 實現身份驗證..6-14
　　　6.2.1　基本概念...6-14
　　　6.2.2　專案準備...6-15
　　　6.2.3　用本地策略實現使用者登入...6-16
　　　6.2.4　用 JWT 策略實現介面驗證...6-19
　　　6.2.5　程式最佳化...6-23
　6.3　基於 RBAC 實現許可權控制..6-26
　　　6.3.1　基本概念...6-27
　　　6.3.2　資料表設計...6-28
　　　6.3.3　專案準備...6-29
　　　6.3.4　建立實體...6-30
　　　6.3.5　啟動服務...6-33
　　　6.3.6　實現角色守衛控制...6-33
　　　6.3.7　生成測試資料...6-37
　　　6.3.8　測試效果...6-39

第 7 章　系統部署與擴充

- 7.1 快速上手 Docker 7-1
  - 7.1.1 初識 Docker 7-1
  - 7.1.2 安裝 Docker 7-2
  - 7.1.3 Docker 的使用 7-4
- 7.2 快速上手 Dockerfile 7-8
  - 7.2.1 Docker 的基本概念 7-8
  - 7.2.2 Dockerfile 的基本語法 7-9
  - 7.2.3 Dockerfile 實踐 7-9

第 3 部分　擴充篇

第 8 章　單元測試與點對點測試

- 8.1 重新認識單元測試 8-1
  - 8.1.1 什麼是單元測試 8-1
  - 8.1.2 為什麼大部分公司沒有進行單元測試 8-2
  - 8.1.3 為什麼要撰寫單元測試 8-3
  - 8.1.4 先撰寫單元測試還是先撰寫程式 8-4
  - 8.1.5 測試驅動開發 8-4
- 8.2 在 Nest 中使用 Jest 撰寫單元測試 8-6
  - 8.2.1 初識 Jest 8-6
  - 8.2.2 專案準備 8-9
  - 8.2.3 撰寫測試用例 8-11
  - 8.2.4 實現業務程式 8-12
  - 8.2.5 重構程式 8-17
- 8.3 整合測試 8-18
  - 8.3.1 撰寫測試用例 8-19
  - 8.3.2 測試效果 8-20
- 8.4 點對點測試 8-21

　　　　8.4.1　撰寫測試用例 8-21
　　　　8.4.2　實現業務程式 8-25

第 9 章　日誌與錯誤處理

　9.1　如何在 Nest 中記錄日誌 9-1
　　　　9.1.1　為什麼要記錄日誌 9-2
　　　　9.1.2　內建日誌器 Logger 9-2
　　　　9.1.3　訂製日誌器 9-4
　　　　9.1.4　記錄日誌的正確姿勢 9-5
　　　　9.1.5　第三方日誌器 Winston 9-6
　9.2　Winston 日誌管理實踐 9-6
　　　　9.2.1　Winston 的基礎使用 9-7
　　　　9.2.2　本地持久化日誌 9-10
　9.3　面向切面日誌統計實踐 9-14
　　　　9.3.1　中介軟體日誌統計 9-14
　　　　9.3.2　攔截器日誌統計 9-16
　　　　9.3.3　篩檢程式日誌統計 9-17

第 4 部分　Nest 專案實戰篇

第 10 章　數字門店管理平臺開發

　10.1　產品需求分析與設計 10-1
　　　　10.1.1　產品需求說明 10-2
　　　　10.1.2　功能原型圖 10-3
　10.2　技術選型與專案準備 10-10
　　　　10.2.1　前端技術選型 10-10
　　　　10.2.2　初始化前端專案 10-10
　　　　10.2.3　前端架構設計 10-11
　　　　10.2.4　後端技術選型 10-12
　　　　10.2.5　初始化後端專案 10-13
　　　　10.2.6　後端架構設計 10-15

10.3	API 介面及資料庫表設計 10-17
	10.3.1　API 介面功能劃分 10-17
	10.3.2　資料庫設計 10-19
10.4	實現註冊登入 10-22
	10.4.1　頁面效果展示 10-22
	10.4.2　介面實現 10-31
10.5	實現使用者與角色模組 10-41
	10.5.1　頁面效果展示 10-41
	10.5.2　表關係設計 10-48
	10.5.3　介面實現 10-52
10.6	實現商品與訂單模組 10-57
	10.6.1　頁面效果展示 10-57
	10.6.2　表關係設計 10-61
	10.6.3　介面實現 10-61
10.7	基於 Redis 實現商品熱銷榜 10-71
	10.7.1　頁面效果展示 10-71
	10.7.2　介面實現 10-74
10.8	實現活動模組與定時任務 10-76
	10.8.1　頁面效果展示 10-77
	10.8.2　表關係設計 10-78
	10.8.3　介面實現 10-79
10.9	使用 Docker Compose 部署專案 10-83
	10.9.1　撰寫後端 Docker Compose 檔案 10-83
	10.9.2　撰寫 Dockerfile 檔案 10-88

完結語：是終點，更是新的起點

第 1 部分

基 礎 篇

　　第 1 部分將深入探討 Nest 框架的基石，涵蓋以下關鍵領域：Nest 的先決條件知識、初步了解 Nest 及其核心概念。前 3 章介紹的核心概念對於理解 Nest 框架與各種中介軟體進行互動至關重要，也是深入學習本書後續章節和進行專案實戰開發的前提條件。

第 1 章
需要提前掌握的知識

本章將引導你學習 Nest 的前置知識：Node 的請求回應物件與 TypeScript 基礎。不必過分擔心，本章不涉及複雜概念，即使你對這兩方面的內容了解不多，也能理解它們的重要性。

首先，Nest 底層預設採用 Express 作為 HTTP 伺服器框架。這可能會引發疑問：原生 HTTP 模組、Express 與 Nest 之間有何聯繫？

其次，TypeScript 在普及過程中受到許多開發者的青睞，但也面臨一些質疑。舉例來說，學習曲線較陡、額外的開發成本、類型系統並非完美等問題讓一些初學者猶豫不決。

因此，在正式學習 Nest 之前，有必要先回顧一些必要的前置知識，以便在後續章節中更進一步地理解和實踐。

1.1 Node 中的請求與回應物件

在 Node 應用中，請求和回應物件是處理 HTTP 請求和發送 HTTP 回應的核心工具，主流框架如 Express、Koa、Nest、Egg 都實現了不同抽象等級的中介軟體。本節將開始學習這些內容。

1.1.1 原生 Node 處理 HTTP 請求

在 Node 的內建 HTTP 模組中有兩個核心物件：請求物件（Request）和回應物件（Response）。請求物件表示用戶端向服務端發送的請求資訊，而回應物件則表示服務端向用戶端傳回的回應資料。

請求物件通常在用戶端發送請求到服務端的過程中使用，攜帶的資訊包含 URL、請求方式、請求標頭、請求本體等。常見的屬性和方法說明如下：

- req.url：包含用戶端請求的 URL，不包括協定、主機名稱和通訊埠編號。
- req.method：包含用戶端使用的 HTTP 請求方法，例如 GET、POST 等。
- req.headers：一個包含所有請求標頭的物件，可以透過屬性名稱存取具體的請求標頭資訊。
- req.params：包含路由中匹配的參數，對於包含參數的路由非常有用。
- req.query：包含 URL 查詢參數的物件，用於解析 URL 中的鍵 - 值對（Key-Value Pair）。
- req.body：對於 POST 請求，包含請求本體的資料。在 Express 中需要使用中介軟體（例如 body-parser）來解析請求本體。

相反，回應物件是向用戶端發送的 HTTP 回應，設置回應狀態、回應標頭和回應本體等資訊。常見的屬性和方法說明如下。

- res.status(code)：設置 HTTP 回應狀態碼。
- res.setHeader(name, value)：設置回應標頭的值。
- res.send(body)：發送回應體內容給用戶端，可以是字串、JSON 物件、緩衝區等。
- res.json(obj)：發送 JSON 格式的回應本體給用戶端。
- res.sendFile(path, options, callback)：發送檔案作為回應。
- res.cookie(name, value, option])：設置 Cookie 資訊。

由此可見，Node 提供了非常豐富的基礎 API 操作物件，但也帶來了一些問題，例如開發者需要關注如何正確處理用戶端發送過來的請求資料。請看下面的範例。

原生 Node 處理請求 URL 程式範例：

```
const http = require("http");
const url = require("url");

const server = http.createServer((req, res) => {
  // 解析請求的 URL
  const parsedUrl = url.parse(req.url, true);

  // 獲取路徑和查詢參數
  const path = parsedUrl.pathname;
  const queryParams = parsedUrl.query;
```

```
  // 設置回應標頭
  res.writeHead(200, { "Content-Type": "text/plain" });
  res.end(`Path: ${path}, Query Parameters: ${JSON.stringify(queryParams)}`);
});

const PORT = 3000;
server.listen(PORT, () => {
  console.log(`Server is listening on port ${PORT}`);
});
```

原生 Node 處理請求本體程式範例：

```
const http = require("http");

const server = http.createServer((req, res) => {
  let body = "";

  // 監聽請求的資料
  req.on("data", (chunk) => {
    body += chunk;
  });

  // 請求結束時處理請求本體
  req.on("end", () => {
    // 解析 JSON 請求本體
    const jsonData = JSON.parse(body);

    res.writeHead(200, { "Content-Type": "application/json" });
    res.end(
      JSON.stringify({ message: "Request Body Processed", data: jsonData })
    );
  });
});

const PORT = 3000;
server.listen(PORT, () => {
  console.log(`Server is listening on port ${PORT}`);
});
```

　　從上述程式可以看出，Node 的原生 HTTP 模組在處理請求 URL 時，需要使用 Node 內建的 url 模組中的 url.parse 方法進行解析。如果是 POST 請求，則需要透過 JSON.parse 方法來解析請求本體。在建構回應時，每個請求都需要設置回應標頭，最終透過 send 方法完成發送操作。顯然，這種方法在實際的生產開發中顯得相當煩瑣。

1.1.2 Express 處理 HTTP 請求

然而，Express 的誕生簡化並抽象了 HTTP 模組，提供了更加靈活的路由系統。結合中介軟體，它使得路由的定義和處理變得更加簡單。請看下面的例子。

Express 處理請求 URL 程式範例：

```
const express = require("express");
const app = express();

// 處理帶有參數的路由
app.get("/users/:id", (req, res) => {
  const userId = req.params.id;
  res.send('User ID: ${userId}');
});

// 處理查詢字串參數
app.get("/search", (req, res) => {
  const query = req.query.q;
  res.send('Search Query: ${query}');
});

const PORT = 3000;
app.listen(PORT, () => {
  console.log('Server is listening on port ${PORT}');
});
```

Express 處理請求本體程式範例：

```
const express = require("express");
const app = express();

// 使用 express.json() 中介軟體
app.use(express.json());
// 使用 express.urlencoded() 中介軟體
app.use(express.urlencoded({ extended: true }));

// 處理 Ajax 發送的 JSON 資料
app.post("/api/data", (req, res) => {
  const data = req.body; // 獲取 JSON 請求本體 res.json(data);
});

// 處理 HTML 表單提交的資料
app.post("/api/data2", (req, res) => {
  const data = req.body; // 獲取 URL 編碼請求本體
  res.json(data);
});

const PORT = 3000;
```

```
app.listen(PORT, () => {
  console.log('Server is listening on port ${PORT}');
});
```

從上述程式可以看出，在 Express 中，GET 請求方式可以直接透過 req 物件獲取對應參數。對於 POST 請求，透過設置路由中介軟體，express.json() 用於解析請求本體中的 JSON 資料，而 express.urlencoded() 則用於處理請求本體中的 URL 編碼參數，從而簡化了煩瑣的資料轉換操作。

此時，Express 的優勢已經顯而易見。然而，這還不是全部。除了提供強大的路由系統和中介軟體之外，它還整合了範本引擎、靜態檔案服務、錯誤處理機制等功能。但這並不是本書的重點，這裡不作過多擴充，感興趣的讀者可以自行了解。

既然 Express 擁有如此多的優勢，Nest 又是憑藉什麼在市場中佔據一席之地呢？答案是它解決了架構問題。

再來看 Express 的路由管理，它可能是這樣的：

```
const express = require("express");
const app = express();

// 定義 /users 路由
app.get("/users", (req, res) => {
  res.send("Welcome to users page");
});

// 定義帶有參數的路由
app.get("/users/:id", (req, res) => {
  const userId = req.params.id;
  res.send(`User ID: ${userId}`);
});

// 使用 Router 物件定義更複雜的路由
const adminRouter = express.Router();

adminRouter.get(«/», (req, res) => {
  res.send("Welcome to admin home");
});

adminRouter.get(«/dashboard», (req, res) => {
  res.send("Welcome to admin dashboard");
});

// 將 /admin 路由掛載到 /app 路徑下
app.use("/app", adminRouter);
```

```
const PORT = 3000;
app.listen(PORT, () => {
  console.log(`Server is listening on port ${PORT}`);
});
```

有經驗的讀者可能會發現，這種方式的可維護性太差了，於是有了下面的路由管理方式。在主應用中使用模組化管理路由，程式如下：

```
// app.js
const express = require('express');
const app = express();

// users 模組
const usersRouter = require('./routes/users');
app.use('/users', usersRouter);

const PORT = 3000;
app.listen(PORT, () => {
  console.log(`Server is listening on port ${PORT}`);
});
```

對 users 路由進行模組化管理，程式如下：

```
// users.js
const express = require('express');
const router = express.Router();

// RESTful 風格的路由
app.get('/', getAllUsers);
app.get('/users/:id', getUserById);
app.post('/users', createUser);
app.put('/users/:id', updateUser);
app.delete('/users/:id', deleteUser);

function getAllUsers(req, res) {}
function getUserById(req, res) {}
function createUser(req, res) {}
function updateUser(req, res) {}
function deleteUser(req, res) {}

module.exports = router;
```

顯然，在對比之下，第二種方式的可維護性和擴充性更高。存在這種差異的原因在於 Express 沒有規定開發者必須遵循哪種方式撰寫程式。開發者之間的水準不一致導致了這種差異化。從設計者的角度來看，這顯然是不可接受的。因此，更高層次的框架—Nest 應運而生。

1.1.3 Nest 處理 HTTP 請求

在 Nest 中是如何處理請求的呢？下面一起來看看。

Nest 處理請求 URL 程式範例：

```
// 透過路由參數
import { Controller, Get, Param } from '@nestjs/common';

@Controller('cats')
export class CatsController {
  // 使用路由參數裝飾器（如 @Param）來獲取請求 URL 中的參數
  @Get(':id')
  findOne(@Param('id') id: string): string {
    return `This action returns a cat with the ID: ${id}`;
  }

  // 使用查詢參數裝飾器（如 @Query）來獲取請求 URL 中的查詢參數
  @Get()
  findAll(@Query('page') page: number): string {
    return `This action returns all cats on page ${page}`;
  }
}
```

Nest 處理請求本體程式範例：

```
import { Controller, Post, Body } from '@nestjs/common';

@Controller('cats')
export class CatsController {
  // 使用 @Body 裝飾器來獲取 POST 請求中的請求本體資料
  @Post()
  create(@Body() data): string {
    return `This action creates a new cat with the name: ${data}`;
  }
}
```

是不是簡單多了？沒錯，這就是在 Nest 中處理請求的常用方式。它提供了強大的裝飾器來操作請求標頭資料，並且自動判斷傳回的資料型態來設置合理的 Content-type，使得請求處理變得非常靈活。

至此，回顧一下開篇的問題：原生 HTTP 模組、Express 與 Nest 之間有何聯繫？相信到這裡讀者已經有了答案，至於 Nest 是如何解決架構問題的，會在後續的章節中詳細講解。

1.2 TypeScript 基礎與應用

Nest 是基於 TypeScript 進行建構的，它提供了類型安全系統，並引入了物件導向程式設計、裝飾器和中繼資料等概念。這些概念貫穿於 Nest 的原始程式設計和專案實踐開發中。本節將從 TypeScript 的編譯配置和類型系統兩大板塊著手，回顧 TypeScript 的基礎知識。

1.2.1 TypeScript 編譯

1. 編譯上下文

TypeScript 被設計為 JavaScript 的超集合，在使用時需要透過特定的編譯器將其編譯為普通的 JavaScript 才能被瀏覽器辨識。編譯器執行的環境被稱為 TypeScript 的編譯上下文，它包含編譯器需要用到的各種配置選項、輸入輸出檔案等資訊。而 tsconfig.json 檔案就是用來指定編譯器執行時期的行為，即編譯選項。

2. 編譯選項

通常用 compilerOptions 來訂製編譯選項，它包含豐富的選項，範例如下：

```
{
  "compilerOptions": {

    /* 基本選項 */
    "target": "es5",                       // 指定 ECMAScript 目標版本：'ES3' (default), 'ES5',
'ES6'/'ES2015', 'ES2016', 'ES2017','ESNEXT'
    "module": "commonjs",                  // 指定使用的模組：'commonjs', 'amd', 'system',
'umd' or 'es2015'
    "lib": [],                             // 指定要包含在編譯中的函數庫檔案
    "allowJs": true,                       // 允許編譯 JavaScript 檔案
    "checkJs": true,                       // 報告 JavaScript 檔案中的錯誤
    "jsx": "preserve",                     // 指定 JSX 程式的生成：'preserve', 'react-
native', or 'react'
    "declaration": true,                   // 生成相應的 '.d.ts' 檔案
    "sourceMap": true,                     // 生成相應的 '.map' 檔案
    "outFile": "./",                       // 將輸出檔案合併為一個檔案
    "outDir": "./",                        // 指定輸出目錄
    "rootDir": "./",                       // 用來控制輸出目錄結構 --outDir.
    "removeComments": true,                // 刪除編譯後的所有註釋
    "noEmit": true,                        // 不生成輸出檔案
    "importHelpers": true,                 // 從 tslib 匯入輔助工具函數
```

```
    "isolatedModules": true,           // 將每個檔案作為單獨的模組 （與 'ts.transpileModule'
類似）

    /* 嚴格的類型檢查選項 */
    "strict": true,                    // 啟用所有嚴格類型的檢查選項
    "noImplicitAny": true,             // 在運算式和宣告上有隱含的 any 類型時顯示出錯
    "strictNullChecks": true,          // 啟用嚴格的 null 檢查
    "noImplicitThis": true,            // 當 this 運算式值為 any 類型時，生成錯誤
    "alwaysStrict": true,              // 以嚴格模式檢查每個模組，並在每個檔案中加入
'use strict'

    /* 額外的檢查 */
    "noUnusedLocals": true,            // 有未使用的區域變數時拋出錯誤
    "noUnusedParameters": true,        // 有未使用的參數時拋出錯誤
    "noImplicitReturns": true,         // 並不是所有函數都有傳回值時拋出錯誤
    "noFallthroughCasesInSwitch": true, // 報告 switch 敘述的 fallthrough 錯誤（即不允許
switch 的 case 敘述貫穿）

    /* 模組解析選項 */
    "moduleResolution": "node",        // 選擇模組解析策略： 'node' (Node.js) or
‹classic› (TypeScript pre-1.6)
    "baseUrl": "./",                   // 用於解析非相對路徑的模組名稱的基目錄
    "paths": {},                       // 模組名稱到基於 baseUrl 的路徑映射的清單
    "rootDirs": [],                    // 根資料夾清單，其組合內容表示專案執行時期的結構內容
    "typeRoots": [],                   // 包含型態宣告的檔案列表
    "types": [],                       // 需要包含的型態宣告檔案名稱清單
    "allowSyntheticDefaultImports": true, // 允許從沒有設置預設匯出的模組中預設匯入

    /* Source Map Options */
    "sourceRoot": "./",                // 指定偵錯器應該找到 TypeScript 檔案而非原始檔案的位置
    "mapRoot": "./",                   // 指定偵錯器應該找到映射檔案而非生成檔案的位置
    "inlineSourceMap": true,           // 生成單一 sourcemaps 檔案，而非將 sourcemaps 生成不同的檔案
    "inlineSources": true,             // 將原始程式碼與 sourcemaps 生成到一個檔案中，要求同時設置了
--inlineSourceMap 或 --sourceMap 屬性

    /* 其他選項 */
    "experimentalDecorators": true,    // 啟用實驗性的裝飾器
    "emitDecoratorMetadata": true      // 為裝飾器提供中繼資料支援
  }
}
```

3. 編譯執行

TypeScript 內建了 TSC 編譯器，可以透過以下方式執行它：

- 執行 tsc，它會在目前的目錄或父級目錄中尋找 tsconfig.json 檔案。

- 執行 tsc -p ./path-to-project-directory。這裡的路徑可以是絕對路徑，也可以是相對於目前的目錄的相對路徑。

4. 指定檔案

可以用 include 和 exclude 選項來指定需要包含的檔案和排除的檔案：

```
{
  "include": [
    "./folder"
  ],
  "exclude": [
    "./folder/**/*.spec.ts",
    "./folder/someSubFolder"
  ]
}
```

1.2.2 TypeScript 類型系統

1. 基本注解

類型注解使用「:TypeAnnotation」語法。在型態宣告空間中，可用的任何內容都可以用作類型注解。

在下面這個例子中，使用了變數、函數參數以及函數傳回值的類型注解：

```
const num: number = 123;
function identity(num: number): number {
  return num;
}
```

2. 原始類型

JavaScript 的原始類型同樣適用於 TypeScript 的類型系統，因此 string、number、boolean 也可以用作類型注解：

```
let num: number;
let str: string;
let bool: boolean;

num = 123;
num = 123.456;
num = '123'; // Error
```

3. 陣列

TypeScript 為陣列提供了專用的類型語法，因此我們可以很容易地為陣列增加類型註解，程式如下：

```
let boolArray: boolean[];

boolArray = [true, false];
console.log(boolArray[0]); // true
```

4. 介面

介面可以將多個型態宣告合併為一個型態宣告：

```
interface Name {
  first: string;
  second: string;
}

let name: Name;
name = {
  first: 'John',
  second: 'Doe'
};

name = {
  // Error: 'Second is missing'
  first: ‹John›
};
```

在這裡，我們把類型註解 first: string + last: string 合併到了一個新的類型註解 Name 中，這樣能夠強制對每個成員進行類型檢查。

5. 特殊類型

除了前面提到的一些原始類型，TypeScript 中還會有一些特殊的類型，它們是 any、null、undefined 以及 void。

1) any

any 類型為我們提供了一個類型系統的「後門」，使用它時，TypeScript 會關閉類型檢查。在類型系統中，any 能夠相容所有的類型（包括它自己），程式如下：

```
let power: any;

// 賦值任意類型
power = '123';
power = 123;
```

2）null 和 undefined

在類型系統中，JavaScript 中的 null 和 undefined 字面量與其他被標注了 any 類型的變數一樣，都可以賦值給任意類型的變數，但不會關閉類型檢查，程式如下：

```
// strictNullChecks: false

let num: number;
let str: string;

// 這些類型可用於賦值
num = null;
str = undefined;
```

3）void

void 表示沒有任何類型，通常用於那些不傳回任何值的函數的傳回類型。

```
function myFunction(): void {
// 函數本體
}
```

6. 泛型

泛型可以應用於 Typescript 中的函數（函數參數、函數傳回值）、介面和類別（類別的實例成員、類別的方法）。

下面我們來建立第一個使用泛型的範例函數：reverse()。這個函數會傳回傳入的任意值。在不使用泛型的情況下，這個函數可能是這樣的：

```
function reverse(arg: any): any {
  return arg;
};
```

使用 any 類型會導致這個函數可以接收任何類型的 arg 參數，但我們希望傳入的類型與傳回的類型相同。

因此，需要一種方法來確保傳回值的類型與傳入參數的類型相同。這裡，我們使用了類型變數，它是一種特殊的變數，只用於表示類型而非值，程式如下：

```
function reverse<T>(arg: T): T {
  return arg;
}
```

我們給 reverse 增加了類型變數 T。T 幫助我們捕捉使用者傳入的類型（比如 string），之後把 T 用作傳回數值型態，這樣就可以追蹤函數中使用的類型資訊了。

我們把這個版本的 reverse() 函數稱為泛型，因為它適用於多個類型。

7. 聯合類型

我們希望 JavaScript 屬性能夠支援多種類型，例如字串或字串陣列。在 TypeScript 中，可以使用聯合類型來滿足這一需求（聯合類型使用「|」作為分隔符號，例如 TypeA | TypeB）。以下是關於聯合類型的例子：

```
function formatCommandline(command: string[] | string) {
  let line = '';
  if (typeof command === 'string') {
    line = command.trim();
  } else {
    line = command.join(' ').trim();
  }
}
```

8. 交叉類型

在 JavaScript 中，extend 是一種非常常見的模式。在這種模式下，我們可以透過兩個物件建立一個新物件，使新物件繼承這兩個物件的所有功能。交叉類型可以讓我們安全地使用這種模式：

```
function extend<T extends object, U extends object>(first: T, second: U): T & U {
  const result = {} as T & U;

  for (let id in first) {
    (result as T)[id] = first[id];
  }

  for (let id in second) {
    (result as U)[id] = second[id];
  }

  return result;
}
```

9. 類型態名

TypeScript 提供了一種便捷的語法來為類型注解設置別名。我們可以使用 type SomeName = someValidTypeAnnotation 來建立別名：

```
type StrOrNum = string | number;

// 使用
let sample: StrOrNum;
```

```
sample = 123;
sample = '123';

// 檢查類型
sample = true; // Error
```

10. 類型斷言

TypeScript 允許覆蓋其類型推斷，並可以按照我們期望的方式重新分析類型。這種機制被稱為「類型斷言」。類型斷言用於告知編譯器，我們對該類型的了解比編譯器的推斷更為準確，因此編譯器不應再發出類型錯誤。以下是類型斷言的常見用法範例：

```
function isArray(o: any): boolean {
  return Object.prototype.toString.call(o) === '[object Array]';
}

function checkAuth(auth: number | number[]) {
  if (isArray(auth)) {
    // 錯誤，Property 'some' does not exist on type 'number'
    return auth.some(auth => auth === 1);
  } else {
    return auth === 1;
  }
}
```

這裡的程式發出了錯誤警告，因為 auth 的類型為 number|number[]，其屬性沒有 some() 方法。因此，我們不能使用 some() 方法。雖然已經知道 auth 的實際類型是 number[]，但 TypeScript 無法推導出這一點。在這種情況下，可以透過類型斷言來避免此問題：

```
function isArray(o: any): boolean {
  return Object.prototype.toString.call(o) === '[object Array]';
}

function checkAuth(auth: number | number[]) {
  if (isArray(auth)) {
    return (auth as number[]).some(auth => auth === 1);
  } else {
    return auth === 1;
  }
}
```

11. 列舉

列舉是一種用於收集相關值的集合的方法,範例如下:

```typescript
enum Status {
  Success,
  Error,
  Warning
}

enum Env {
  Null = '',
  Dev = 'development',
  Prod = 'production'
}

// 簡單使用
let status = Status.Success;

let env = Env.Dev;
```

12. 多載

TypeScript 允許我們宣告函數多載。這對於文件 + 類型安全非常實用。請看以下程式:

```typescript
function padding(top: number, right?: number, bottom?: number, left?: number) {
  if (right === undefined && bottom === undefined && left === undefined) {
    right = bottom = left = top;
  } else if (bottom === undefined && left === undefined) {
    left = right;
    bottom = top;
  }

  return { top, right, bottom, left };
}
```

我們可以透過函數多載來強制執行並記錄這些約束,只需多次宣告函數標頭即可。最後一個函數標頭在函數體內實際處於活動狀態,但這函數標頭不可在外部直接使用,如下所示:

```typescript
// 多載
function padding(all: number);
function padding(topBottom: number, leftRight: number);
function padding(top: number, right: number, bottom: number, left: number);
// 函數主體的具體實現包含需要處理的所有情況
function padding(top: number, right?: number, bottom?: number, left?: number) {
```

```
  if (right === undefined && bottom === undefined && left === undefined) {
    right = bottom = left = top;
  } else if (bottom === undefined && left === undefined) {
    left = right;
    bottom = top;
  }

  return { top, right, bottom, left };
}
```

本節透過範例介紹了 TypeScript 的基本概念和用法，這些類型在 Nest 開發中都會頻繁使用。掌握本節內容可以為我們以後撰寫 Nest 應用打下堅實的基礎。

第 2 章

Nest 初識

本章主要介紹 Nest 框架的基礎概念和主要特點，講解如何透過 Nest 命令列介面（CLI）快速建立你的應用程式。此外，本章還將結合前端 React 框架，演示一個完整的前後端專案請求閉環流程。接下來，將介紹 Nest 中兩個非常重要的軟體設計思想：面向切面程式設計（Aspect-Oriented Programming，AOP）和控制反轉（Inversion of Control，IoC），以及它們在 Nest 中的應用。最後，本章將介紹如何透過 Chrome 開發者工具（DevTools）和 VS Code 高效偵錯 Nest 專案。

2.1 什麼是 Nest

在第 1 章中，我們了解到在 Nest 中可以更優雅地處理 HTTP 請求方式。它的優勢遠不止於此，還包括強大的相依注入系統，以及對物件導向程式設計和函數式程式設計等多種範式的支援。雖然這聽起來可能有些複雜，但不必擔心。我們先來掌握 Nest 的基本概念，這有助在後續的章節中更進一步地理解它。

2.1.1 Nest 概述

Nest 是一個用於建構高效、可擴充的 Node.js 伺服器端應用的框架。它預設使用 Express 作為 HTTP 服務端框架，同時也支援 Fastify 作為替代選項。然而，Nest 並不依賴於這兩種框架，而是採用了優雅的轉接器模式來實現其可擴充性。簡言之，即使未來出現更優秀的第三方模組，開發者仍然擁有自由選擇的權利。

2.1.2 Nest 的主要特點

在架構設計方面，Nest 屬於 MVC（Model-View-Controller，模型 - 視圖 - 控

制器）架構系統，具有模組化和鬆散耦合的特性。它引入了相依注入來管理物件之間的關係，使得 Model、View、Controller 之間的互動變得簡單且高效。

在專案管理方面，模組化結構將各個功能模組拆分為更細的粒度進行管理。每個模組中都包含獨立的控制器、路由、服務、攔截器等，最後將這些模組聚合在一起，形成一個類似搭積木的過程。

在開發風格方面，Nest 支援 JavaScript 與 TypeScript 兩種語言。使用 TypeScript 具備更多優勢，例如靜態類型檢測、程式自動補全、智慧提示以及函數多載等高級語法支援。

在擴充性方面，Nest 支持各種中介軟體和外掛程式，包括身份驗證、日誌記錄、快取等。此外，Nest 還允許與各種資料庫、通訊協定以及其他第三方服務無縫整合，使開發者能夠輕鬆擴充應用功能。

2.1.3 Nest 的應用場景

Nest 是一個多功能的 Node 後端框架。其靈活性使得它能夠滿足不同需求，從小型專案到大型應用等各種場景，都可以用 Nest 來建構高品質、可維護的後端服務。應用場景包括但不限於：

- Web 應用程式：Nest 提供強大的路由系統和 HTTP 模組，能夠與各種範本引擎良好配合，用於建構部落格系統、企業官網、電子商務網站等各種 Web 應用。
- 微服務架構：Nest 支援微服務架構，可建立多個微服務，並輕鬆實現不同微服務之間的互動和通訊，適用於建構複雜的分散式系統。
- 服務端著色（Server-Side Rendering，SSR）：Nest 支援與前端主流的服務端著色框架（如 Next.js、Nuxt.js）配合使用，用於建構單頁面應用（Single-Page Application，SPA）。
- 企業級應用：Nest 擁有強大的相依注入和模組系統，與各種資料庫、日誌系統集成緊密，能夠滿足大型企業級應用的各種需求。

無論你需要建構小型應用還是大型複雜的企業級應用，Nest 都是一個值得選擇的框架。

2.2 快速上手 Nest CLI

相信 CLI 對讀者來說並不陌生，它廣泛應用於各個工程化領域，用於快速建立專案、執行編譯、執行建構等操作，如前端的 Vue CLI、Vite、Create React App，後端的 Spring Boot CLI、express-generator 等。Nest 也不例外，提供了強大的 CLI 命令列工具，它被放在 @nestjs/cli 套件中。本節首先介紹 CLI 的安裝並建立一個實際的專案，接著透過專案演示來介紹它生成各種程式部分的命令，幫助讀者徹底掌握 CLI 的使用。

2.2.1 Nest CLI 的安裝

執行以下命令可將 Nest CLI 安裝為全域工具，方便日後使用：

```
npm install -g @Nestjs/cli
```

需要注意的是，這種方式需要不定時更新，以獲取最新的範本程式：

```
npm update -g @Nestjs/cli
```

安裝完成之後，執行「nest –h」命令，結果如圖 2-1 所示。

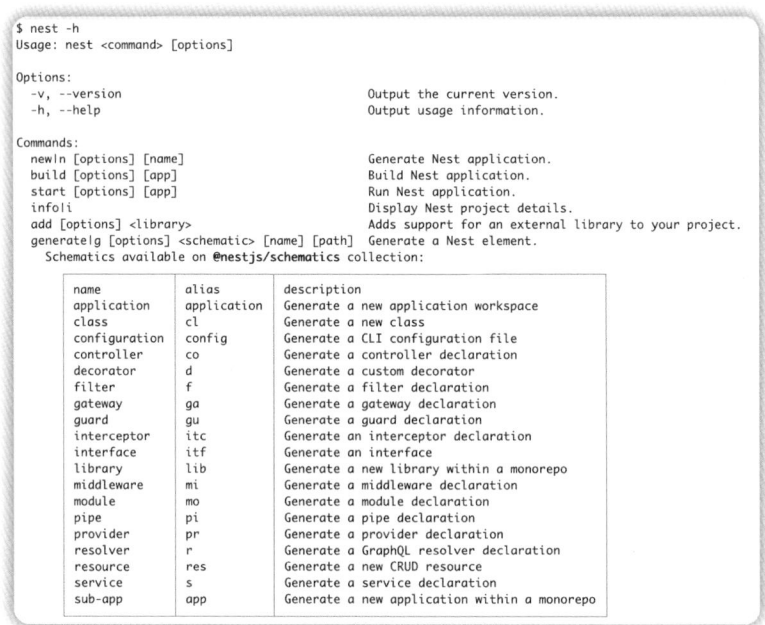

▲ 圖 2-1 「nest –h」命令的執行結果

由圖 2-1 可見，Nest 提供了非常豐富的命令。

- nest new：用於建立專案。
- nest build：用於建構生產環境程式。
- nest start：用於啟動本地開發服務。
- nest info：用於查看當前專案中的 Nest 套件資訊。
- nest add：用於增加官方外掛程式或第三方模組。
- nest generate：用於生成各種模組程式，如 Module、Controller、Service、Pipe、Middleware 等。

同時，部分命令也支援別名，如「nest n」「nest i」「nest g」，後續將分別進行測試。

2.2.2 建立專案

使用者可以用「nest new」命令來建立專案，具體參數如圖 2-2 所示。

```
$ nest new -h
Usage: nest new|n [options] [name]

Generate Nest application.

Options:
  --directory [directory]              Specify the destination directory
  -d, --dry-run                        Report actions that would be performed without writing out results. (default: false)
  -g, --skip-git                       Skip git repository initialization. (default: false)
  -s, --skip-install                   Skip package installation. (default: false)
  -p, --package-manager [packageManager]  Specify package manager.
  -l, --language [language]            Programming language to be used (TypeScript or JavaScript) (default: "TypeScript")
  -c, --collection [collectionName]    Schematics collection to use (default: "@nestjs/schematics")
  --strict                             Enables strict mode in TypeScript. (default: false)
  -h, --help                           Output usage information.
```

▲ 圖 2-2　建立專案時可以使用的參數選項

下面挑選幾個常用的參數說明。

- --skip-git 和 --skip-install：這些參數用於跳過 Git 初始化和 npm 套件安裝步驟。
- --package-manager：此參數用於指定專案使用的套件管理器（npm、yarn、pnpm）。推薦使用 pnpm，它作為繼 npm、yarn 之後推出的套件管理器，因其速度快、節省磁碟空間而受到青睞。
- --language：此參數決定使用 TypeScript 還是 JavaScript 進行撰寫。推薦使用預設的 TypeScript。

- --collection：用於指定工作流集合。預設是 @nestjs/schematics，用於快速建立模組、控制器、服務等，與「nest generate」命令相關。通常使用預設值，後續將詳細講解。
- --strict：此參數用於指定 TypeScript 是否以嚴格模式執行。

執行「nest n cli-test」命令後，CLI 提供了互動式命令讓你選擇套件管理器，如圖 2-3 所示。

```
$ nest n cli-test
⚡  We will scaffold your app in a few seconds..

? Which package manager would you ❤ to use? (Use arrow keys)
❯ npm
  yarn
  pnpm
```

▲ 圖 2-3 選擇套件管理器

當然，我們可以直接使用「nest n cli-test -p pnpm」命令來指定使用 pnpm，如圖 2-4 所示。

```
$ nest n cli-test -p pnpm
⚡  We will scaffold your app in a few seconds..

CREATE cli-test/.eslintrc.js (663 bytes)
CREATE cli-test/.prettierrc (51 bytes)
CREATE cli-test/README.md (3347 bytes)
CREATE cli-test/nest-cli.json (171 bytes)
CREATE cli-test/package.json (1949 bytes)
CREATE cli-test/tsconfig.build.json (97 bytes)
CREATE cli-test/tsconfig.json (546 bytes)
CREATE cli-test/src/app.controller.spec.ts (617 bytes)
CREATE cli-test/src/app.controller.ts (274 bytes)
CREATE cli-test/src/app.module.ts (249 bytes)
CREATE cli-test/src/app.service.ts (142 bytes)
CREATE cli-test/src/main.ts (208 bytes)
CREATE cli-test/test/app.e2e-spec.ts (630 bytes)
CREATE cli-test/test/jest-e2e.json (183 bytes)

▶▶▶▶▶ Installation in progress... 🍺
```

▲ 圖 2-4 指定 pnpm

由圖 2-4 可見，CLI 直接建立預設的專案範本，然後執行安裝相依，完成後自動建立一個 pnpm-lock.yaml 檔案，如圖 2-5 所示。

▲ 圖 2-5　自動建立 pnpm-lock.yaml 檔案

「nest generate」命令預設執行的是 @nestjs/schematics 套件中的命令。而 Schematics 是一種用於建立、刪除和更新 Angular 應用程式碼的工具。Nest 在此基礎上進行了擴充和訂製，以適應其框架的需求。

2.2.3　生成指定的程式部分

執行「nest generate -h」命令來生成指定的程式部分，可以查看其中包含的內容，如圖 2-6 所示。

```
$ nest g -h
Usage: nest generate|g [options] <schematic> [name] [path]

Generate a Nest element.
  Schematics available on @nestjs/schematics collection:
  ┌─────────────────┬─────────────┬──────────────────────────────────────────────────┐
  │ name            │ alias       │ description                                      │
  │ application     │ application │ Generate a new application workspace             │
  │ class           │ cl          │ Generate a new class                             │
  │ configuration   │ config      │ Generate a CLI configuration file                │
  │ controller      │ co          │ Generate a controller declaration                │
  │ decorator       │ d           │ Generate a custom decorator                      │
  │ filter          │ f           │ Generate a filter declaration                    │
  │ gateway         │ ga          │ Generate a gateway declaration                   │
  │ guard           │ gu          │ Generate a guard declaration                     │
  │ interceptor     │ itc         │ Generate an interceptor declaration              │
  │ interface       │ itf         │ Generate an interface                            │
  │ library         │ lib         │ Generate a new library within a monorepo         │
  │ middleware      │ mi          │ Generate a middleware declaration                │
  │ module          │ mo          │ Generate a module declaration                    │
  │ pipe            │ pi          │ Generate a pipe declaration                      │
  │ provider        │ pr          │ Generate a provider declaration                  │
  │ resolver        │ r           │ Generate a GraphQL resolver declaration          │
  │ resource        │ res         │ Generate a new CRUD resource                     │
  │ service         │ s           │ Generate a service declaration                   │
  │ sub-app         │ app         │ Generate a new application within a monorepo     │
  └─────────────────┴─────────────┴──────────────────────────────────────────────────┘

Options:
  -d, --dry-run                    Report actions that would be taken without writing out results.
  -p, --project [project]          Project in which to generate files.
  --flat                           Enforce flat structure of generated element.
  --no-flat                        Enforce that directories are generated.
  --spec                           Enforce spec files generation. (default: true)
  --spec-file-suffix [suffix]      Use a custom suffix for spec files.
  --skip-import                    Skip importing (default: false)
  --no-spec                        Disable spec files generation.
  -c, --collection [collectionName] Schematics collection to use.
  -h, --help                       Output usage information.
```

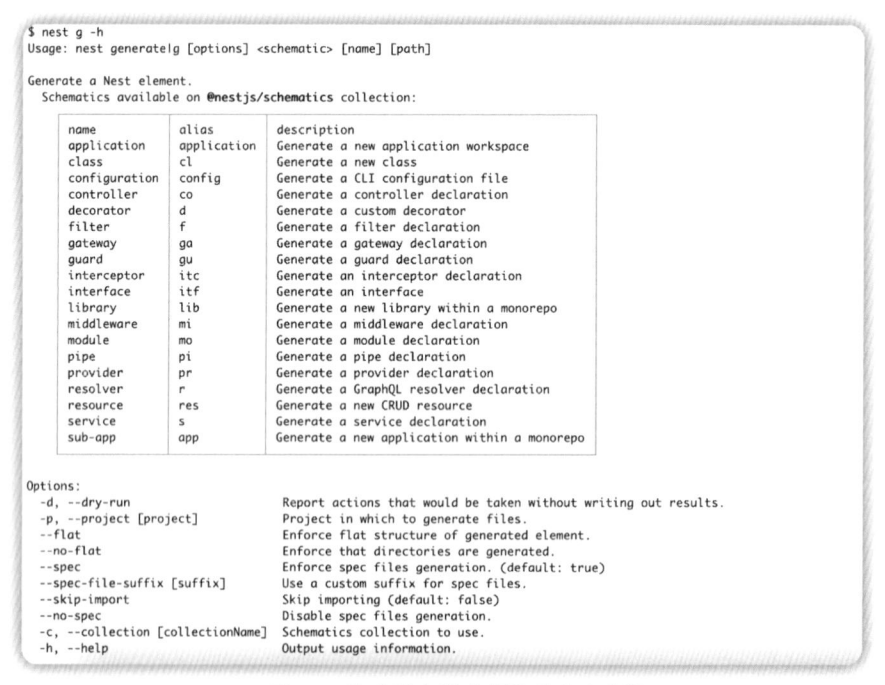

▲ 圖 2-6　生成指定程式部分的命令參數

其中包含豐富的 Nest 元素（如 Controller、Decorator、Filter 等），同時提供了靈活的參數控制。

我們來試一下，進入剛剛建立的專案 cli-test，執行「nest g controller」命令，如圖 2-7 所示。

```
$ nest g controller user
CREATE src/user/user.controller.spec.ts (478 bytes)
CREATE src/user/user.controller.ts (97 bytes)
UPDATE src/app.module.ts (446 bytes)
```

▲ 圖 2-7 使用命令生成控制器（controller）

接著可以看見在 src 目錄下建立了一個 user 控制器，如圖 2-8 所示。

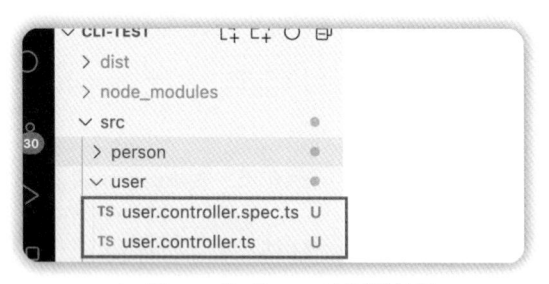

▲ 圖 2-8 生成 user 控制器結果

其中，.spec.ts 是單元測試檔案，可以透過設置 --no-spec 參數表明不生成測試檔案。

- --flat 和 --not-flat 這兩個參數表示是否使用扁平化結構。我們接著建立一個 user 篩檢程式來測試。
- --flat：參數表示扁平化，會將生成的檔案放到 src 目錄下，而不生成對應的目錄，如圖 2-9 所示。

```
$ nest g filter user --flat
CREATE src/user.filter.spec.ts (164 bytes)
CREATE src/user.filter.ts (186 bytes)
```

▲ 圖 2-9 使用扁平化參數對應的執行結果

- --not-flat：參數表示非扁平化，會生成對應的目錄，如圖 2-10 所示。

```
● $ nest g filter user --no-flat
  CREATE src/user/user.filter.spec.ts (164 bytes)
  CREATE src/user/user.filter.ts (186 bytes)
```

▲ 圖 2-10 非扁平化參數執行結果

- --skip-import 表示是否跳過自動匯入相依，預設情況下會自動匯入，以 user controller 為例，如圖 2-11 所示。

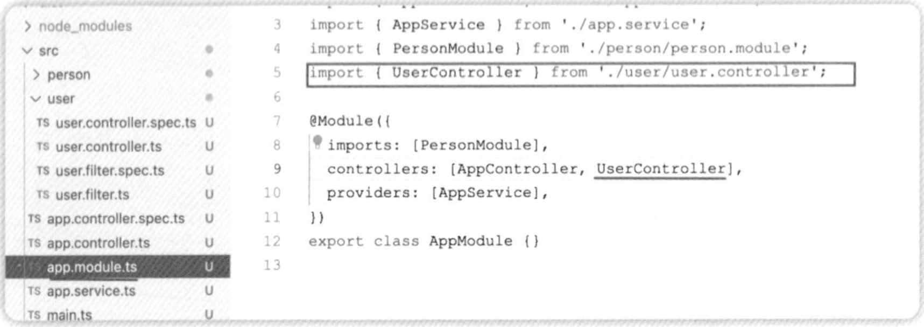

```
3   import { AppService } from './app.service';
4   import { PersonModule } from './person/person.module';
5   import { UserController } from './user/user.controller';
6
7   @Module({
8     imports: [PersonModule],
9     controllers: [AppController, UserController],
10    providers: [AppService],
11  })
12  export class AppModule {}
13
```

▲ 圖 2-11 預設自動匯入相依的效果圖

由圖 2-11 可見，UserController 會自動匯入應用的主模組中，並自動將其增加到 Controllers 相依項列表中。同樣的流程也適用於生成 Service 服務和 module 模組。

然而，一個一個建立 Controller 或 Filters 可能不夠方便。能否一次性生成所需的範本呢？當然可以。Nest 提供了 nest generate resource 命令，可以一鍵生成程式範本。執行該命令後，CLI 會詢問我們選擇使用哪種程式風格。我們選擇 REST 風格的 API，如圖 2-12 所示。

```
$ nest g resource person
? What transport layer do you use? (Use arrow keys)
> REST API
  GraphQL (code first)
  GraphQL (schema first)
  Microservice (non-HTTP)
  WebSockets
```

▲ 圖 2-12 選擇不同風格的 API

選擇 yes 後，系統將自動生成與 CRUD（建立、讀取、更新、刪除）相關的程式，如圖 2-13 所示。

```
$ nest g resource person
? What transport layer do you use? REST API
? Would you like to generate CRUD entry points? Yes
CREATE src/person/person.controller.spec.ts (576 bytes)
CREATE src/person/person.controller.ts (925 bytes)
CREATE src/person/person.module.ts (254 bytes)
CREATE src/person/person.service.spec.ts (460 bytes)
CREATE src/person/person.service.ts (635 bytes)
CREATE src/person/dto/create-person.dto.ts (32 bytes)
CREATE src/person/dto/update-person.dto.ts (177 bytes)
CREATE src/person/entities/person.entity.ts (23 bytes)
UPDATE src/app.module.ts (444 bytes)
```

▲ 圖 2-13 選擇 CRUD 入口節點

建立完成後，生成 REST 風格的 API，內容如下：

```typescript
import { Controller, Get, Post, Body, Patch, Param, Delete } from '@nestjs/common';
import { PersonService } from './person.service';
import { CreatePersonDto } from './dto/create-person.dto';
import { UpdatePersonDto } from './dto/update-person.dto';

@Controller('person')
export class PersonController {
  constructor(private readonly personService: PersonService) {}

  @Post()
  create(@Body() createPersonDto: CreatePersonDto) {
    return this.personService.create(createPersonDto);
  }

  @Get()
  findAll() {
    return this.personService.findAll();
  }

  @Get(':id')
  findOne(@Param('id') id: string) {
    return this.personService.findOne(+id);
  }

  @Patch(':id')
  update(@Param('id') id: string, @Body() updatePersonDto: UpdatePersonDto) {
    return this.personService.update(+id, updatePersonDto);
  }

  @Delete(':id')
  remove(@Param('id') id: string) {
    return this.personService.remove(+id);
  }
}
```

dto 和 entities 是 CRUD 相關的程式，最終整合到 PersonModule 中，並被自動匯入 AppModule，如圖 2-14 所示。

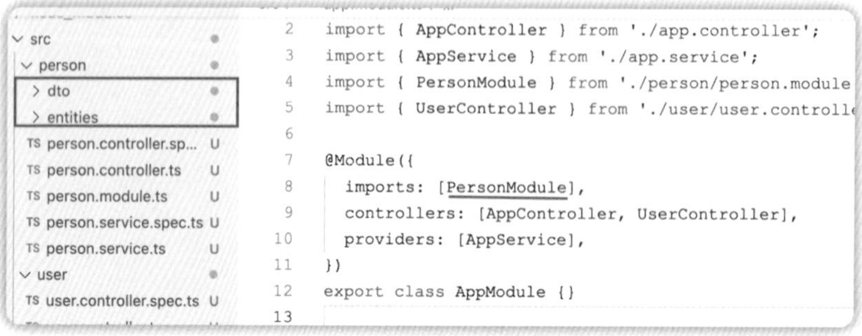

▲ 圖 2-14 CRUD 程式展示

以上就是 Nest CLI提供的用於快速建立專案程式的工具。

2.2.4 建構應用

前面介紹了 nest generate 命令，接下來使用 new build 命令來建構應用。

執行 nest build –h 命令，可以看到 build 命令提供了一些可選參數，如圖 2-15 所示。

```
$ nest build -h
Usage: nest build [options] [app]

Build Nest application.

Options:
  -c, --config [path]      Path to nest-cli configuration file.
  -p, --path [path]        Path to tsconfig file.
  -w, --watch              Run in watch mode (live-reload).
  -b, --builder [name]     Builder to be used (tsc, webpack, swc).
  --watchAssets            Watch non-ts (e.g., .graphql) files mode.
  --webpack                Use webpack for compilation (deprecated option, use --builder instead).
  --type-check             Enable type checking (when SWC is used).
  --webpackPath [path]     Path to webpack configuration.
  --tsc                    Use typescript compiler for compilation.
  --preserveWatchOutput    Use "preserveWatchOutput" option when using tsc watch mode.
  -h, --help               Output usage information.
```

▲ 圖 2-15 build 命令提供的參數

其中，各選項說明如下。

- --path：用於指定 tsconfig 檔案的路徑。
- --watch：開啟即時監聽模式，在檔案發生變化時自動執行建構操作。

- --builder：選擇使用指定的工具進行建構，可選的工具包括 tsc、webpack、swc 等。

預設情況下，Nest 使用 tsc 進行編譯，執行 nest build 命令的效果如圖 2-16 所示。

若要切換為使用 webpack 進行打包，可以執行 nest build -b webpack 命令，效果如圖 2-17 所示。

▲ 圖 2-16 建構之後的目錄檔案　　▲ 圖 2-17 webpack 打包後的目錄檔案

--webpack 和 --tsc 分別指定了不同的編譯器，webpack 用於編譯和打包，而 tsc 只用於編譯，它們的執行效果與前文介紹的類似。

每次都需要在命令後面增加參數，這樣有點麻煩，是否可以將這些參數寫入設定檔中進行管理呢？答案是可以的。

- --config：指定 nest-cli 的設定檔路徑，即 nest-cli.json 檔案，可以用來配置打包參數，如圖 2-18 所示。

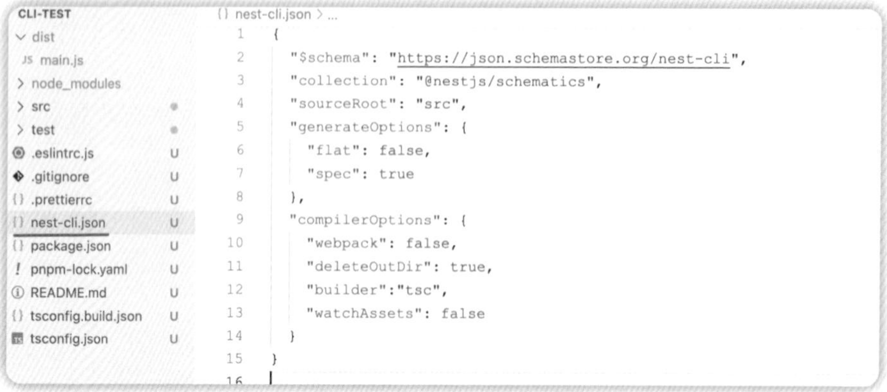

```
CLI-TEST                      {} nest-cli.json > ...
∨ dist                    1   {
  JS main.js              2     "$schema": "https://json.schemastore.org/nest-cli",
> node_modules            3     "collection": "@nestjs/schematics",
> src                     4     "sourceRoot": "src",
> test                    5     "generateOptions": {
  .eslintrc.js      U     6       "flat": false,
  .gitignore        U     7       "spec": true
  {} .prettierrc    U     8     },
  {} nest-cli.json  U     9     "compilerOptions": {
  {} package.json   U    10       "webpack": false,
  ! pnpm-lock.yaml  U    11       "deleteOutDir": true,
  ① README.md       U    12       "builder":"tsc",
  {} tsconfig.build.json U 13      "watchAssets": false
  {} tsconfig.json  U    14     }
                         15   }
                         16
```

▲ 圖 2-18 nest-cli 的設定檔

由圖 2-18 可以看到，前面提到的 flat、spec 都可以在這裡進行配置。編譯成功選項也可以指定是否使用 webpack 進行建構，以及是否使用 builder 指定選擇的編譯器等。

2.2.5 啟動開發偵錯

本小節將用 nest start 命令來啟動開發偵錯，它的可選參數如圖 2-19 所示。

```
$ nest start -h
Usage: nest start [options] [app]

Run Nest application.

Options:
  -c, --config [path]         Path to nest-cli configuration file.
  -p, --path [path]           Path to tsconfig file.
  -w, --watch                 Run in watch mode (live-reload).
  -b, --builder [name]        Builder to be used (tsc, webpack, swc).
  --watchAssets               Watch non-ts (e.g., .graphql) files mode.
  -d, --debug [hostport]      Run in debug mode (with --inspect flag).
  --webpack                   Use webpack for compilation (deprecated option, use --builder instead).
  --webpackPath [path]        Path to webpack configuration.
  --type-check                Enable type checking (when SWC is used).
  --tsc                       Use typescript compiler for compilation.
  --sourceRoot [sourceRoot]   Points at the root of the source code for the single project in standard mode str
  --entryFile [entryFile]     Path to the entry file where this command will work with. Defaults to the one def
  -e, --exec [binary]         Binary to run (default: "node").
  --preserveWatchOutput       Use "preserveWatchOutput" option when using tsc watch mode.
  -h, --help                  Output usage information.
```

▲ 圖 2-19 start 命令提供的參數

在開發階段，nest start 命令用於開啟本機服務。執行該命令的結果如圖 2-20 所示。

```
$ nest start
[Nest] 86023  - 2023/12/29 16:35:58     LOG [NestFactory] Starting Nest application...
[Nest] 86023  - 2023/12/29 16:35:58     LOG [InstanceLoader] AppModule dependencies initialized +7ms
[Nest] 86023  - 2023/12/29 16:35:58     LOG [InstanceLoader] PersonModule dependencies initialized +0ms
[Nest] 86023  - 2023/12/29 16:35:58     LOG [RoutesResolver] AppController {/}: +5ms
[Nest] 86023  - 2023/12/29 16:35:58     LOG [RouterExplorer] Mapped {/, GET} route +1ms
[Nest] 86023  - 2023/12/29 16:35:58     LOG [RoutesResolver] UserController {/user}: +0ms
[Nest] 86023  - 2023/12/29 16:35:58     LOG [RoutesResolver] PersonController {/person}: +0ms
[Nest] 86023  - 2023/12/29 16:35:58     LOG [RouterExplorer] Mapped {/person, POST} route +0ms
[Nest] 86023  - 2023/12/29 16:35:58     LOG [RouterExplorer] Mapped {/person, GET} route +1ms
[Nest] 86023  - 2023/12/29 16:35:58     LOG [RouterExplorer] Mapped {/person/:id, GET} route +0ms
[Nest] 86023  - 2023/12/29 16:35:58     LOG [RouterExplorer] Mapped {/person/:id, PATCH} route +0ms
[Nest] 86023  - 2023/12/29 16:35:58     LOG [RouterExplorer] Mapped {/person/:id, DELETE} route +0ms
[Nest] 86023  - 2023/12/29 16:35:58     LOG [NestApplication] Nest application successfully started +1ms
```

▲ 圖 2-20 nest start 執行結果

--debug 參數用來偵錯。執行 nest start-d 命令後會啟動一個 WebSocket 偵錯服務，透過偵錯工具連結到這個通訊埠即可進行偵錯，如圖 2-21 所示。

```
$ nest start -d
Debugger listening on ws://127.0.0.1:9229/c40562b2-0028-45a2-9fa6-7cee4518e195
For help, see: https://nodejs.org/en/docs/inspector
Debugger attached.
[Nest] 86113  - 2023/12/29 16:38:50     LOG [NestFactory] Starting Nest application...
[Nest] 86113  - 2023/12/29 16:38:50     LOG [InstanceLoader] AppModule dependencies initialized +9ms
[Nest] 86113  - 2023/12/29 16:38:50     LOG [InstanceLoader] PersonModule dependencies initialized +0ms
[Nest] 86113  - 2023/12/29 16:38:52     LOG [RoutesResolver] AppController {/}: +2710ms
[Nest] 86113  - 2023/12/29 16:38:52     LOG [RouterExplorer] Mapped {/, GET} route +3ms
[Nest] 86113  - 2023/12/29 16:38:52     LOG [RoutesResolver] UserController {/user}: +0ms
[Nest] 86113  - 2023/12/29 16:38:52     LOG [RoutesResolver] PersonController {/person}: +0ms
[Nest] 86113  - 2023/12/29 16:38:52     LOG [RouterExplorer] Mapped {/person, POST} route +1ms
[Nest] 86113  - 2023/12/29 16:38:52     LOG [RouterExplorer] Mapped {/person, GET} route +0ms
[Nest] 86113  - 2023/12/29 16:38:52     LOG [RouterExplorer] Mapped {/person/:id, GET} route +1ms
[Nest] 86113  - 2023/12/29 16:38:52     LOG [RouterExplorer] Mapped {/person/:id, PATCH} route +0ms
[Nest] 86113  - 2023/12/29 16:38:52     LOG [RouterExplorer] Mapped {/person/:id, DELETE} route +0ms
[Nest] 86113  - 2023/12/29 16:38:52     LOG [NestApplication] Nest application successfully started +3ms
```

▲ 圖 2-21 帶 --debug 參數執行 nest 命令的執行結果

詳細的偵錯內容及技巧將在 2.6 節介紹。其他配置與 build 命令類似，這裡不再贅述。

2.2.6 查看專案資訊

nest info 命令用於查看 Node.js、npm 以及 Nest 相依套件的相關版本資訊，如圖 2-22 所示。

```
$ nest info
 _   _             _      ___  _     ___
| \ | |           | |    / __|| |   |_ _|
|  \| | ___  ___ | |_  | |   | |    | |
| . ` |/ _ \/ __|| __| | |   | |    | |
| |\  |  __/\__ \| |_  | |__ | |___ | |
|_| \_|\___||___/ \__|  \___||_____||___|

[System Information]
OS Version     : macOS Big Sur
NodeJS Version : v16.20.1
PNPM Version   : 8.6.12

[Nest CLI]
Nest CLI Version : 10.0.0

[Nest Platform Information]
platform-express version : 10.0.0
mapped-types version     : 0.0.1
schematics version       : 10.0.0
passport version         : 10.0.2
typeorm version          : 10.0.0
testing version          : 10.0.0
common version           : 10.0.0
core version             : 10.0.0
cli version              : 10.0.0
```

▲ 圖 2-22 nest info 命令的執行結果

2.3 建立第一個 Nest 應用

了解了 Nest CLI 的使用之後，本節我們來小試牛刀，建立一個服務端應用，並使用 React 建構一個用戶端應用。我們將實現用戶端發送請求後，服務端接收請求、進行資料處理，並傳回新的資料給用戶端。如果你喜歡使用 Vue，也可以選擇透過 Vue 來建立專案；但請注意，本案例不涉及複雜的前端互動。

2.3.1 生成後端專案

首先，執行 nest n web-app -p pnpm 命令生成服務端專案，如圖 2-23 所示。

```
$ nest n web-app -p pnpm
⚡ We will scaffold your app in a few seconds..

CREATE web-app/.eslintrc.js (663 bytes)
CREATE web-app/.prettierrc (51 bytes)
CREATE web-app/README.md (3347 bytes)
CREATE web-app/nest-cli.json (171 bytes)
CREATE web-app/package.json (1948 bytes)
CREATE web-app/tsconfig.build.json (97 bytes)
CREATE web-app/tsconfig.json (546 bytes)
CREATE web-app/src/app.controller.spec.ts (617 bytes)
CREATE web-app/src/app.controller.ts (274 bytes)
CREATE web-app/src/app.module.ts (249 bytes)
CREATE web-app/src/app.service.ts (142 bytes)
CREATE web-app/src/main.ts (208 bytes)
CREATE web-app/test/app.e2e-spec.ts (630 bytes)
CREATE web-app/test/jest-e2e.json (183 bytes)

✔ Installation in progress... 🐈

✔ Successfully created project web-app
```

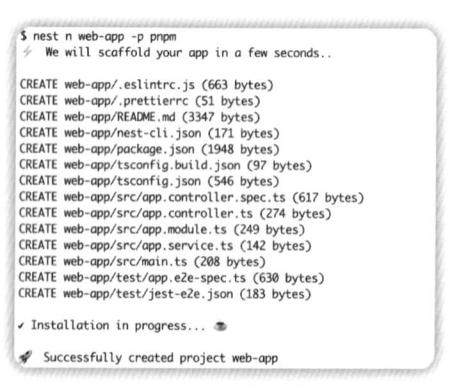

▲ 圖 2-23 建立服務端專案

然後，啟動該專案並保持熱多載，如圖 2-24 所示。

```
$ nest start --watch
[16:59:47] Starting compilation in watch mode...

[16:59:48] Found 0 errors. Watching for file changes.

[Nest] 86452  - 2023/12/29 16:59:48     LOG [NestFactory] Starting Nest application...
[Nest] 86452  - 2023/12/29 16:59:48     LOG [InstanceLoader] AppModule dependencies initialized +5ms
[Nest] 86452  - 2023/12/29 16:59:48     LOG [RoutesResolver] AppController {/}: +6ms
[Nest] 86452  - 2023/12/29 16:59:48     LOG [RouterExplorer] Mapped {/, GET} route +1ms
[Nest] 86452  - 2023/12/29 16:59:48     LOG [NestApplication] Nest application successfully started +1ms
```

▲ 圖 2-24 執行服務端專案

2.3.2 生成前端專案

使用 React 提供的鷹架 create-react-app 生成專案，先全域安裝一下：

```
npm install -g create-react-app
```

查看是否安裝成功：

```
create-react-app --version
```

結果如圖 2-25 所示。

```
$ create-react-app --version
5.0.1
```

▲ 圖 2-25 檢查是否安裝成功

執行 create-react-app web-app-front 命令，建立名為 web-app-front 的專案。建立成功後的效果如圖 2-26 所示。

```
Inside that directory, you can run several commands:

  npm start
    Starts the development server.

  npm run build
    Bundles the app into static files for production.

  npm test
    Starts the test runner.

  npm run eject
    Removes this tool and copies build dependencies, configuration files
    and scripts into the app directory. If you do this, you can't go back!

We suggest that you begin by typing:

  cd web-app-front
  npm start

Happy hacking!
```

▲ 圖 2-26 建立用戶端專案

同樣，啟動該專案，執行 npm start 命令，結果如圖 2-27 所示。

```
Compiled successfully!

You can now view web-app-front in the browser.

  Local:            http://localhost:3000
  On Your Network:  http://192.168.6.49:3000

Note that the development build is not optimized.
To create a production build, use npm run build.

webpack compiled successfully
```

▲ 圖 2-27　行用戶端專案

2.3.3 準備工作

在發送請求之前，我們需要進行一些準備工作。首先，在前端設置請求代理，以便能夠存取 8088 通訊埠下的服務。接著，透過 axios 發送一個 GET 請求，成功後將資料顯示在頁面上。

為了設置請求代理，在目錄 web-app-front/src 下新建 setupProxy.js 檔案，如圖 2-28 所示。

```
∨ WEB-APP-F...
  > node_modules
  > public
  ∨ src
    # App.css
    JS App.js                    M
    JS App.test.js
    # index.css
    JS index.js
    🖼 logo.svg
    JS reportWebVitals.js
    JS setupProxy.js             U
    JS setupTests.js
    ♦ .gitignore
    {} package-lock.json
```

▲ 圖 2-28　新建 setupProxy 檔案

設置代理邏輯，這裡將通訊埠自訂為 8088，程式如下：

```js
// 引入 http-proxy-middleware，react 鷹架已經安裝
const proxy = require("http-proxy-middleware");

module.exports = function (app) {
  app.use(
    // 遇見 /api 首碼的請求就會觸發該代理配置
    proxy.createProxyMiddleware("/api", {
```

```
      // 請求轉發給誰
      target: "http://localhost:8088",
      // 控制伺服器收到的請求標頭中 Host 的值
      changeOrigin: true,
      // 重寫入請求路徑
      pathRewrite: { "^/api": "" },
    })
  );
};
```

執行 npm install axios 命令來安裝 axios 函數庫。然後在 App.js 檔案中引入 axios，並定義一個名為 getHello 的 API 介面，用於請求 /123 路徑的資源。以下是範例程式：

```
import axios from 'axios';

function getHello() {
  return axios.get('/api/123');
}
```

透過點擊按鈕觸發 AJAX 請求，並回顯接收到的資料。以下是部分範例程式：

```
// 頭部匯入 useState
import { useState } from "react";

// App 方法
function App() {
  const [nestData, setNestData] = useState({});

  const handleGetData = async () => {
    const { data } = await getHello();
    setNestData(data);
  };

  const handleCreateUser = async () => {
    const params = {
      name: "mouse",
      age: "22",
      address: "廣州",
    };
    const { data } = await createUser(params);
    setNestData(data);
  };

  return (
    <div className="App">
      <header className="App-header">
        <img src={logo} className="App-logo" alt="logo" />
        <button onClick={handleGetData}>發送請求給 Nest</button>
```

```
      <br />
      <button onClick={handleCreateUser}>建立使用者</button>
      <br />
      後端傳回資料：{JSON.stringify(nestData)}
    </header>
  </div>
);
}
```

切換到 web-app 服務端，在 main.ts 檔案中將通訊埠修改為 8088，以匹配前端代理通訊埠，程式如下：

```
async function bootstrap() {
  const app = await NestFactory.create(AppModule);
  // 修改通訊埠為 8088
  await app.listen(8088);
}
bootstrap();
```

接下來，在 app.controller.ts 檔案的 getHello 方法中接收參數 id，同時呼叫 Service 服務的 getHello() 方法並傳遞參數 id，程式如下：

```
@Controller()
export class AppController {
  constructor(private readonly appService: AppService) {}

  @Get(':id')
  getHello(@Param('id') id: string): Person {
    return this.appService.getHello(id);
  }
}
```

最終 getHello() 方法會傳回新的 JSON 資料：

```
@Injectable()
export class AppService {
  getHello(id: string): Person {
    return {
      name: 'mouse',
      age: 22,
      hello: '你好啊，mouse',
      desc: '我是用戶端請求的 id:${id}'
    };
  }
}
```

2.3.4 執行結果

前後端互動邏輯完成之後，在用戶端點擊按鈕發送請求，可以從網路面板看到正常傳回的資料並在頁面上顯示，如圖 2-29 所示。

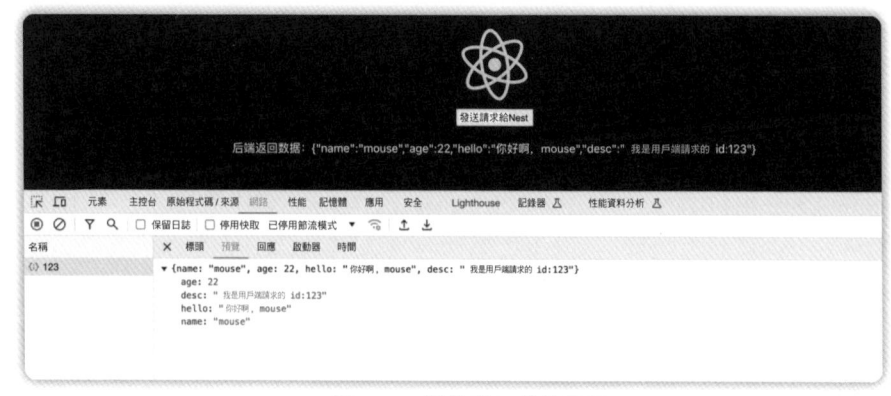

▲ 圖 2-29 前後端互動結果圖

至此，一個簡易的前後端互動流程就完成了。

2.3.5 模組化開發

本小節將把請求方式改為 POST，並最佳化專案結構。透過模組化來管理請求 API。以 user 模組為例，執行「nest g resource user」命令來生成 user 模組。接下來，在 user.controller.ts 中修改請求路徑，範例程式如下：

```
@Controller('user')
export class UserController {
  constructor(private readonly userService: UserService) {}

  @Post('/create-user')
  create(@Body() createUserDto: CreateUserDto) {
    return this.userService.create(createUserDto);
  }
}
```

同時，在 user.service.ts 中列印並傳回提示訊息：

```
@Injectable()
export class UserService {
  create(createUserDto: CreateUserDto) {
    console.log('createUserDto', createUserDto);
    return {
      status: true,
      msg: '建立成功'
    };
  }
}
```

接下來，同步修改 web-app-front 前端專案的 App.js，新增 createUser 請求介面和 handleCreateUser 使用者事件，程式如下：

```js
function getHello() {
  return axios.get('/api/123');
}
// 新增 createUser 請求介面函數
function createUser(data) {
  return axios.post('/api/user/create-user', data)
}

function App() {
  const [ nestData, setNestData] = useState({})

  const handleGetData = async() => {
    const { data } = await getHello()
    setNestData(data)
  }
  // 新增建立使用者頁面事件
  const handleCreateUser = async() => {
    const params = {
      name: 'mouse',
      age: '22',
      address: '廣州'
    }
    const { data } = await createUser(params)
    setNestData(data)
  }

  return (
    <div className="App">
      <header className="App-header">
        <img src={logo} className="App-logo" alt="logo" />
        <button onClick={handleGetData}>發送請求給 Nest</button>
        <br />
        <button onClick={handleCreateUser}>建立使用者</button>
        <br />
        後端傳回資料：{ JSON.stringify(nestData) }
      </header>
    </div>
  );
}
```

點擊頁面上的「建立使用者」按鈕，後端會正常接收到請求參數，並向用戶端發送回應，如圖 2-30 所示。

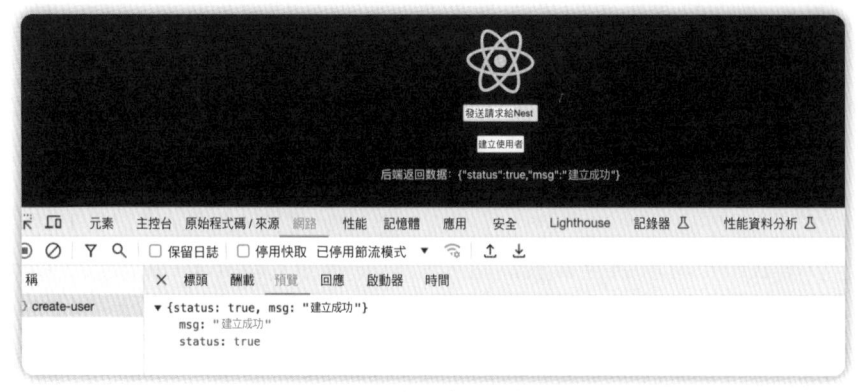

▲ 圖 2-30 建立使用者傳回結果

值得注意的是，在請求介面中，/user/* 後面的路由對應於 User 控制器下的各類介面方法。這樣的設計具有以下優點：它提供了清晰的路由層次結構，並且可以方便地實現對控制器層面的許可權管理和攔截器操作。這些內容將在後續章節中詳細講解。

透過本節的學習，一個完整的前後端分離的 Nest 應用已經建構完成。現在，你已經具備了建立自己的 Nest 應用的能力。趕快去實踐，將所學知識應用到實際專案中吧！

2.4 Nest 的 AOP 架構理念

Nest 採用了多種優秀的設計模式和軟體設計思想，其中之一就是切面程式設計（AOP）。AOP 導向的引入解決了傳統物件導向程式設計中的一些問題，比如程式重複和業務邏輯的混雜。作為 Nest 最核心的設計理念之一，本節將深入學習 AOP 的工作機制及其應用。

2.4.1 MVC 架構概述

Nest 屬於 MVC（Model-View-Controller，模型 - 視圖 - 控制器）架構系統，實際上，大多數後端框架都是基於這一架構設計的。在 MVC 架構中：

- Model 層負責業務邏輯處理，包括資料的獲取、儲存、驗證以及資料庫操作。

- Controller 層通常用於處理使用者的輸入，排程 Service 服務，以及進行 API 的路由管理。
- View 層在傳統的伺服器端著色中，可能使用如 ejs、hbs 等範本引擎。在前後端分離的系統中，通常指的是用戶端框架（如 Vue 或 React）負責的部分。

當一個 HTTP 請求到達伺服器時，它首先會被 Controller 層接收。Controller 層會根據請求呼叫 Model 層中的相應模組來處理業務邏輯，並將處理結果傳回給 View 層以進行展示。整個基礎流程的示意圖如圖 2-31 所示。

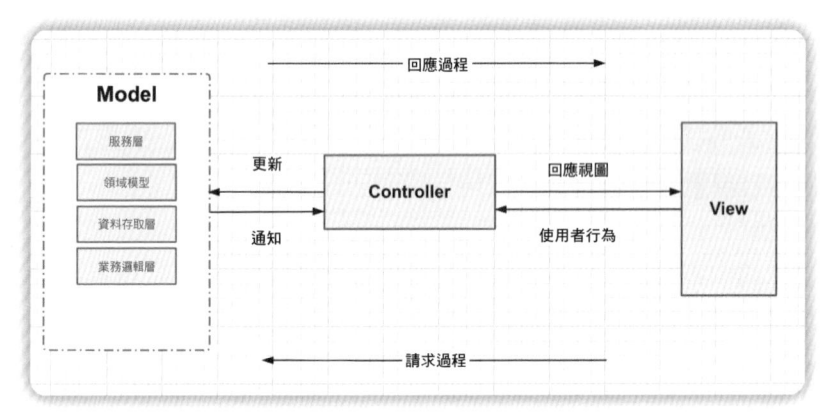

▲ 圖 2-31 HTTP 請求基本流程

在 MVC 架構的基礎上，Nest 還引入了 AOP 的思想，從而具備了切面程式設計導向的能力。我們經常聽到後端開發者提到 AOP 切面，那麼究竟什麼是面向切面程式設計呢？接下來將舉出答案。

2.4.2 AOP 解決的問題

以一個 HTTP 請求為例，用戶端發送請求時首先會經過 Controller（控制器）、Service（服務）、DB（資料存取或操作）等模組。如果想要在這些模組中加入一些操作，例如資料驗證、許可權驗證或日誌統計，應該怎麼辦呢？

首先，我們可能會想到在 Controller 中加入參數驗證邏輯或許可權驗證，如果不通過驗證，就直接傳回錯誤。這樣看起來似乎沒什麼問題。但是，如果有多個功能模組都需要進行驗證，並且許可權驗證的邏輯相同，那麼是不是表示需要在多個 Controller 中重複加入這段邏輯？顯然，這樣做會導致公共邏輯與業務邏輯耦合。有沒有辦法可以做到統一管理呢？

答案是有的，辦法如圖 2-32 所示。

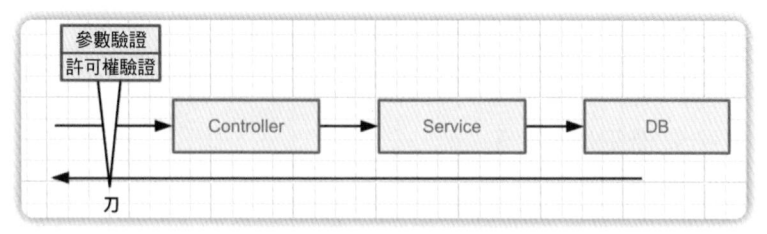

▲ 圖 2-32 切面工作原理圖

由圖 2-32 可見，在 Controller 的前後都可以「切一刀」，用來統一處理公共邏輯，這樣就不會侵入 Controller、Service 等業務程式。

事實上，在 Nest 中，請求流程可以換一種角度來看，如圖 2-33 所示。

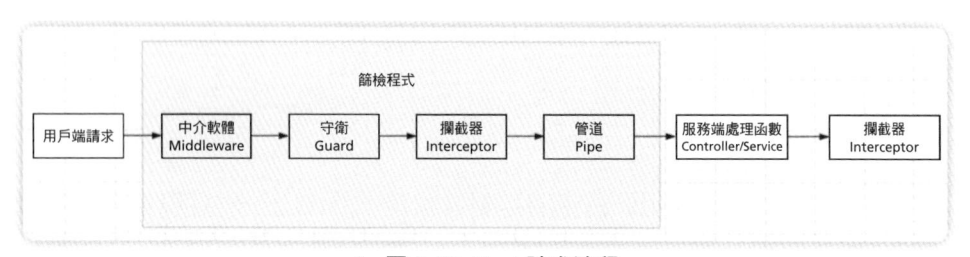

▲ 圖 2-33 Nest 請求流程

中間的灰色區域屬於 AOP 切面部分，包含 Middleware（中介軟體）、Guard（守衛）、Interceptor（攔截器）、Pipe（管道）和 Filter（篩檢程式）。它們都是 AOP 思想的具體實現。

2.4.3 AOP 在 Nest 中的應用

前面我們了解了 AOP 思想的優勢，接下來看看 AOP 在 Nest 中的應用。這裡將以 store-web 專案作為演示，程式已經放到 GitHub 對應章節的目錄下。

1. 中介軟體

Nest 的中介軟體預設是基於 Express 的，它在請求流程中的位置如圖 2-34 所示。

▲ 圖 2-34 中介軟體的位置

　　中介軟體可以在路由處理常式之前或之後插入執行任務，它們分為全域中介軟體和局部中介軟體兩種類型。

　　全域中介軟體透過 use 方法呼叫，與 Express 中的使用方式類似。所有進入應用的請求都會經過全域中介軟體，通常用於執行日誌統計、監控、安全性處理等任務。舉例來說，在 main.ts 檔案中使用全域中介軟體的方式如下，其中 LoggerMiddleware 是一個用於日誌統計的中介軟體。範例程式如下：

```
async function bootstrap() {
  const app = await NestFactory.create(AppModule);
  // 中介軟體
  app.use(new LoggerMiddleware().use)
  // 啟動服務
  await app.listen(8088);
}
bootstrap();
```

　　局部中介軟體通常應用於特定的控制器或單一路由上，以實現更細粒度的邏輯控制。舉例來說，可以將 LoggerMiddleware 綁定到 /person/create-person 路由，並指定其適用的 HTTP 請求方法。以下是如何在控制器中應用局部中介軟體的範例程式：

```
@Injectable()
// person 模組
export class PersonModule implements NestModule{
  configure(consumer: MiddlewareConsumer) {
      consumer.apply(LoggerMiddleware).forRoutes({
          // 指定路由路徑
          path: '/person/create-Person',
          // 指定 HTTP 請求方式
       method: RequestMethod.Post
        })
    }
}
```

　　對於中介軟體的更深入的知識，我們將在後續章節中進行詳細介紹。

2. 守衛

守衛的職責很明確，通常用於許可權、角色等授權操作。守衛所在的位置與中介軟體類似，可以對請求進行攔截和過濾，如圖 2-35 所示。

▲ 圖 2-35 守衛的位置

守衛在呼叫路由程式之前傳回 true 或 false 來判斷是否通行，分為全域守衛和局部守衛。

守衛必須實現 CanActivate 介面中的 canActivate() 方法，程式如下：

```
@Injectable()
export class PersonGuard implements CanActivate {
  canActivate(
    context: ExecutionContext,
  ): boolean | Promise<boolean> | Observable<boolean> {
    console.log(' 進入守衛 ');
    // 通常根據 ExecutionContext 資訊來判斷許可權，傳回 true/false，表示放行或禁止通行
    return true;
  }
}
```

全域守衛在 main.ts 中透過 useGlobalGuards 來呼叫，每個路由程式都會經過它進行許可權驗證才能夠通行。

```
async function bootstrap() {
  const app = await NestFactory.create(AppModule);
  // 守衛
  app.useGlobalGuards(new PersonGuard())
  // 啟動服務
  await app.listen(8088);
}
```

同樣，與中介軟體類似，作為局部守衛，可以縮小控制範圍，從而實現更加精細的許可權控制。控制器中的範例程式如下：

```
@Controller('person')
// 宣告守衛
@UseGuards(new PersonGuard())
// 控制器
export class PersonController {}
```

3. 攔截器

攔截器不同於中介軟體和守衛，它在路由請求之前和之後都可以進行邏輯處理，能夠充分操作 request 和 response 物件，如圖 2-36 所示。攔截器通常用於記錄請求日誌、轉換或格式化回應資料等。

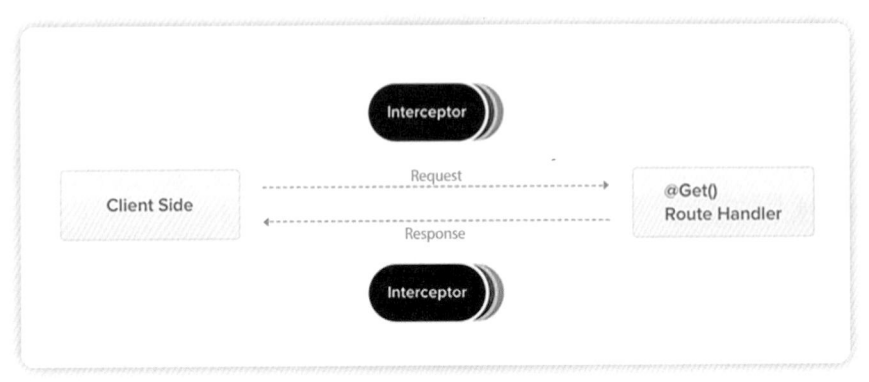

▲ 圖 2-36 攔截器的位置

為了更進一步地說明攔截器的作用，下面的程式定義了一個用於統計介面逾時的攔截器：

```
import { CallHandler, ExecutionContext, Injectable, NestInterceptor } from "@nestjs/common";
import { log } from "console";
import { Observable, tap, timeout } from "rxjs";

// 統計介面逾時的攔截器
@Injectable()
export class TimeoutInterceptor implements NestInterceptor{
  intercept(
    context: ExecutionContext,
    next: CallHandler
  ): Observable<any>{
    log(' 進入攔截器 ', context.getClass());
    const now = Date.now();
    return next.handle().pipe(
      tap(() => {
        log('Timeout: ', Date.now() - now)
      }),
      timeout({
        each: 10
      })
    );
  }
}
```

在上述程式中，攔截器必須實現 NestInterceptor 介面中的 intercept 方法。該方法包含與守衛相同的 ExecutionContext 上下文物件作為第一個參數。第二個參數是 CallHandler，它代表每個路由處理常式的方法。只有當攔截器呼叫了 handle 方法後，控制權才會交給路由處理常式。在 Nest 中，攔截器通常與 RxJS 非同步處理函數庫一起使用，以執行一些非同步邏輯，例如上例中的逾時（timeout）統計。

類似於守衛，攔截器可以設置為控制器作用域、方法作用域或全域作用域。

（1）控制器作用域允許攔截器只作用於某個控制器。當程式執行到控制器時觸發攔截器邏輯，透過 @UseInterceptors 裝飾器將 TimeoutInterceptor 綁定到控制器類別，程式如下：

```
@Controller('person')
// 為控制器綁定逾時攔截器
@UseInterceptors(new TimeoutInterceptor())

export class PersonController {}
```

（2）方法作用域把攔截器的作用範圍限制在某個方法上，比全域或控制器等級的範圍更加精確。當程式執行到該方法時，觸發攔截器的邏輯。綁定方式如下：

```
@Get()
// 為單獨的方法綁定逾時攔截器
@UseInterceptors(new TimeoutInterceptor())
findAll() {
  return this.personService.findAll();
}
```

（3）全域作用域允許攔截器應用到整個應用中。在 main.ts 檔案中，可以透過 app.useGlobalInterceptors() 方法進行綁定，程式如下：

```
async function bootstrap() {
  const app = await NestFactory.create(AppModule);
  // 全域逾時攔截器
  app.useGlobalInterceptors(new TimeoutInterceptor());
  // 啟動服務
  await app.listen(8088);
}
bootstrap();
```

4. 管道

管道用於處理通用邏輯，其中兩個典型的用例是處理請求參數的驗證

（validation）和轉換（transformation）。在執行路由方法之前，會首先執行管道邏輯，並將經過管道轉換後的參數傳遞給路由方法。它的位置如圖 2-37 所示。

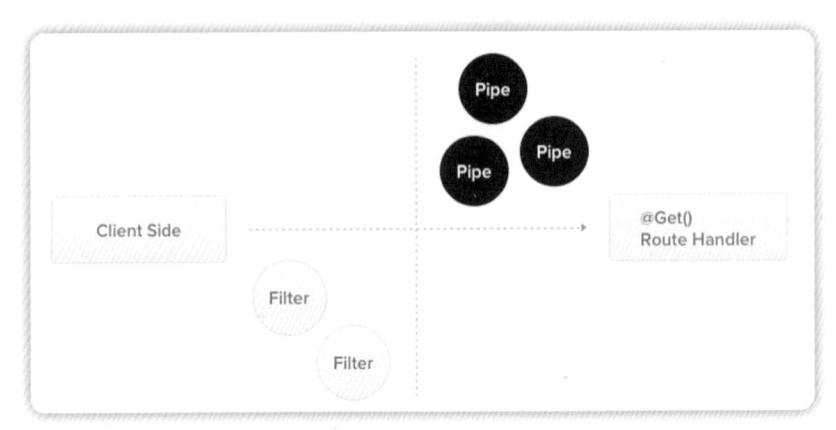

▲ 圖 2-37 管道的位置

儘管 Nest 框架中已經內建了一些管道（Pipes），例如 ParseIntPipe，但有時我們可能需要實現特定的轉換功能，比如將數字轉為八進制。在這種情況下，我們需要建立自訂管道。以下是一個自訂管道的範例程式：

```
import { ArgumentMetadata, BadRequestException, Injectable, PipeTransform } from "@nestjs/common";

@Injectable()
export class ParseIntPipe implements PipeTransform<string, number> {
  transform(value: string, metadata: ArgumentMetadata): number {
    console.log('進入自訂管道');
    // 轉為八進制
    const val = parseInt(value, 8);
    if (isNaN(val)) {
      throw new BadRequestException('Validation failed');
    }
    return val;
  }
}
```

自訂管道需要實現 PipeTransform 介面的 transform() 方法。其中，value 表示需要處理的方法參數，而 metadata 則是描述該參數的中繼資料，用於標識其屬於哪種類型，如 'body'、'query'、'param' 和 'custom'。

除了自訂管道和內建管道外，通常我們還會結合第三方驗證函數庫，例如 class-validator。在後面的章節中將詳細介紹這部分內容。

如果管道驗證器驗證失敗，則需要拋出例外並回應給用戶端。這時就需要使用異常篩檢程式來處理。

5. 篩檢程式

Nest 中最為常見的是 HTTP 異常篩檢程式，通常用於在後端服務發生異常時向用戶端報告異常的類型。目前內建的 HTTP 異常包含：

- BadRequestException
- UnauthorizedException
- NotFoundException
- ForbiddenException
- NotAcceptableException
- RequestTimeoutException
- ConflictException
- GoneException
- HttpVersionNotSupportedException
- PayloadTooLargeException
- UnsupportedMediaTypeException
- ...

它們都繼承自 HttpException 類別。當然，我們也可以自訂異常篩檢程式，並向前端傳回統一的資料格式：

```typescript
import { ArgumentsHost, Catch, ExceptionFilter, HttpException } from "@nestjs/common";
import { Request, Response } from "express";

@Catch(HttpException)
export class HttpExceptionFilter implements ExceptionFilter {
    catch(exception: HttpException, host: ArgumentsHost) {
        // 指定傳輸協定
        const ctx = host.switchToHttp();
        // 獲取請求回應物件
        const response = ctx.getResponse<Response>();
        const request = ctx.getRequest<Request>();
        // 獲取狀態碼和異常訊息
        const status = exception.getStatus();
        const message = exception.message;
        let resMessage: string | Record<string, any> = exception.getResponse();
        console.log(' 進入異常篩檢程式 ');
```

```
    if (typeof resMessage === 'object') {
        resMessage = resMessage.message
    }
    // 統一組裝傳回格式
    response.status(status).json({
        message: resMessage || message,
        success: false,
        path: request.url,
        status
    });
  }
}
```

　　自訂異常篩檢程式需要實現 ExceptionFilter 介面的 catch() 方法來攔截異常。ArgumentsHost 能夠獲取不同平臺的傳輸協定上下文，用於存取 request 和 response 物件。

　　另外，@Catch() 裝飾器用於宣告要攔截的異常類型，這裡使用的是 HttpException。異常篩檢程式可應用於控制器作用域、方法作用域和全域作用域，我們都會嘗試一遍。

（1）將篩檢程式綁定到控制器，程式如下：

```
@Controller('person')
// 綁定到控制器
@UseFilters(new HttpExceptionFilter())
export class PersonController {}
```

（2）將篩檢程式綁定到某個路由方法中，程式如下：

```
@Get()
// 綁定到路由方法
@UseFilters(new HttpExceptionFilter())
findAll() {
    return this.personService.findAll();
}
```

（3）將篩檢程式綁定到全域中，程式如下：

```
async function bootstrap() {
    const app = await NestFactory.create(AppModule);
    // 全域異常篩檢程式
    app.useGlobalFilters(new HttpExceptionFilter());
    // 啟動服務
    await app.listen(8088);
}
bootstrap();
```

以上就是 Nest 實現 AOP 架構思想的幾種方式。不同的切面解決不同場景下的通用邏輯抽離問題，從而實現了更加靈活和可維護的應用程式。

2.5 IoC 思想解決了什麼問題

在 2.4 節中，我們介紹了 AOP（切面程式設計）架構理念導向的基本概念及其在 Nest 中的應用方式。本節將介紹 Nest 中的另一種核心設計思想—IoC（控制反轉）思想。IoC 思想貫穿於整個 Nest 應用的開發過程。接下來，讓我們一起學習 IoC 如何解決問題，並學習在 Nest 中如何應用 IoC。

2.5.1 IoC 核心思想概述

在後端系統中，通常包含以下幾個組件。

- Controller（控制器）：負責接收用戶端的請求。
- Service（服務）：處理業務邏輯。
- Dao（資料存取物件）或 Repository（倉庫）：負責對資料執行增刪改查（CRUD）操作。
- DataSource（資料來源）：根據配置資訊連接和管理資料庫。

這表示在開發過程中，你需要按照合適的順序建立這些元件。例如：

```
const dataSource = new DataSource(config)
const dao = new Dao(dataSource)
const Service = new Service(dao)
const Controller = new Controller(Service)
```

在後端架構中，當多個 Controller 模組呼叫同一個 Service 類別時，我們希望確保它們使用的是同一個實例，即維持單例模式。在大型應用中，手動管理這種依賴關係可能會變得複雜，這是傳統後端開發經常面臨的挑戰。

幸運的是，IoC（Inverse Of Control，控制反轉）提供了一種解決方案。IoC 容器在應用初始化時，會查詢每個類別上宣告的相依，並按順序建立相應的實例，然後管理這些實例。當需要使用某個相依時，IoC 容器會提供相應的物件實例。

相依注入（DI）是實現 IoC 的一種常見方式。為什麼稱之為「控制反轉」呢？讓我們透過一個生活化的比喻來說明：

想像一下，通常我們做飯之前需要準備各種食材，有時還要在市場上討價還價，回家後要清洗、切割食材，烹飪時還要考慮食材的下鍋順序等，這個過程相當煩瑣。那麼，有沒有更簡單的方法呢？

答案是肯定的。如果我們的目的是為了解決饑餓問題，為什麼不選擇去餐廳就餐呢？我們只需告訴服務員：「請給我來一份番茄炒蛋飯。」

此時，後廚接到通知，控制權已經轉移。廚師們會根據選單要求，有序地處理後廚的所有事務，並最終為我們提供一份美味的「番茄炒蛋飯」。

在這個比喻中，後廚相當於 IoC 容器，選單相當於在類別上宣告的相依，服務員則相當於相依注入的過程。這樣，IoC 容器會根據類別上宣告的相依來建立和管理物件。

透過 IoC，我們從主動建立和維護物件轉變為被動等待相依注入，實現了從主動下廚到等待服務員上菜的轉變，這就是 IoC 控制反轉的精髓。

2.5.2 IoC 在 Nest 中的應用

本小節將介紹在 Nest 中如何實現 IoC。首先，透過執行 nest n ioc-test 命令新建一個專案，並在建立過程中選擇 pnpm 作為套件管理器，如圖 2-38 所示。

▲ 圖 2-38 建立 ioc-test 專案

進入該目錄，執行 pnpm start:dev 命令啟動服務。服務啟動成功後的介面如圖 2-39 所示。

```
$ pnpm start:dev

> ioc-test@0.0.1 start:dev /Users/jmin/ioc-test
> nest start --watch
[16:56:27] Starting compilation in watch mode...

[16:56:28] Found 0 errors. Watching for file changes.

[Nest] 98889  - 2023/12/31 16:56:28     LOG [NestFactory] Starting Nest application...
[Nest] 98889  - 2023/12/31 16:56:28     LOG [InstanceLoader] AppModule dependencies initialized +6ms
[Nest] 98889  - 2023/12/31 16:56:28     LOG [RoutesResolver] AppController {/}: +5ms
[Nest] 98889  - 2023/12/31 16:56:28     LOG [RouterExplorer] Mapped {/, GET} route +1ms
[Nest] 98889  - 2023/12/31 16:56:28     LOG [NestApplication] Nest application successfully started +1ms
```

▲ 圖 2-39 啟動服務

下面來看在 Nest 中如何組織程式。開啟 app.controller.ts 檔案，核心程式如下：

```
@Controller()
export class AppController {
  constructor(private readonly appService: AppService) {}

  @Get()
  getHello(): string {
    return this.appService.getHello();
  }
}
```

可見，AppController 類別透過 @Controller 裝飾器來修飾，表示它可以進行相依注入，由 Nest 內建的 IoC 容器接管。

接著開啟 app.service.ts，程式如下：

```
import { Injectable } from '@nestjs/common';

@Injectable()
export class AppService {
  getHello(): string {
    return 'Hello World!';
  }
}
```

AppService 類別透過 @Injectable 進行裝飾，表示這個類別（class）可以被注入，同時也可以注入其他物件中。我們再建立一個 user 中介軟體，如圖 2-40 所示。

```
$ nest g middleware user
CREATE src/user/user.middleware.spec.ts (180 bytes)
CREATE src/user/user.middleware.ts (196 bytes)
```

▲ 圖 2-40 建立 user 中介軟體

開啟 user.middleware.ts 檔案，程式如下：

```
import { Injectable, NestMiddleware } from '@nestjs/common';

@Injectable()
export class UserMiddleware implements NestMiddleware {
  use(req: any, res: any, next: () => void) {
    next();
  }
}
```

我們發現 UserMiddleware 也是透過 @Injectable 裝飾器來標記的。這可能會讓讀者產生一些疑問：

為什麼控制器是單獨使用 @Controller 來裝飾的？

除了已知的服務（Service）和中介軟體（Middleware）使用 @Injectable 裝飾，還有哪些元件也使用它？

對於第一個問題，控制器（Controller）只用於處理請求，不作為相依物件被其他物件元件注入。我們可以把控制器看作是消費者，而服務（Service）和中介軟體（Middleware）則是提供者。

通常使用 @Injectable 進行裝飾的還包括篩檢程式（Filter）、攔截器（Interceptor）、提供者（Provider）、閘道（Gateway）等。在沒有特殊情況下，Nest 中所有需要相依注入的模組都可以使用 @Injectable 進行標記，以便 IoC 容器進行收集和管理。

回到程式中，這些元件會在 AppModule 中進行引入，如圖 2-41 所示。

```
1   import { Module } from '@nestjs/common';
2   import { AppController } from './app.controller';
3   import { AppService } from './app.service';
4
5   @Module({
6     imports: [],
7     controllers: [AppController],
8     providers: [AppService],
9   })
10  export class AppModule {}
```

▲ 圖 2-41 相依注入

@Module 裝飾器在 Nest 中用於定義模組，這些模組包含需要注入的元件，例如控制器（Controllers）。控制器僅能作為消費者被注入，而提供者（Providers），如服務（Services），既可以作為相依被注入，也可以注入其他相依物件中。

除此之外,在 Nest 中實現模組化管理非常簡單。imports 屬性用於引入其他模組,這有助實現功能邏輯的分組和重用。舉例來說,我們可以建立一個 UserModule,如圖 2-42 所示。

```
$ nest g resource user
? What transport layer do you use? REST API
? Would you like to generate CRUD entry points? Yes
CREATE src/user/user.controller.spec.ts (556 bytes)
CREATE src/user/user.controller.ts (883 bytes)
CREATE src/user/user.module.ts (240 bytes)
CREATE src/user/user.service.spec.ts (446 bytes)
CREATE src/user/user.service.ts (607 bytes)
CREATE src/user/dto/create-user.dto.ts (30 bytes)
CREATE src/user/dto/update-user.dto.ts (169 bytes)
CREATE src/user/entities/user.entity.ts (21 bytes)
UPDATE src/app.module.ts (357 bytes)
```

▲ 圖 2-42 建立 user 模組

可見,在 AppModule 中自動引入了 UserModule,如圖 2-43 所示。

```
1  import { Module } from '@nestjs/common';
2  import { AppController } from './app.controller';
3  import { AppService } from './app.service';
4  import { UserModule } from './user/user.module';
5
6  @Module({
7    imports: [UserModule],
8    controllers: [AppController],
9    providers: [AppService],
10 })
11 export class AppModule {}
```

▲ 圖 2-43 相依自動引入

另外,如果還有 aaaModule、bbbModule 等模組,它們也會自動在 AppModule 中引入,最終交給 Nest 工廠方法完成應用的建立,如圖 2-44 所示。

```
1  import { NestFactory } from '@nestjs/core';
2  import { AppModule } from './app.module';
3
4  async function bootstrap() {
5    const app = await NestFactory.create(AppModule);
6    await app.listen(3000);
7  }
8  bootstrap();
```

▲ 圖 2-44 應用主入口

有了這些步驟，在初始化過程中，Nest 就可以輕鬆透過模組之間的相依關係找到 UserModule 模組。根據 @module 中宣告的相依關係，UserController 只需在建構函數中宣告對 UserService 的相依，而不需要顯式建立實例就可以呼叫 UserService 的實例方法，如圖 2-45 所示。

```
4    import { UpdateUserDto } from './dto/update-user.dto';
5
6    @Controller('user')
7    export class UserController {
8      constructor(private readonly userService: UserService) {}
9
10     @Post()
11     create(@Body() createUserDto: CreateUserDto) {
12       return this.userService.create(createUserDto);
13     }
14
```

▲ 圖 2-45 呼叫服務方法

這就是 Nest 的相依注入與模組機制。在後續章節中，我們將頻繁使用這些機制，並享受 IoC 帶來的便捷性。

2.6 學會偵錯 Nest 應用

偵錯能力是衡量一個開發者水準的重要指標。在撰寫應用時，首先應該了解如何偵錯應用程式。透過偵錯，我們可以清晰地看到程式在執行過程中的具體執行狀態。

本節將透過偵錯請求過程和異常資訊來演示如何在 VS Code 中偵錯 Nest。

```
$ nest n nest-debug
⚡ We will scaffold your app in a few seconds..

? Which package manager would you ❤ to use? pnpm
CREATE nest-debug/.eslintrc.js (663 bytes)
CREATE nest-debug/.prettierrc (51 bytes)
CREATE nest-debug/README.md (3347 bytes)
CREATE nest-debug/nest-cli.json (171 bytes)
CREATE nest-debug/package.json (1951 bytes)
CREATE nest-debug/tsconfig.build.json (97 bytes)
CREATE nest-debug/tsconfig.json (546 bytes)
CREATE nest-debug/src/app.controller.spec.ts (617 bytes)
CREATE nest-debug/src/app.controller.ts (274 bytes)
CREATE nest-debug/src/app.module.ts (249 bytes)
CREATE nest-debug/src/app.service.ts (142 bytes)
CREATE nest-debug/src/main.ts (208 bytes)
CREATE nest-debug/test/app.e2e-spec.ts (630 bytes)
CREATE nest-debug/test/jest-e2e.json (183 bytes)

▶▶▶▶▶ Installation in progress... 🍺
```

▲ 圖 2-46 建立 nest-debug 專案

2.6.1 Chrome DevTools 偵錯

首先，執行 nest n nest-debug 命令建立一個專案，如圖 2-46 所示。

在專案中執行 nest start--watch–debug 命令，可以啟動應用並開啟熱多載以及偵錯模式，執行成功後，效果如圖 2-47 所示。

▲ 圖 2-47 主控台上的執行結果

存取 http://localhost:3000/，可以看到正常傳回的內容，如圖 2-48 所示。

▲ 圖 2-48 瀏覽器上的執行結果

除此之外，Nest 啟動了一個如圖 2-47 所示的通訊埠為 9229 的 WebSocket 服務。我們只需要連接這個通訊埠就可以進行偵錯了。在 Chrome 中輸入 chrome://inspect 後按 Enter 鍵，顯示已經連接上遠端偵錯目標，連接的正是 nest-debug 這個專案，如圖 2-49 所示。

▲ 圖 2-49 連接偵錯目標

其中，Configure 是用來配置通訊埠編號的。我們可以設置偵錯服務的通訊埠，Nest 中預設是 9229，如圖 2-50 所示。

▲ 圖 2-50 預設 9229 通訊埠

當然，可以在啟動服務時修改通訊埠編號，比如輸入 nest start --watch --debug 8083，此時啟動的就是 8083 通訊埠，如圖 2-51 所示。然後在 Configure 中配置即可連接。

▲ 圖 2-51 自訂偵錯通訊埠

在圖 2-49 中點擊 inspect，匯入專案檔案夾，在 app.Controller.ts 的 getHello() 方法中增加一個中斷點。刷新頁面之後，可以看到程式自動在中斷點處停下，如圖 2-52 所示。

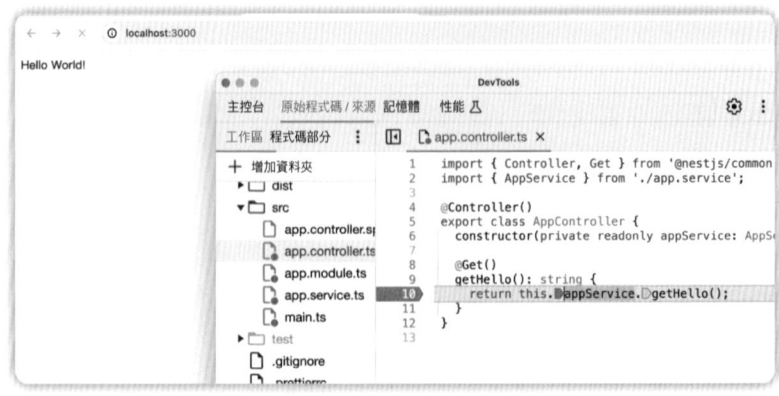

▲ 圖 2-52 中斷點演示效果

這樣就完成了使用 Chrome 瀏覽器偵錯 Nest 應用的過程。然而，這種方法可能不太方便。我們通常會使用 VS Code 等整合式開發環境（IDE）進行開發。那麼，能否直接透過這些工具進行偵錯呢？答案是肯定的。接下來將詳細講解。

2.6.2　VS Code 偵錯

在終端視窗透過命令「nest start --watch –debug」啟動偵錯服務，如圖 2-53 所示。

```
[18:00:10] Starting compilation in watch mode...

[18:00:11] Found 0 errors. Watching for file changes.
Debugger listening on ws://127.0.0.1:9229/6e9600c9-60ae-4815-89fc-b6f62af843f5
For help, see: https://nodejs.org/en/docs/inspector
Debugger attached.
[Nest] 1519  - 2023/12/31 18:00:12     LOG [NestFactory] Starting Nest application...
[Nest] 1519  - 2023/12/31 18:00:12     LOG [InstanceLoader] AppModule dependencies initialized +9ms
[Nest] 1519  - 2023/12/31 18:00:12     LOG [RoutesResolver] AppController {/}: +8ms
[Nest] 1519  - 2023/12/31 18:00:12     LOG [RouterExplorer] Mapped {/, GET} route +1ms
[Nest] 1519  - 2023/12/31 18:00:12     LOG [NestApplication] Nest application successfully started +1ms
```

▲ 圖 2-53　啟動偵錯服務

接著在需要偵錯的地方打個中斷點，如圖 2-54 所示。

▲ 圖 2-54　在指定行打斷點

然後在 http://localhost:3000/ 位址刷新瀏覽器，此時程式會直接在中斷點處停下，如圖 2-55 所示。

```
TS app.service.ts U ×
  TS app.service.ts > ⚛ AppService > ⊗ getHello
1  import { Injectable } from '@nestjs/common';
2
3  @Injectable()
4  export class AppService {
5    getHello(): string {
6      return 'Hello World!';
7    }
8  }
9
```

▲ 圖 2-55 中斷點效果

這種方式是否比之前的方法更簡單呢？確實如此。無論是 Chrome DevTools 還是 VS Code 等第三方整合式開發環境（IDE），它們都透過 CDP（Chrome DevTools Protocol）協定實現了偵錯服務的功能。

有些讀者可能會提出這樣的問題：我不想每次都手動輸入 --debug 參數，能否在執行「yarn start:dev」命令後自動進入偵錯模式？

當然可以。你只需在 package.json 檔案中增加一個指令稿（script）命令即可實現這一點。在 Nest 專案中，這個指令稿預設就已經存在了，如圖 2-56 所示。

```
6    "private": true,
7    "license": "UNLICENSED",
     ▷ 偵錯
8    "scripts": {
9      "build": "nest build",
10     "format": "prettier --write \"src/**/*.ts\" \"test/**/*.ts\"",
11     "start": "nest start",
12     "start:dev": "nest start --watch",
13     "start:debug": "nest start --debug --watch",
14     "start:prod": "node dist/main",
15     "lint": "eslint \"{src,apps,libs,test}/**/*.ts\" --fix",
16     "test": "jest",
17     "test:watch": "jest --watch",
```

▲ 圖 2-56 package.json 中的 scripts 配置

確實，使用 Nest CLI 的 --debug 參數可以方便地啟動偵錯模式。但如果是其他 Node.js 指令稿，情況就不同了。這正是筆者想要提到的：VS Code 提供了一個便捷的自動附加模式。

要使用這個功能，你可以透過按快速鍵 Command+Shift+P（在 Windows 中）或 Ctrl+Shift+P（在 Linux 中），然後在彈出的命令面板中輸入「Toggle Auto

Attach」，接著選擇「總是」選項，以自動附加偵錯器到 Node.js 處理程序，如圖 2-57 所示。

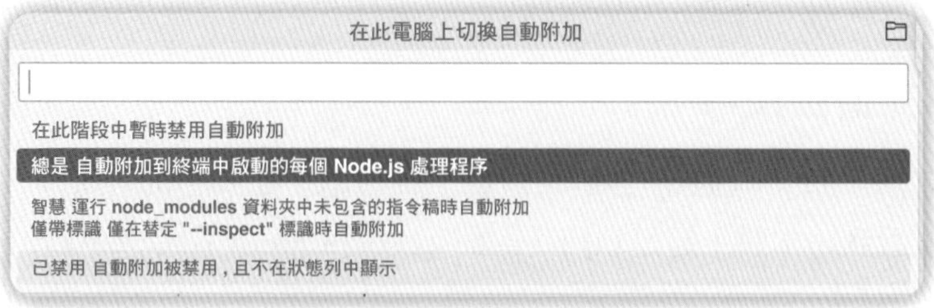

▲ 圖 2-57 VS Code 自動附加

重新開啟主控台並執行「yarn start:dev」命令，此時 VS Code 在啟動應用的同時，也會自動建立一個偵錯器（debugger）服務。該偵錯服務的預設通訊埠編號為 61174，如圖 2-58 所示。

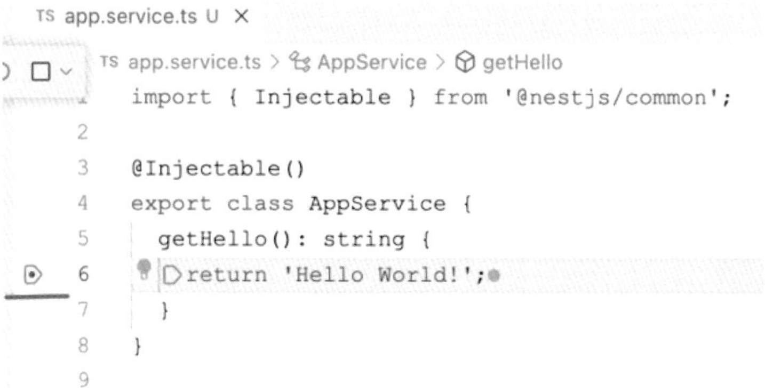

▲ 圖 2-58 附加模式啟動效果

重新在瀏覽器中存取 http://localhost:3000/，依然可以進入中斷點偵錯狀態，如圖 2-59 所示。

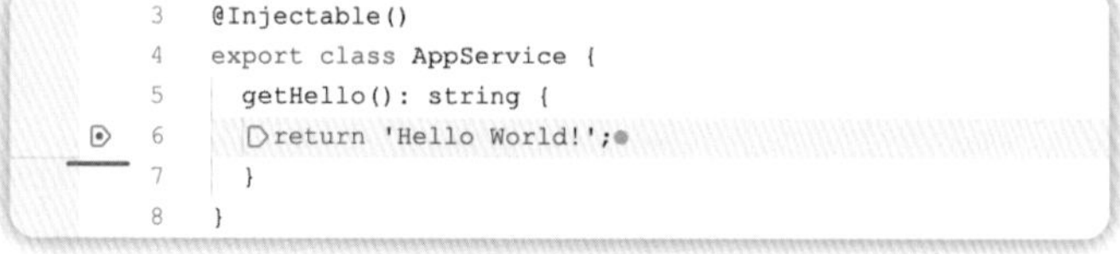

▲ 圖 2-59 偵錯效果

以上兩種方式會在每次執行命令時建立一個新的 WebSocket 偵錯服務。然而，對一些開發者來說，這種方法可能並不適合他們的需求。接下來，在 2.6.3 節中，我們將探討一些擴充的偵錯技巧。

2.6.3 擴充偵錯技巧

前面講解了基於 --debug 參數來啟動偵錯服務，如圖 2-60 所示。從 CLI 原始程式中可以看到，本質上還是執行「node xxxx –inspect」命令來執行的。

▲ 圖 2-60 --debug 參數的定義

因此，我們也可以選擇在 VS Code 中使用這種方式進行偵錯。首先，在偵錯面板中點擊建立 launch.json 檔案的選項，如圖 2-61 所示。

▲ 圖 2-61 建立 launch.json 偵錯檔案

然後選擇 Node.js 作為偵錯器，如圖 2-62 所示。

2.6 學會偵錯 Nest 應用 | 2-43

▲ 圖 2-62 選擇 Node.js 偵錯器

接著根據需要配置相關的偵錯資訊。在 launch.json 檔案中，將 program 欄位的值修改為你的應用程式的進入點，例如 main.ts：

```
{
  // 移過以查看現有屬性的描述
  "version": "0.2.0",
  "configurations": [
    {
      "type": "node",
      "request": "launch",
      "name": " 啟動偵錯 ",
      "skipFiles": [
        "<node_internals>/**"
      ],
      "program": "${workspaceFolder}/src/main.ts",
      "outFiles": [
        "${workspaceFolder}/dist/**/*.js"
      ]
    }
  ]
}
```

配置完成之後，在 app.service.ts 的 getHello() 方法中增加中斷點，點擊左上角的「啟動偵錯」按鈕，如圖 2-63 所示。

▲ 圖 2-63 打斷點並啟動偵錯

可以看到在「呼叫堆疊」中新增了一個啟動程式，這就是 VS Code 偵錯器。我們依舊在 Chrome 中造訪 http://localhost:3000/，此時程式自動在中斷點處停住，並且在左側可以看到呼叫堆疊資訊和變數情況，如圖 2-64 所示。

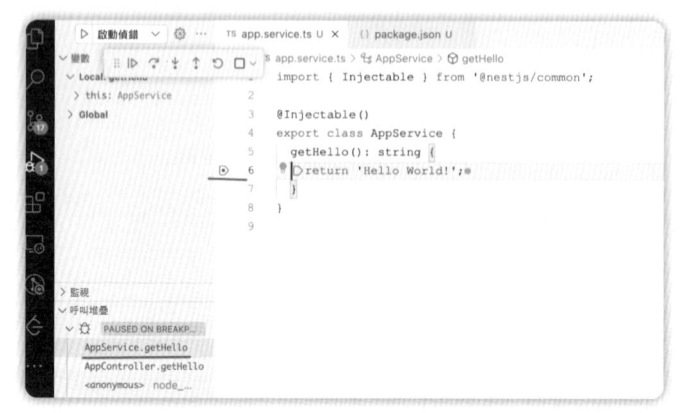

▲ 圖 2-64　偵錯呼叫堆疊情況

這種方法的優勢在於它避免了每次啟動都需要建立一個新的 WebSocket 偵錯服務，從而節省了執行時期的記憶體消耗。偵錯工具僅在需要時啟動，這正是我們最常用的偵錯方式。它適用於前端應用，如 Vue、React 以及 Node.js 等多種應用的偵錯。

第 3 章
Nest 核心概念介紹

本章將介紹 Nest 核心概念及其應用場景。首先，探討貫穿 Nest 開發過程的裝飾器，以及實現功能模組化的模組概念。接著，介紹管理路由的控制器，以及實現複雜業務邏輯並與資料層互動的服務。最後，介紹 AOP 架構下的中介軟體機制和攔截器功能。讀者可結合實際程式範例，更加了解這些概念在實際開發中的應用。

3.1 貫穿全書的裝飾器

本節首先介紹裝飾器（Decorator）的基本概念和常見的 5 種裝飾器類型，包括類別裝飾器、方法裝飾器、屬性裝飾器、參數裝飾器和存取器裝飾器。接著介紹 Nest 中常用的裝飾器，我們先初步認識它們。

3.1.1 基本概念

顧名思義，裝飾器（Java 中有個類似的概念叫注解）是用來裝飾和擴充物件功能的。它能夠在不改變原有物件結構的前提下，增加額外的功能，以滿足更多的實際需求。

舉個例子：一間房子裡放了一張床，就滿足了基本的居住需求。在此基礎上，如果新增了沙發、紅酒杯、電視機，那麼就不僅能夠滿足休息的需求，還能提供娛樂和休閒的體驗。在這個比喻中，沙發、紅酒杯、電視機就像是裝飾器，它們可以在不影響基本功能的前提下，根據需要隨意增加或移除，從而實現功能的擴充和鬆散耦合。

3.1.2 裝飾器的種類

上例中，沙發、紅酒杯、電視機屬於不同的裝飾器，它們實現了不同的功能。同樣地，常用的裝飾器也分為以下幾種：

（1）類別裝飾器。
（2）方法裝飾器。
（3）屬性裝飾器。
（4）參數裝飾器。
（5）存取器裝飾器。

儘管裝飾器仍處於提案（stage-3）階段，但它們已經廣泛應用於 TypeScript 和 Babel 編譯中。以方法裝飾器為例，在 stage-3 提案中的實現如下：

```
/**
 * 方法裝飾器
 */
const Log: MethodDecorator = (
  target: Function,
  context: {
    kind: "method";
    name: string | symbol;
    access: { get(): unknown };
    static: boolean;
    private: boolean;
    addInitializer(initializer: () => void): void;
  }
) => {
  // 此處撰寫你的邏輯
};
```

上述程式中定義了一個 Log 方法裝飾器，用於修飾類別方法，其中 target 是被修飾的物件，context 提供上下文資訊。在基於 TypeScript 建構的 Nest 專案中，採用的是 stage-1 提案的裝飾器版本，具體如下：

```
/**
 * 方法裝飾器
 */
const log: MethodDecorator = (
  target: Function,
  propertyKey: string | Symbol,
  descriptor: PropertyDescriptor
) => {
```

```
  // 此處撰寫你的邏輯
};
```

上述程式中，Log 方法裝飾器接收 3 個參數：target 是被修飾的值，propertyKey 是被修飾的屬性鍵（方法名稱），description 是被裝飾的方法的屬性描述符號物件。

下面透過簡單的範例分別介紹。

1. 類別的裝飾

裝飾器可以用來裝飾整個類別，以此增強類別的功能，範例程式如下：

```
/**
 * 類別裝飾器
 */
const doc: ClassDecorator = (target: Function) => {
  console.log("--------------- 類別裝飾器 -----------------");
  target.prototype.name = "mouse";
  console.log(target);
  console.log("--------------- 類別裝飾器 -----------------");
};

@doc
class App {
  constructor() {}
}

const app: Record<string, any> = new App();
console.log(«app name: « + app.name);
```

在上面的程式中，@doc 用來裝飾 App 類別，同時往原型鏈上增加一個屬性 name，target 是被修飾的類別。

我們透過 CodePen 來編譯 ES6 程式。CodePen 是一個社區驅動的線上程式編輯器，支持編輯 HTML、CSS 和 JavaScript。CodePen 還支持 Babel，這使得它能夠將 ES6 程式轉換成 ES5 並進行即時預覽。

由於本節案例中我們使用 TypeScript 開發，因此需要在 CodePen 中找到 Settings 下的 JS 選項，並將 JavaScript Preprocessor（JavaScript 前置處理器）設置為 TypeScript，如圖 3-1 所示。

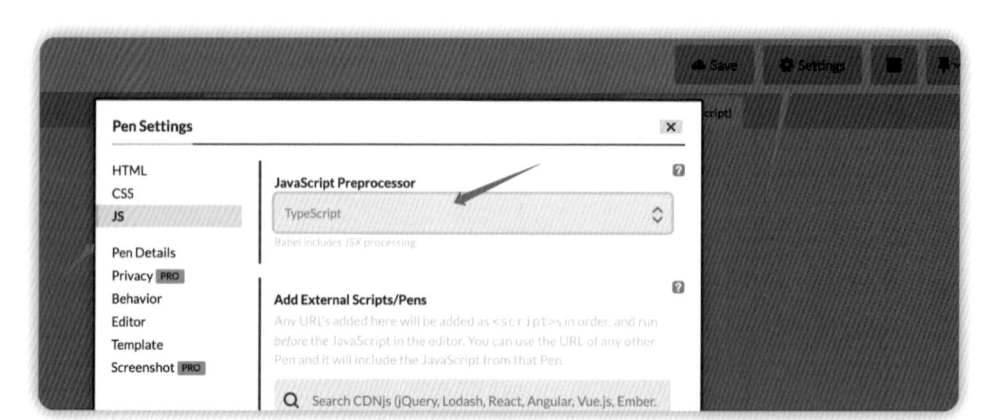

▲ 圖 3-1 設置 CodePen 前置處理器

儲存後，類別裝飾器執行結果如圖 3-2 所示。

▲ 圖 3-2 類別裝飾器的執行結果

上面列印了程式預期的結果，成功地給實例增加了屬性 name，並獲取了 App 的建構函數，這樣我們就可以動態地修改類別的屬性和行為了。

如果覺得一個 target 參數不夠用，怎麼辦呢？可以透過工廠函數來增強裝飾器的能力。程式如下：

```typescript
/**
 * 類別裝飾器
 */
const doc: ClassDecorator = (module) => {
  return (target: Function) => {
    console.log("--------------- 類別裝飾器 ----------------");
    target.prototype.module = module;
    console.log(target);
    console.log("--------------- 類別裝飾器 ----------------");
  };
};

@doc(«user»)
class App {
  constructor() {}
}

const app: Record<string, any> = new App();
console.log(«app module: « + app.module);
```

透過工廠函數增強裝飾器的能力後，@doc 可以接收外部參數，使得 App 被實例化時能夠獲取更多的資訊來組織程式，具體效果如圖 3-3 所示。

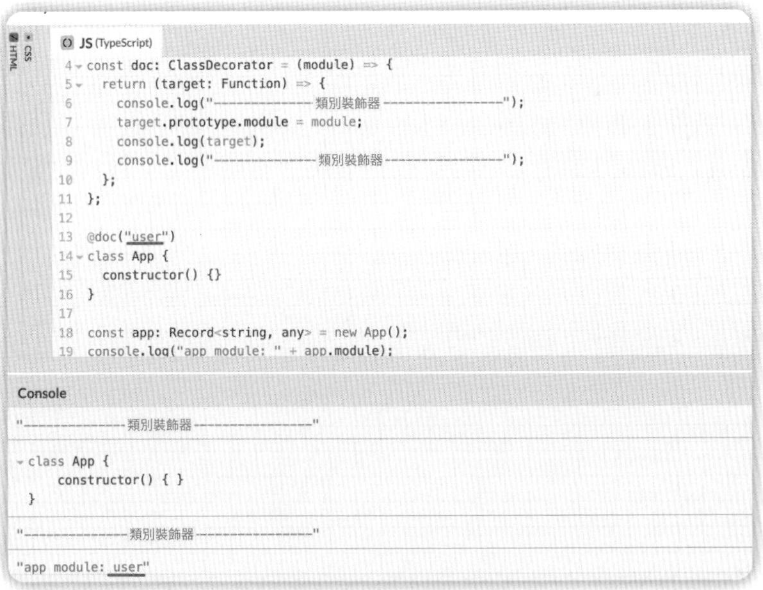

▲ 圖 3-3 裝飾器工廠

至此，你應該已經理解了 Nest 中的裝飾器的實現方式，它在相依注入中扮演著重要的角色。然而，請不要急於下結論，讓我們繼續深入探討。

2. 方法的裝飾

裝飾器可以用來裝飾類別的方法，程式如下：

```
/**
 * 方法裝飾器
 */
const Log: MethodDecorator = (
  target: Object,
  propertyKey: string | Symbol,
  descriptor: PropertyDescriptor
) => {
  console.log("--------------- 方法裝飾器 -----------------");
  console.log(target);
  console.log(propertyKey);
  console.log(descriptor);
  console.log("--------------- 方法裝飾器 -----------------");
};

class User {
  @Log
  getName() {
    return "mouse";
  }
}
```

在上面的程式中，@Log 裝飾器函數修飾了 User 類別的 getName 方法。其中，target 是被「裝飾」的類別的原型，即 User.prototype。由於此時類別還沒有被實例化，因此只能裝飾原型物件。這與類別裝飾器有所不同。執行結果如圖 3-4 所示。

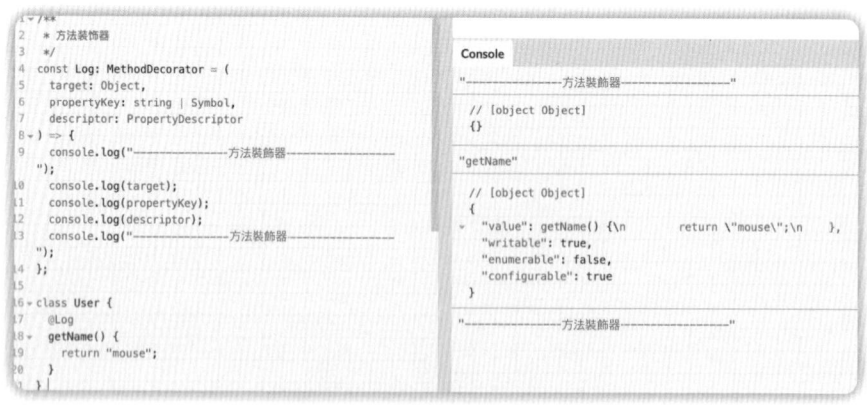

▲ 圖 3-4 方法裝飾器的執行結果

由於 User 類別的原型尚未附加屬性方法，因此是空白物件 {}。

3. 屬性的裝飾

裝飾器也可以用來裝飾類別的屬性，程式如下：

```
/**
 * 屬性裝飾器
 */
const prop: PropertyDecorator = (
  target: Object,
  propertyKey: string | Symbol
) => {
  console.log("------屬性裝飾器-------");
  console.log(target);
  console.log(propertyKey);
  console.log("------屬性裝飾器-------");
};

class User {
  @prop
  name: string = "mouse";
}
```

在上面的程式中，@prop 用於裝飾 User 類別的 name 屬性。此時，target 是被裝飾的類別的原型物件，propertyKey 是被修飾的屬性名稱。具體的執行結果如圖 3-5 所示。

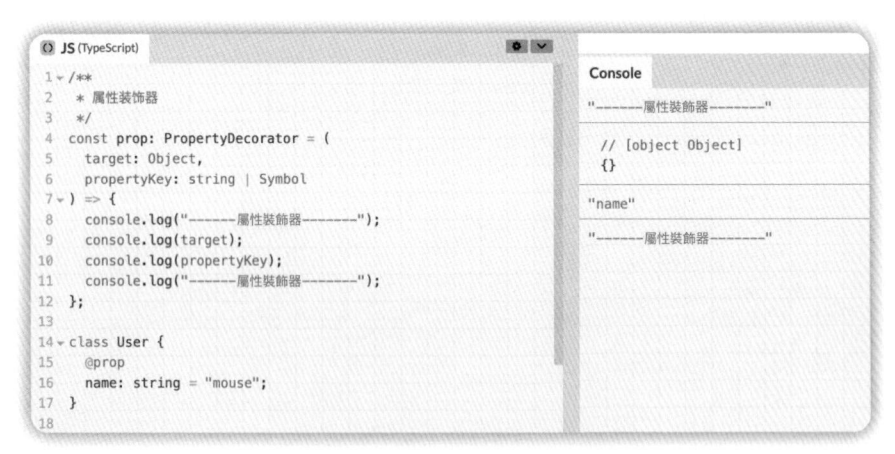

▲ 圖 3-5 屬性裝飾器的執行結果

4. 參數的裝飾

除了裝飾類別、類別方法和類別屬性外，裝飾器還能裝飾類別方法的參數，範例程式如下：

```
/**
 * 參數裝飾器
 */
const Param: ParameterDecorator = (
  target: Object,
  propertyKey: string | Symbol | undefined,
  index: number
) => {
  console.log("------------- 參數裝飾器 ------------------");
  console.log(target);
  console.log(propertyKey);
  console.log(index);
  console.log("------------- 參數裝飾器 ------------------");
};
class User {
  getName(@Param name: string) {
    return name;
  }
}
```

在上述程式中，@Param 用於裝飾類別方法 getName 的參數，其中 target 是被裝飾類別的原型，propertyKey 是被裝飾的類別方法，index 是裝飾方法參數的索引位置。具體的執行結果如圖 3-6 所示。

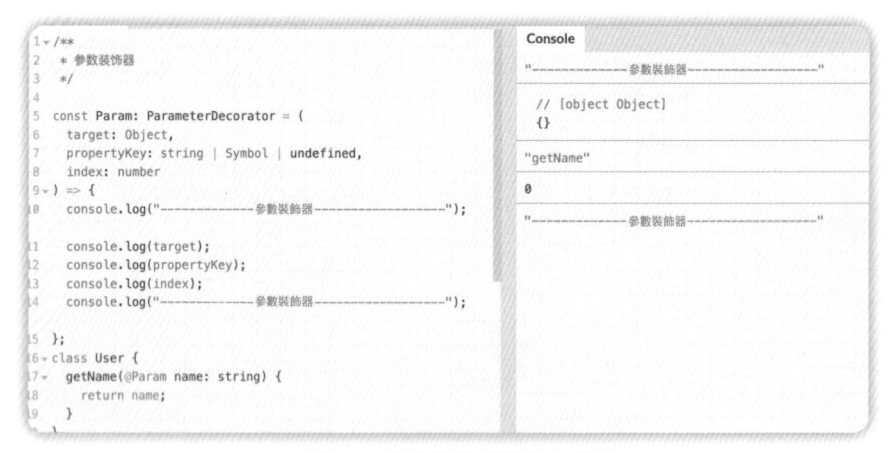

▲ 圖 3-6 參數裝飾器的執行結果

5. 存取器的裝飾

裝飾器還可以用來修飾類別的存取器，即屬性的 getter 方法。範例程式如下：

```
/**
 * 存取器裝飾器
 */
const Immutable: ParameterDecorator = (
  target: any,
  propertyKey: string | Symbol,
  descriptor: PropertyDescriptor
) => {
  console.log("------------ 存取器裝飾器 ------------------");
  console.log(target);
  console.log(propertyKey);
  console.log(descriptor);
  console.log("------------ 存取器裝飾器 ------------------");
};
class User {
  private _name = "mouse";
  @Immutable
  get name() {
    return this._name;
  }
}
```

存取器裝飾器與類別方法裝飾器類似，唯一的區別在於它們的描述符號（descriptor）中某些鍵（key）不同。

方法裝飾器的描述器鍵包括：

- value
- writable
- enumerable
- configurable

存取器裝飾器的描述器的 key 為：

- get
- set
- enumerable
- configurable

執行結果如圖 3-7 所示。

▲ 圖 3-7 存取器裝飾器的執行結果

了解了這幾個裝飾器的概念後，接下來學習 Nest 中的裝飾器，你應該能夠輕鬆理解。

3.1.3 Nest 中的裝飾器

在 Nest 中實現了前面列舉的前 4 種裝飾器，它們包含但不限於：

- 類別裝飾器：@Controller、@Injectable、@Module、@UseInterceptors。
- 方法裝飾器：@Get、@Post、@UseInterceptors。
- 屬性裝飾器：@IsNotEmpty、@IsString、@IsNumber。
- 參數裝飾器：@Body、@Param、@Query。

下面分別對它們進行解釋。

- @Controller()：用於裝飾控制器類別，使之能夠管理應用中的路由程式，並透過設置路由路徑首碼來模組化管理路由。
- @Injectable()：裝飾後成為服務提供者，可以被其他物件進行依賴注入。
- @Module()：模組裝飾器，用於在 Nest 中劃分功能模組並限制相依注入的範圍。
- @UseInterceptors()：用於綁定攔截器，將攔截器的作用範圍限制在控制器類別範圍中（當然也可以作用在類別方法上）。
- @Get、@Post：用於定義路由方法的 HTTP 請求方式。
- @IsNotEmpty、@IsString、@IsNumber：用於在參數的驗證場景中驗證 HTTP 請求的參數是否符合預期。

- @Body、@Param、@Query：用於接收 HTTP 請求發送的資料，不同的請求方式對應不同的參數接收方式。

Nest 中的裝飾器主要分為這幾類別。理解了它們將有助我們在使用過程中更加得心應手。關於這些裝飾器的具體使用方法，我們將在後續章節中逐一進行詳細介紹。

3.2 井然有序的模組化

當前，模組化思想已廣泛應用於軟體開發領域。它透過將大型軟體應用劃分為相互獨立且功能明確的模組或元件，實現了類似搭積木的過程。同樣，Nest 框架也提供了一種結構化和模組化的方法來管理應用程式中的不同部分。本節將學習如何建立並使用這些模組。

3.2.1 基本概念

模組透過 @Module 裝飾器來宣告。每個應用都會有一個根模組，Nest 框架會從根模組開始收集各個模組之間的依賴關係，形成相依樹狀結構。在應用初始化時，根據相依樹狀結構實例化不同的模組物件，具體如圖 3-8 所示。

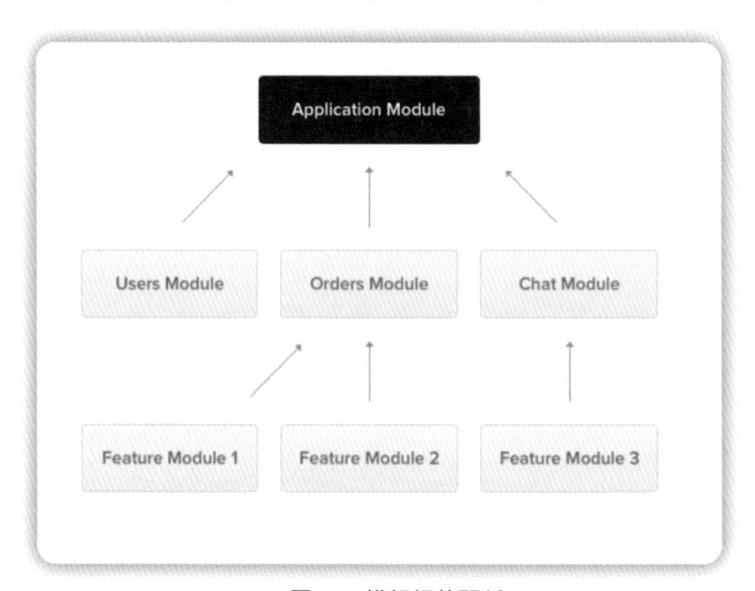

▲ 圖 3-8 模組相依關係

在模組樹中，每個模組都有自己獨立的作用域。它們之間的程式是相互隔離的，各自擁有自己的控制器（Controllers）、服務提供者（Providers）、中介軟體（Middlewares）和其他組件。當然，它們也可以透過一定的方式進行相互共用。

3.2.2 建立模組

下面透過建立一個名為 nest-module 的專案來演示。執行「nest n nest-module -p pnpm」命令，具體的執行結果如圖 3-9 所示。

```
$ nest n nest-module -p pnpm
⚡  We will scaffold your app in a few seconds..

CREATE nest-module/.eslintrc.js (663 bytes)
CREATE nest-module/.prettierrc (51 bytes)
CREATE nest-module/README.md (3347 bytes)
CREATE nest-module/nest-cli.json (171 bytes)
CREATE nest-module/package.json (1952 bytes)
CREATE nest-module/tsconfig.build.json (97 bytes)
CREATE nest-module/tsconfig.json (546 bytes)
CREATE nest-module/src/app.controller.spec.ts (617 bytes)
CREATE nest-module/src/app.controller.ts (274 bytes)
CREATE nest-module/src/app.module.ts (249 bytes)
CREATE nest-module/src/app.service.ts (142 bytes)
CREATE nest-module/src/main.ts (208 bytes)
CREATE nest-module/test/app.e2e-spec.ts (630 bytes)
CREATE nest-module/test/jest-e2e.json (183 bytes)

✔ Installation in progress... ☕

🚀  Successfully created project nest-module
```

▲ 圖 3-9　建立專案

首先來看 app.module.ts 檔案，程式如下：

```
import { Module } from '@nestjs/common';
import { AppController } from './app.controller';
import { AppService } from './app.service';

@Module({
  imports: [],
  controllers: [AppController],
  providers: [AppService],
})
export class AppModule {}
```

其中，AppModule 是預設的根模組。類別裝飾器 @Module 的參數中，controllers 用於注入該模組的控制器集合，而 providers 用於注入該模組的服務提供者，這些服務提供者將在該模組中共用。

Imports 用於匯入應用中的其他模組，預設為空。下面以建立新的 User 和 Order 模組為例，執行以下命令：

```
// 生成 User 模組
nest g resource User --no-spec
// 生成 Order 模組
nest g resource Order --no-spec
```

--no-spec 表示不生成單元測試檔案，執行結果如圖 3-10 所示。

```
$ nest g resource User --no-spec
nest g resource Order --no-spec
? What transport layer do you use? REST API
? Would you like to generate CRUD entry points? No
CREATE src/user/user.controller.ts (204 bytes)
CREATE src/user/user.module.ts (240 bytes)
CREATE src/user/user.service.ts (88 bytes)
UPDATE src/app.module.ts (310 bytes)
? What transport layer do you use? REST API
? Would you like to generate CRUD entry points? No
CREATE src/order/order.controller.ts (210 bytes)
CREATE src/order/order.module.ts (247 bytes)
CREATE src/order/order.service.ts (89 bytes)
UPDATE src/app.module.ts (375 bytes)
```

▲ 圖 3-10 生成 User 和 Order 模組

回到 AppModule 根模組，可以看到 UserModule 和 OrderModule 被自動匯入根模組，成為 AppModule 的子模組。具體程式如下：

```
import { Module } from '@nestjs/common';
import { AppController } from './app.controller';
import { AppService } from './app.service';
import { UserModule } from './user/user.module';
import { OrderModule } from './order/order.module';

@Module({
  // 模組被自動匯入了
  imports: [UserModule, OrderModule],
  controllers: [AppController],
  providers: [AppService],
})
export class AppModule {}
```

3.2.3 共用模組

既然有 imports，那麼必然也有 exports。假設存在這樣的需求，Order 模組需要相依 User 模組中的 UserService。這時可以將 UserService 增加到 UserModule 的

exports 中，使之成為共用服務。這樣，Order 模組只需匯入 UserModule 即可存取 UserService。下面透過一個案例來演示。

在 User 模組中匯出 User 服務：

```
@Module({
  controllers: [UserController],
  providers: [UserService],
  // 匯出 UserService 服務
  exports: [UserService]
})
export class UserModule {}
```

接著在 Order 模組中匯入 User 模組，程式如下：

```
@Module({
  // 匯入 UserModule
  imports: [UserModule],
  controllers: [OrderController],
  providers: [OrderService]
})
export class OrderModule {}
```

此時，在 Order 模組的任何地方，都可以共用 UserService 服務了。下面來演示一下。

在 order.controller.ts 中定義路由方法 getOrder，然後呼叫 order.service.ts 中的 getOrderDesc 方法：

```
// order.controller.ts
import { Controller, Get } from '@nestjs/common';
import { OrderService } from './order.service';

@Controller('order')
export class OrderController {
  constructor(private readonly orderService: OrderService) {}

  @Get()
  getOrderDesc(): string {
    return this.orderService.getOrderDesc()
  }
}
```

在 order.service.ts 中透過屬性注入 UserService 相依，同時呼叫 UserService 的 getUserHello 方法，最終傳回一個問候語字串，程式如下：

```
// order.service.ts
import { Inject, Injectable } from '@nestjs/common';
import { UserService } from 'src/user/user.service';
```

```typescript
@Injectable()
export class OrderService {
  // 相依注入之屬性注入共用的服務
  @Inject(UserService)
  private userService: UserService;
  getOrderDesc(): string {
    let name = this.userService.getUserName()
    return '訂單 ID：xxxx，下單人：${name}'
  }
}

// user.service.ts
import { Injectable } from '@nestjs/common';

@Injectable()
export class UserService {
  getUserName(): string {
    return 'mouse'
  }
}
```

除使用屬性注入相依外，還可以使用建構函數注入：

```typescript
import { Inject, Injectable } from '@nestjs/common';
import { UserService } from 'src/user/user.service';

@Injectable()
export class OrderService {
  // 相依注入之建構函數注入
  constructor(private userService: UserService) {}
  getOrderDesc(): string {
    let name = this.userService.getUserName()
    return '訂單 ID：xxxx，下單人：${name}'
  }
}
```

最後，在瀏覽器中造訪 http://localhost:3000/order，頁面成功傳回了內容，這表明實現了模組間的資料互動，如圖 3-11 所示。

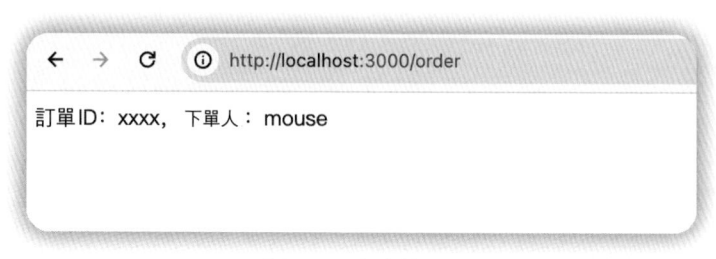

▲ 圖 3-11 程式執行結果

3.2.4 全域模組

如果某個模組在多個地方被引用，為了簡化管理，可以使用 @Global 裝飾器將其宣告為全域模組。這樣，便可以直接注入透過 exports 匯出的 providers，而無須在每個模組的 imports 中重複宣告它們。程式如下：

```
// user.module.ts
// 宣告為全域模組
@Global()
@Module({
  controllers: [UserController],
  providers: [UserService],
  exports: [UserService]
})
export class UserModule {}

// order.module.ts
@Module({
  // 這裡無須再匯入
  // imports: [UserModule],
  controllers: [OrderController],
  providers: [OrderService]
})
export class OrderModule {}
```

需要注意的是，使用全域模組之前，需要確保你的模組確實需要全域使用，以避免不必要地增加模組之間的耦合性。

3.2.5 動態模組

前面介紹的都是靜態模組的綁定和使用。Nest 中還提供了動態載入模組的功能，使得應用可以在執行時期建立模組，通常用於動態讀取配置或根據許可權判斷來載入模組。

建立動態模組的第一步是在需要使用動態模組的地方，例如某個服務或控制器中，使用 forFeature 方法來動態載入模組。程式如下：

```
// dynamic.module.ts
import { Module, DynamicModule } from '@nestjs/common';

@Module({})
export class DynamicModule {
  static forFeature(entities: Function[]): DynamicModule {
    // 在此方法中根據需要建立模組
    return {
```

```
    module: DynamicModule,
      providers: [],
      exports: [],
    };
  }
}
```

動態載入模組時,呼叫 loadModule 方法即可載入相應的模組。程式如下:

```
// some.service.ts
import { Injectable } from '@nestjs/common';
import { DynamicModule } from './dynamic.module'

@Injectable()
export class SomeService {
  constructor(private readonly dynamicModule: DynamicModule) {}
  // 在某個方法中動態載入模組
  loadModule(entities: Function[]) {
    return this.dynamicModule.forFeature(entities);
  }
}
```

除 loadModule 方法之外,還有 register 和 forRoot 方法可用於載入動態模組。其中,register 方法通常用於與外部模組或第三方函數庫整合時,它能將外部模組動態載入到 Nest.js 模組中。

```
import { Module } from '@nestjs/common';

@Module({})
export class DynamicModule {
  static register(options: SomeOptions) {
    // 在此方法中註冊外部模組
    return {
      module: DynamicModule,
      providers: [],
      exports: [],
    };
  }
}
```

而 forRoot 方法在 Nest 中通常用於註冊根模組,例如配置根模組的全域服務或中介軟體等。程式如下:

```
import { Module } from '@nestjs/common';

@Module({})
export class CoreModule {
  static forRoot(options: SomeOptions) {
    // 在此方法中配置根模組的全域服務或特性
```

```
    return {
      module: CoreModule,
      providers: [],
      exports: [],
    };
  }
}
```

以上介紹了載入靜態和動態模組的幾種方式。動態模組的載入方式在後續章節中會頻繁出現，特別是與第三方服務整合時。

3.3 控制器與服務的默契配合

在了解模組的基礎知識之後，本節將深入學習模組中的核心部分：控制器和服務。首先介紹它們各自在模組中承擔的職責。透過請求流程解釋控制器如何管理應用路由和處理請求參數，以及服務是如何進行資料處理的。此外，還將介紹服務與服務提供者之間的區別，以幫助讀者理清想法，從而更進一步地理解相依注入的核心概念。

3.3.1 基本概念

Nest 提供了分層結構，其中 Controller 和 Service 分別扮演著不同但緊密相關的角色。它們各自承擔的任務也不相同，你必須清楚地了解它們的職責，才能確保你的應用既清晰又可維護。

1. Controller 的職責

（1）處理請求和路由：接收 HTTP 請求並確定路由，一個控制器中通常存在多個路由，不同的路由執行不同的操作。

（2）解析和驗證輸入資料：控制器通常負責解析並驗證請求的資料。舉例來說，透過使用管道或資料傳輸物件（Data Transfer Object，DTO），控制器可以定義和實施資料驗證、格式化或轉換操作。

（3）呼叫 Service：Controller 層用於排程 Service 層執行業務邏輯處理，是 Service 層的入口。

2. Service 的職責

（1）處理業務邏輯：包含對資料的處理、外部系統的呼叫和複雜業務邏輯處理。

（2）資料持久化：與資料庫、Redis 快取等儲存庫進行互動，執行持久化操作。通常在 Service 中使用模型或實體來進行資料的增刪改查操作。

（3）由此可見，Controller 負責與 HTTP 互動相關事項，如路由管理和參數處理等，而 Service 則負責業務資料處理。

下面我們透過實際案例來演示說明。

3.3.2 Controller 管理請求路由

建立一個新的 Nest 專案，執行「nest n nest-controller-service -p pnpm」命令，結果如圖 3-12 所示。

```
$ nest n nest-controller-service -p pnpm
⚡  We will scaffold your app in a few seconds..

CREATE nest-controller-service/.eslintrc.js (663 bytes)
CREATE nest-controller-service/.prettierrc (51 bytes)
CREATE nest-controller-service/README.md (3347 bytes)
CREATE nest-controller-service/nest-cli.json (171 bytes)
CREATE nest-controller-service/package.json (1964 bytes)
CREATE nest-controller-service/tsconfig.build.json (97 bytes)
CREATE nest-controller-service/tsconfig.json (546 bytes)
CREATE nest-controller-service/src/app.controller.spec.ts (617 bytes)
CREATE nest-controller-service/src/app.controller.ts (274 bytes)
CREATE nest-controller-service/src/app.module.ts (249 bytes)
CREATE nest-controller-service/src/app.service.ts (142 bytes)
CREATE nest-controller-service/src/main.ts (208 bytes)
CREATE nest-controller-service/test/app.e2e-spec.ts (630 bytes)
CREATE nest-controller-service/test/jest-e2e.json (183 bytes)

✓ Installation in progress... 🍺

🚀  Successfully created project nest-controller-service
```

▲ 圖 3-12 建立專案

接著透過 CLI 快速建立 User 模組的 CRUD 控制器，執行「nest g resource User」命令，選擇 REST API 和 CRUD 入口，如圖 3-13 所示。

```
$ nest g resource User
? What transport layer do you use? REST API
? Would you like to generate CRUD entry points? Yes
CREATE src/user/user.controller.spec.ts (556 bytes)
CREATE src/user/user.controller.ts (883 bytes)
CREATE src/user/user.module.ts (240 bytes)
CREATE src/user/user.service.spec.ts (446 bytes)
CREATE src/user/user.service.ts (607 bytes)
CREATE src/user/dto/create-user.dto.ts (30 bytes)
CREATE src/user/dto/update-user.dto.ts (169 bytes)
CREATE src/user/entities/user.entity.ts (21 bytes)
UPDATE package.json (1997 bytes)
UPDATE src/app.module.ts (308 bytes)
✔ Packages installed successfully.
```

▲ 圖 3-13 生成 User 模組

建立完成後，開啟 user.controller.ts 檔案，核心程式如下：

```
// 定義 user 路徑首碼
@Controller('user')
export class UserController {
  constructor(private readonly userService: UserService) {}

  @Post()
  create(@Body() createUserDto: CreateUserDto) {
    return this.userService.create(createUserDto);
  }

  @Get()
  findAll() {
    return this.userService.findAll();
  }
}
```

在 Nest 應用程式中，User 控制器會自動生成對應的路由 API，並在 Controller() 裝飾器中指定一個路徑首碼，例如 user。這樣做的好處是可以實現路由的分組，同時最大限度地減少程式重複。在大型應用系統中，這種路由分組方式使得開發者能夠透過路徑快速辨識路由所屬的控制器。舉例來說，透過 /orders/* 路徑可以清楚地知道這些路由是由 OrderController 管理的，而 /products/* 路由則屬於 ProductController。

Create 方法透過 @Post() 裝飾器宣告了請求方式，告訴 HTTP 用戶端以 /user 路徑和 POST 方法進行請求存取。如果使用 /user 路徑和 GET 方法請求 Nest 應用，則會映射到 findAll 這個路由處理常式中。

如果一個控制器中存在多個 POST 請求或 GET 請求,可以透過向裝飾器傳遞參數來區分它們。舉例來說,設置 @Post("create-user") 會將請求路徑組合為 /user/create-user,並將該路徑映射到 create 方法。範例程式如下:

```
// 定義 user 路徑首碼
@Controller('user')
export class UserController {
  constructor(private readonly userService: UserService) {}
  // 定義方法路徑
  @Post('create-user')
  create(@Body() createUserDto: CreateUserDto) {
    return this.userService.create(createUserDto);
  }
}
```

另外,Nest 支援所有標準的 HTTP 方法,如 @Get()、@Post()、@Put()、@Delete()、@Patch()、@Options() 和 @Head() 等,讀者可以根據需要自由組合路徑。

3.3.3 Controller 處理請求參數與請求本體

在 Express 中,通常需要透過 Request、Response 請求物件來獲取 header、query、body 等屬性。而 Nest 基於 Express 作為底層的 HTTP 框架,自然也提供了類似的物件,如 @Req、@Res,但一般情況下不需要那麼麻煩。Nest 提供了開箱即用的裝飾器 @Headers、@Body、@Param、@Query 等,範例程式如下:

```
@Get(':id')
findOne(@Param('id') id: string) {
  return this.userService.findOne(+id);
}

@Patch(':id')
update(@Param('id') id: string, @Body() updateUserDto: UpdateUserDto) {
  return this.userService.update(+id, updateUserDto);
}

@Delete(':id')
remove(@Param('id') id: string) {
  return this.userService.remove(+id);
}
```

首先,我們來探討路由中攜帶有參數的情況。使用 @Get(':id') 裝飾器可以建構出像 /user/123 這樣的請求 URL,而 @Param('id') 用於從 URL 中提取參數 id,其值為 123。如果是查詢參數形式的請求 URL,如 /user?id=123,則可以透過

@Query('id') 來獲取參數 id，其值為 123。

接下來是 POST 請求的處理。我們通常透過 @Body() 來獲取請求本體，並利用最常見的資料傳輸物件（Data Transfer Object，DTO）進行資料驗證。舉例來說，如上文程式所示的 updateUserDto，它將用於驗證資料。更詳細的 DTO 介紹將在本章後續部分進行。

完成一系列操作後，我們將呼叫服務層（Service）進行資料持久化。服務層的方法通常與控制器（Controller）中的路由方法名稱一一對應。舉例來說，在 use.service.ts 中定義的方法，其程式如下：

```
import { Injectable } from '@nestjs/common';
import { CreateUserDto } from './dto/create-user.dto';
import { UpdateUserDto } from './dto/update-user.dto';

@Injectable()
export class UserService {
  create(createUserDto: CreateUserDto) {
    return 'This action adds a new user';
  }

  findAll() {
    return `This action returns all user`;
  }

  findOne(id: number) {
    return `This action returns a #${id} user`;
  }

  update(id: number, updateUserDto: UpdateUserDto) {
    return `This action updates a #${id} user`;
  }

  remove(id: number) {
    return `This action removes a #${id} user`;
  }
}
```

3.3.4 Service 處理資料層

至此，我們已經了解到，當用戶端發出一個請求時，首先由控制器（Controller）處理，然後控制器會排程服務層（Service）的特定方法來執行業務邏輯，例如與資料庫互動以執行資料的增加、刪除、更新和查詢操作。那麼，在服務層中，資料又是如何被處理的呢？

3.3 控制器與服務的默契配合

先來看下面的虛擬程式碼：

```
import { Injectable } from '@nestjs/common';
import { CreateXxxxDto } from './dto/create-xxx.dto';
import { UpdateXxxDto } from './dto/update-xxx.dto';
import { InjectRepository } from '@nestjs/typeorm';
import { Repository } from 'typeorm';
import { Xxx } from './entities/xxx.entity';

@Injectable()
export class XxxService {
  // 注入 Xxx 實體
  @InjectRepository(Xxx) private xxxRepository: Repository<User>
  // 新增
  async create(createXxxDto: CreateXxxxDto) {
    createXxxDto.createTime = createXxxDto.updateTime = new Date()
    return await this.xxxRepository.save(createXxxDto);
  }
  // 查詢全部
  async findAll() {
    return await this.xxxRepository.find();
  }
  // 查詢單一
  async findOne(id: number) {
    return this.xxxRepository.findBy;
  }
  // 修改
  async update(id: number, updateXxxDto: UpdateXxxDto) {
    updateXxxDto.updateTime = new Date()
    return await this.xxxRepository.update(id, updateXxxDto);
  }
  // 刪除
  async remove(id: number) {
    return await this.xxxRepository.delete(id);
  }
}
```

由此可見，服務層（Service）會呼叫 Repository 實例方法對物理資料進行 CRUD 操作，而實體物件資料的改變將同步映射到資料庫中對應欄位的更新。這就是物件關係映射（Object-Relational Mapping，ORM）的作用，後面會更加詳細地介紹 ORM 的概念。

在 Nest 框架中，用戶端發出的請求會經歷一個分層處理過程：首先由控制器（Controller）接收請求，然後傳遞給服務層（Service），接著服務層可能會呼叫倉庫層（Repository）進行資料操作，最終將處理結果傳回給用戶端。這一流程正是 Nest 分層架構的典型表現。

3.3.5 服務與服務提供者

Service 服務是 Nest 分層架構中的組成部分。透過使用 @Injectable 裝飾器宣告，它可以被控制器（Controller）排程使用，或被其他服務共用。這時，一些讀者可能會有疑問：服務與服務提供者之間是什麼關係？實際上，所有透過 @Injectable 裝飾器裝飾的類別都是服務提供者。這是一個抽象概念，它不僅包括服務類別，還可能包括中介軟體、攔截器、管道等，它們可以是工廠函數或常數。這些服務提供者被相依注入系統注入其他元件中（例如控制器、服務等），以提供特定的功能、資料或操作。

3.4 耳熟能詳的中介軟體

後端系統中有很多中介軟體，比如訊息中介軟體如 Kafka 和 MQ，快取中介軟體如 Redis，資料庫中介軟體如 Hibernate 和 MyBatis 等。在 Node.js 生態系統中，Express 和 Koa 等第三方框架都實現了中介軟體功能。同樣地，Nest 預設基於 Express 框架，自然也支援中介軟體，但實際上它們並不完全相同。本節將探討 Nest 的中介軟體。

Nest 實現切面程式設計（AOP）思想導向的方式之一是中介軟體。根據實現方式，可以分為函數式中介軟體和類別中介軟體；根據作用域，又可以分為局部中介軟體和全域中介軟體。

建立一個 Nest 專案，執行「nest n nest-middleware -p pnpm」命令，如圖 3-14 所示。

```
$ nest n nest-middleware -p pnpm
⚡  We will scaffold your app in a few seconds..

CREATE nest-middleware/.eslintrc.js (663 bytes)
CREATE nest-middleware/.prettierrc (51 bytes)
CREATE nest-middleware/README.md (3347 bytes)
CREATE nest-middleware/nest-cli.json (171 bytes)
CREATE nest-middleware/package.json (1956 bytes)
CREATE nest-middleware/tsconfig.build.json (97 bytes)
CREATE nest-middleware/tsconfig.json (546 bytes)
CREATE nest-middleware/src/app.controller.spec.ts (617 bytes)
CREATE nest-middleware/src/app.controller.ts (274 bytes)
CREATE nest-middleware/src/app.module.ts (249 bytes)
CREATE nest-middleware/src/app.service.ts (142 bytes)
CREATE nest-middleware/src/main.ts (208 bytes)
CREATE nest-middleware/test/app.e2e-spec.ts (630 bytes)
CREATE nest-middleware/test/jest-e2e.json (183 bytes)

✔ Installation in progress... ☕

🚀  Successfully created project nest-middleware
```

▲ 圖 3-14 建立專案

建立成功後,接下來演示這幾種中介軟體的使用及它們之間的區別。

3.4.1 類別中介軟體

類別中介軟體透過使用 @Injectable() 裝飾器來宣告,並需要實現 NestMiddleware 介面的 use 方法。在預設情況下,如果使用 Express 作為 HTTP 框架,中介軟體可以操作 Express 提供的 Request 和 Response 物件。而當使用 Fastify 作為 HTTP 框架時,對應的請求和回應物件分別是 FastifyRequest 和 FastifyReply 物件。

執行「nest g middleware logger」命令生成一個中介軟體,logger.middleware.ts 程式如下:

```
import { Injectable, NestMiddleware } from '@nestjs/common';
import { Request, Response, NextFunction } from 'express';

@Injectable()
export class LoggerMiddleware implements NestMiddleware {
  use(req: Request, res: Response, next: NextFunction) {
    console.log('before 中介軟體');
    next();
    console.log('after 中介軟體');
  }
}
```

接著執行「nest g resource user」命令建立 User 模組,並在 user.module.ts 檔案中使用中介軟體,範例程式如下:

```
import { MiddlewareConsumer, Module, NestModule } from '@nestjs/common';
import { UserService } from './user.service';
import { UserController } from './user.controller';
import { LoggerMiddleware } from 'src/logger/logger.middleware';

@Module({
  controllers: [UserController],
  providers: [UserService]
})
export class UserModule implements NestModule {
  configure(consumer: MiddlewareConsumer) {
    // 針對此模組的所有路由綁定中介軟體
    consumer.apply(LoggerMiddleware).forRoutes('*');
  }
}
```

可見,@Module 裝飾器中並沒有直接提供註冊中介軟體的位置。使用中介軟體的模組必須實現 NestModule 介面的 configure 方法,並透過 consumer 輔助類別對中介軟體進行配置。forRoutes 方法用於指定應用中間的路由。該中介軟體可以應用到 UserController 中的所有路由程式中,當然,我們也可以指定更精確的路由路徑或請求方式,例如:

```typescript
import { MiddlewareConsumer, Module, NestModule, RequestMethod } from '@nestjs/common';
import { UserService } from './user.service';
import { UserController } from './user.controller';
import { LoggerMiddleware } from 'src/logger/logger.middleware';

@Module({
  controllers: [UserController],
  providers: [UserService]
})
export class UserModule implements NestModule {
  configure(consumer: MiddlewareConsumer) {
      // consumer.apply(LoggerMiddleware).forRoutes('*')
      // 指定應用中介軟體的路由和請求方式
      consumer.apply(LoggerMiddleware).forRoutes({
        path: '/user',
        method: RequestMethod.GET
      })
  }
}
```

我們指定中介軟體應用在 /person 路徑,並指定了 GET 請求方式,完成後在瀏覽器存取 localhost:3000,主控台列印效果如圖 3-15 所示。

```
[Nest] 53714   - 2024/01/03 15:18:43    LOG [RouterExplorer] Mapped {/user/:id, GET} route +0ms
[Nest] 53714   - 2024/01/03 15:18:43    LOG [RouterExplorer] Mapped {/user/:id, PATCH} route +0ms
[Nest] 53714   - 2024/01/03 15:18:43    LOG [RouterExplorer] Mapped {/user/:id, DELETE} route +0ms
[Nest] 53714   - 2024/01/03 15:18:43    LOG [NestApplication] Nest application successfully started +1ms
before中介軟體
after中介軟體
```

▲ 圖 3-15 中介軟體的執行結果

在 Nest 中,類別中介軟體的作用不僅限於處理 HTTP 請求和回應,更重要的是它能夠實現相依注入。這表示我們可以在中介軟體中注入特定的相依項,並呼叫這些相依項內部的方法,例如 UserService 服務。範例程式如下:

```typescript
import { Injectable, NestMiddleware } from '@nestjs/common';
import { Inject } from '@nestjs/common/decorators';
import { Request, Response, NextFunction } from 'express';
import { UserService } from 'src/user/user.service';

@Injectable()
export class LoggerMiddleware implements NestMiddleware {
  // 注入 UserService 相依
```

```
  @Inject(UserService)
  private userService: UserService;
  use(req: Request, res: Response, next: NextFunction) {
    console.log('before 中介軟體 ');
    console.log(' 中介軟體作用的方法執行結果：', this.userService.findAll());
    next();
    console.log('after 中介軟體 ');
  }
}
```

刷新瀏覽器，執行結果如圖 3-16 所示。

```
[Nest] 53533  - 2024/01/03 15:04:21    LOG [RouterExplorer] Mapped {/user, GET} route +0ms
[Nest] 53533  - 2024/01/03 15:04:21    LOG [RouterExplorer] Mapped {/user/:id, GET} route +0ms
[Nest] 53533  - 2024/01/03 15:04:21    LOG [RouterExplorer] Mapped {/user/:id, PATCH} route +1ms
[Nest] 53533  - 2024/01/03 15:04:21    LOG [RouterExplorer] Mapped {/user/:id, DELETE} route +0ms
[Nest] 53533  - 2024/01/03 15:04:21    LOG [NestApplication] Nest application successfully started +1ms
before中介軟體
中介軟體的作用的方法執行結果：  This action returns all user
after中介軟體
```

▲ 圖 3-16 中介軟體注入外部相依

這就是 Nest 中介軟體的相依注入。當然，如果不需要相依，可以用輕量的函數式中介軟體。這就好比在 React/Vue 中，當我們的元件不需要狀態管理時，優先選擇函數式元件是類似的道理。

3.4.2 函數式中介軟體

Nest 中提供了函數式中介軟體，又稱為功能中介軟體。當中介軟體不需要成員、沒有額外的方法和相依時，應優先考慮使用它。

下面以 LoggerMiddleware 為例：

```
import { Request, Response, NextFunction } from 'express';

export function LoggerMiddleware(
  req: Request,
  res: Response,
  next: NextFunction,
) {
  console.log(' 中介軟體 before');
  next();
  console.log(' 中介軟體 after');
}
```

簡單吧？與定義普通的函數沒什麼區別。

如果管道驗證器驗證失敗，則需要拋出例外並回應給用戶端。這時就需要使用異常篩檢程式來處理。

5. 篩檢程式

Nest 中最為常見的是 HTTP 異常篩檢程式，通常用於在後端服務發生異常時向用戶端報告異常的類型。目前內建的 HTTP 異常包含：

- BadRequestException
- UnauthorizedException
- NotFoundException
- ForbiddenException
- NotAcceptableException
- RequestTimeoutException
- ConflictException
- GoneException
- HttpVersionNotSupportedException
- PayloadTooLargeException
- UnsupportedMediaTypeException
- …

它們都繼承自 HttpException 類別。當然，我們也可以自訂異常篩檢程式，並向前端傳回統一的資料格式：

```
import { ArgumentsHost, Catch, ExceptionFilter, HttpException } from "@nestjs/common";
import { Request, Response } from "express";

@Catch(HttpException)
export class HttpExceptionFilter implements ExceptionFilter {
    catch(exception: HttpException, host: ArgumentsHost) {
        // 指定傳輸協定
        const ctx = host.switchToHttp();
        // 獲取請求回應物件
        const response = ctx.getResponse<Response>();
        const request = ctx.getRequest<Request>();
        // 獲取狀態碼和異常訊息
        const status = exception.getStatus();
        const message = exception.message;
        let resMessage: string | Record<string, any> = exception.getResponse();
        console.log(' 進入異常篩檢程式 ');
```

在中介軟體中強制注入 UserService，可能會導致錯誤。因此，在 Express 中，通常推薦在應用模組（AppModule）中透過 configure 方法來註冊中介軟體。

Nest 框架中的中介軟體分為類別中介軟體和函數式中介軟體，它們各自有不同的使用場景和作用域。類別中介軟體支援相依注入，可以透過 forRoutes 方法綁定到一個或多個具體的路由處理器或控制器。此外，也可以透過 app.use 方法將中介軟體註冊為全域中介軟體。

在 Nest 應用的初始化過程中，中介軟體的執行順序是首先執行全域中介軟體，然後執行局部（模組層級）中介軟體。

3.5　攔截器與 RxJS 知多少

攔截器是 AOP 程式設計思想的第二種實現方式，第一種是 3.4 節介紹的中介軟體。本節首先介紹攔截器的基本概念和常見的應用場景，然後透過程式演示如何實現一個介面逾時的攔截器。這個攔截器需要與非同步事件處理函數庫 RxJS 配合來使用，最後還會介紹 RxJS 常用的 API，並且練習如何使用它。

3.5.1　基本概念

攔截器的請求和回應流程遵循先進後出的順序。請求首先透過全域攔截器，然後是控制器等級的攔截器，最後是路由等級的攔截器進行處理。回應流程則相反，即從路由等級的攔截器開始，經過控制器等級的攔截器，最終到達全域攔截器。這樣的設計允許在請求處理的任何階段，包括由管道、控制器或服務拋出的錯誤，都能夠透過攔截器進行捕捉和處理。

每個攔截器透過 @Injectable() 來宣告，並且需要實現 NestInterceptor 介面的 intercept 方法，接收兩個參數：ExecutionContext 上下文物件和 CallHandler 處理常式。

ExecutionContext 能夠存取當前請求的詳細資訊，包括路由資訊、HTTP 方法、請求本體以及回應本體資料。它主要應用於以下幾個場景：

（1）記錄請求和回應的日誌，用於追蹤、監控和偵錯。

（2）進行身份驗證和許可權檢查。

（3）根據請求標頭或路由資訊來設置快取策略。

（4）修改或轉換回應資料，例如對回應進行包裝、格式化、加密操作。

CallHandler 實現了 handle 方法，必須在攔截器中呼叫 handle 方法才能執行路由處理方法。範例程式如下：

```
import {
  CallHandler,
  ExecutionContext,
  Injectable,
  NestInterceptor,
} from '@nestjs/common';

@Injectable()
export class TimeoutInterceptor implements NestInterceptor {
  intercept(context: ExecutionContext, next: CallHandler): Observable<any> {
    // 呼叫 handle
    return next.handle();
  }
}
```

通常情況下，intercept 方法傳回 RxJS 的 Observable 物件。那麼，什麼是 RxJS 呢？

RxJS 是一個用於處理非同步資料流程的 JavaScript 函數庫。你可以把它理解為一個「管道」，它可以幫你更方便地處理各種事件和資料流程。比如，當你需要處理使用者的點擊事件、網路請求傳回的資料、計時器觸發的事件等，RxJS 能夠幫助你更加優雅地管理這些複雜的非同步作業。

然而，我們不必擔心這會增加額外的認知負擔，可以簡單地將其視為前端的 Lodash 工具集。接下來，我們透過新建專案來演示說明這一點。

3.5.2 建立專案

執行「nest n nest-interceptor -p pnpm」命令建立專案，如圖 3-17 所示。

```
$ nest n nest-interceptor -p pnpm
⚡ We will scaffold your app in a few seconds..
CREATE nest-interceptor/.eslintrc.js (663 bytes)
CREATE nest-interceptor/.prettierrc (51 bytes)
CREATE nest-interceptor/README.md (3347 bytes)
CREATE nest-interceptor/nest-cli.json (171 bytes)
CREATE nest-interceptor/package.json (1957 bytes)
CREATE nest-interceptor/tsconfig.build.json (97 bytes)
CREATE nest-interceptor/tsconfig.json (546 bytes)
CREATE nest-interceptor/src/app.controller.spec.ts (617 bytes)
CREATE nest-interceptor/src/app.controller.ts (274 bytes)
CREATE nest-interceptor/src/app.module.ts (249 bytes)
CREATE nest-interceptor/src/app.service.ts (142 bytes)
CREATE nest-interceptor/src/main.ts (208 bytes)
CREATE nest-interceptor/test/app.e2e-spec.ts (630 bytes)
CREATE nest-interceptor/test/jest-e2e.json (183 bytes)

✔ Installation in progress... 🍺

🚀 Successfully created project nest-interceptor
```

▲ 圖 3-17 建立專案

3.5.3 攔截器的基本使用方法

在 src 目錄下新建一個名為 interceptor 的子目錄，專門用於存放應用攔截器。接下來，我們將實現一個用於統計介面逾時的攔截器。範例程式如下：

```
import {
  CallHandler,
  ExecutionContext,
  Injectable,
  NestInterceptor,
} from '@nestjs/common';
import { log } from 'console';
import { Observable, tap, timeout } from 'rxjs';

@Injectable()
export class TimeoutInterceptor implements NestInterceptor {
  intercept(context: ExecutionContext, next: CallHandler): Observable<any> {
    console.log('進入攔截器', context.getClass());
    const now = Date.now();
    // 呼叫 handle
    return next.handle().pipe(
      tap(() => {
        // 統計耗時
        log('Timeout: ', Date.now() - now);
      }),
      // 逾時時間
      timeout(1000),
    );
```

```
    }
}
```

在上面的程式中，tap 和 timeout 操作符號不會改變資料或干擾回應週期，而是用於執行額外的邏輯處理。

將這些操作符號綁定到 app.controller.ts 檔案中的 getHello 方法上，具體範例如圖 3-18 所示。

```
@Controller()
export class AppController {
  constructor(private readonly appService: AppService) {}

  @Get()
  @UseInterceptors(TimeoutInterceptor)
  getHello(): string {
    return this.appService.getHello();
  }
}
```

▲ 圖 3-18 綁定攔截器

接著在瀏覽器輸入 localhost:3000，執行結果如圖 3-19 所示。

```
[Nest] 65604  - 2024/01/03 20:17:28     LOG [RouterExplorer] Mapped {/, GET} route +1ms
[Nest] 65604  - 2024/01/03 20:17:28     LOG [NestApplication] Nest application successful
進入攔截器 [class AppController]
Timeout: 2
```

▲ 圖 3-19 攔截器的執行結果

除此之外，還有其他操作符號，如 map、filter、delay、from、toArray、catchError 等。我們接下來定義一個稍微複雜一些的攔截器，用於轉換並過濾指定使用者的名稱。範例程式如下：

```
import {
  Injectable,
  NestInterceptor,
  ExecutionContext,
  CallHandler,
} from '@nestjs/common';
import { Observable, of } from 'rxjs';
import { tap, map, catchError, filter, toArray } from 'rxjs/operators';

@Injectable()
export class AuthInterceptor implements NestInterceptor {
  intercept(context: ExecutionContext, next: CallHandler): Observable<any> {
    // 在請求處理之前進行日誌記錄
    console.log('Request received...');
```

```
    return next.handle().pipe(
      // 使用 map 操作符號將每個使用者名稱轉為大寫
      map((name) => name.toUpperCase()),

      // 使用 filter 操作符號過濾出名字長度大於 2 的使用者
      filter((name) => name.length > 2),

      // 使用 tap 操作符號進行簡單的日誌記錄
      tap((arr) => console.log('Filtered Name: ${arr}')),

      // 轉為陣列
      toArray(),

      // 使用 catchError 操作符號捕捉並處理任何錯誤
      catchError((error) => {
        console.error('Error occurred:', error);
        // 傳回新的 Observable
        return of('Fallback Value');
      }),
    );
  }
}
```

在上面的程式中，透過 map 操作符號把結果轉為大寫，然後使用 filter 操縱符號過濾長度大於 2 的名稱，接著列印日誌並最後轉為陣列。如果出現異常，使用 catchError 來捕捉並傳回新的資料。

將這些操作符號綁定到 app.controller.ts 檔案中的 getNameList 方法上，具體範例如圖 3-20 所示。

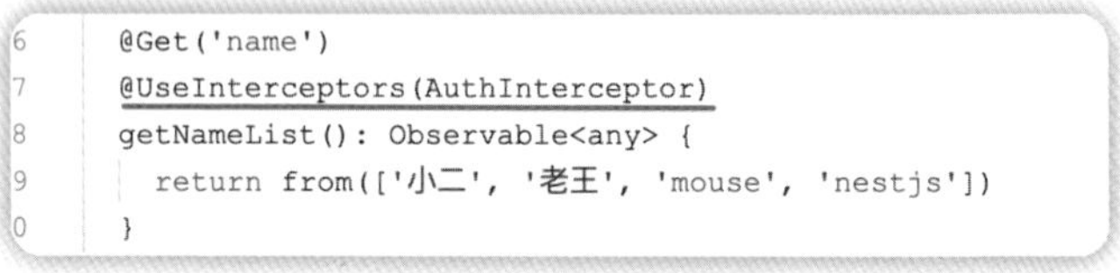

```
 6    @Get('name')
 7    @UseInterceptors(AuthInterceptor)
 8    getNameList(): Observable<any> {
 9      return from(['小二', '老王', 'mouse', 'nestjs'])
10    }
```

▲ 圖 3-20 綁定攔截器

在瀏覽器中輸入 localhost:3000/name，執行結果如圖 3-21 所示。

```
[
    "MOUSE",
    "NESTJS"
]
```

▲ 圖 3-21 攔截器的執行結果

如果介面在執行過程中拋出例外，catchError 操作符號將介入並傳回資料，作為錯誤處理的最後手段，如圖 3-22 所示。

```
15    }
16    @Get('name')
17    @UseInterceptors(AuthInterceptor)
18    getNameList(): Observable<any> {
19      throw Error('34343')
20    }
21  }
```

▲ 圖 3-22 介面拋出例外

刷新瀏覽器，傳回「Fallback Value」，效果如圖 3-23 所示。

```
← → C  ⓘ http://localhost:3000/name
Fallback Value
```

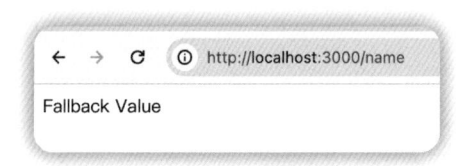

▲ 圖 3-23 傳回資料

除此之外，在前面的 2.4 節中，我們學習了 HTTP 框架擁有許多內建的異常篩檢程式。這些異常篩檢程式不僅可以捕捉和處理異常，還可以作為「Fallback Value」（備用傳回值）使用。

3.6 資料之源守護者：管道

前面講過，控制器層（Controllers）的職責之一是解析和驗證請求資料。這項工作是由什麼來完成的呢？正是本節將要介紹的管道。本節首先介紹管道的基本概念和用法，包括 Nest 框架中內建的 9 種開箱即用的管道驗證器。接下來，將演示這些驗證器如何應用於處理 GET 和 POST 請求參數。最後，透過實現自訂管道，將進一步加深對這一概念的理解。

3.6.1 基本概念

在後端開發中，資料庫表的欄位類型在建立時就已經被明確定義，任何不符合預期類型的資料儲存操作都會導致錯誤。為了確保傳入的資料滿足預期的格式和標準，Nest 框架會在用戶端發起請求時，將請求資料傳遞給管道進行前置處理。這些前置處理操作包括資料驗證、轉換或過濾等，以確保資料的準確性。處理後的資料隨後會被傳遞給路由處理常式。

如果在驗證過程中遇到任何異常，Nest 框架可以使用異常篩檢程式來統一捕捉和處理這些異常。這包括記錄錯誤日誌、傳回統一格式的錯誤回應等操作。

通常情況下，我們會將管道綁定在方法參數上，這樣管道就會與特定的路由方法連結起來，這種方式稱為參數等級管道。除此之外，Nest 還允許將管道綁定在全域作用域，使其適用於每個控制器和路由方法。

3.6.2 內建管道

Nest 內建了 9 種開箱即用的管道驗證器，它們分別是：

- ParseIntPipe
- ParseFloatPipe
- ParseBoolPipe
- ParseUUIDPipe
- ParseEnumPipe
- DefaultValuePipe
- ValidationPipe
- ParseArrayPipe
- ParseFilePipe

接下來建立一個 Nest 專案，用來分別演示 GET 和 POST 請求中常用的幾個管道。執行「nest n nest-pipe -p pnpm」命令，結果如圖 3-24 所示。

```
$ nest n nest-pipe -p pnpm
⚡  We will scaffold your app in a few seconds..

CREATE nest-pipe/.eslintrc.js (663 bytes)
CREATE nest-pipe/.prettierrc (51 bytes)
CREATE nest-pipe/README.md (3347 bytes)
CREATE nest-pipe/nest-cli.json (171 bytes)
CREATE nest-pipe/package.json (1950 bytes)
CREATE nest-pipe/tsconfig.build.json (97 bytes)
CREATE nest-pipe/tsconfig.json (546 bytes)
CREATE nest-pipe/src/app.controller.spec.ts (617 bytes)
CREATE nest-pipe/src/app.controller.ts (274 bytes)
CREATE nest-pipe/src/app.module.ts (249 bytes)
CREATE nest-pipe/src/app.service.ts (142 bytes)
CREATE nest-pipe/src/main.ts (208 bytes)
CREATE nest-pipe/test/app.e2e-spec.ts (630 bytes)
CREATE nest-pipe/test/jest-e2e.json (183 bytes)

✔ Installation in progress... ☕

🚀  Successfully created project nest-pipe
```

▲ 圖 3-24 建立專案

執行「pnpm start:dev」命令以啟動服務，然後在瀏覽器中存取 localhost:3000。如果執行成功，結果如圖 3-25 所示。

▲ 圖 3-25　成功執行結果

1. ParseIntPipe

ParseIntPipe 接收的參數必須能夠被解析為整數（parseInt），否則將拋出例外。範例程式如下：

```
@Get('int/:id')
getHello(@Param('id', ParseIntPipe) id: number): string {
  log('id:', id);
  return this.appService.getHello();
}
```

@Param 裝飾器用於接收 URL 中的參數，例如在瀏覽器中存取 localhost:3000/int/123，即可成功獲取 id 參數，如圖 3-26 所示。

▲ 圖 3-26　ParseIntPipe 驗證參數

把位址改為 localhost:3000/xxx 並刷新，此時請求不能透過管道驗證，會拋出例外，如圖 3-27 所示。

▲ 圖 3-27　驗證異常效果

這個錯誤回應是 Nest 預設定義的，也可以自訂友善的錯誤訊息資訊，驗證器允許傳遞下面兩個參數。

（1）errorHttpStatusCode：驗證器失敗時拋出的 HTTP 狀態碼，預設為 400（錯誤請求）。

（2）exceptionFactory：工廠函數，用於接收錯誤訊息並傳回相應的錯誤物件。

修改後的程式如下：

```
@Get(':id')
getHello(
  @Param(
    'id',
    new ParseIntPipe({
      exceptionFactory: () => {
        // 拋出 HTTP 例外
        throw new HttpException(' 參數 id 類型錯誤 ', HttpStatus.BAD_REQUEST);
      },
    }),
  )
  id: number,
): string {
  log('id:', id);
  return this.appService.getHello();
}
```

刷新瀏覽器，傳回結果如圖 3-28 所示。

▲ 圖 3-28 自訂異常資訊

同樣地，接下來要介紹的驗證器也允許我們透過這兩個參數來自訂錯誤訊息資訊。

2. ParseFloatPipe

接下來，我們來看 ParseFloatPipe。此管道的作用是將接收到的資料轉為浮點數（Float）類型。範例程式如下：

```
@Get('float/:id')
getHello2(@Param('id', ParseFloatPipe) id: Number): string {
  log('float id:', id);
  return this.appService.getHello()
}
```

在瀏覽器中輸入 localhost:3000/float/123.45，執行結果如圖 3-29 所示。

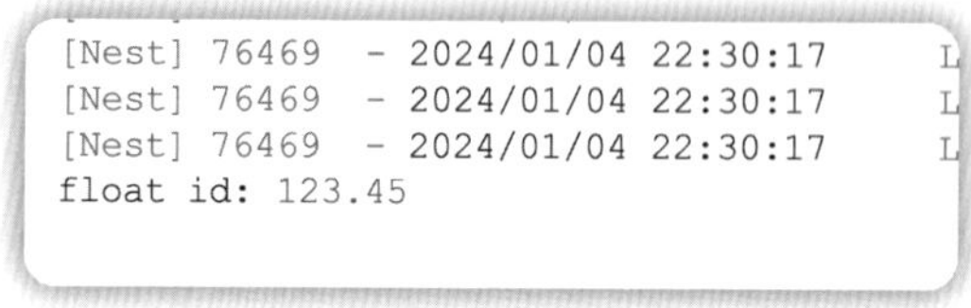

```
[Nest] 76469    - 2024/01/04 22:30:17      L
[Nest] 76469    - 2024/01/04 22:30:17      L
[Nest] 76469    - 2024/01/04 22:30:17      L
float id: 123.45
```

▲ 圖 3-29 ParseFloatPipe 驗證參數

3. ParseBoolPipe

ParseBoolPipe 負責將參數轉為布林（Boolean）類型。我們可以透過 @Query() 裝飾器來接收查詢參數，並利用此管道進行轉換。範例程式如下：

```
@Get('bool')
getHello3(@Query('flag', ParseBoolPipe) flag: boolean): string {
  log('bool flag:', flag)
  return this.appService.getHello()
}
```

瀏覽器中輸入 localhost:3000/bool?flag=true，執行結果如圖 3-30 所示。

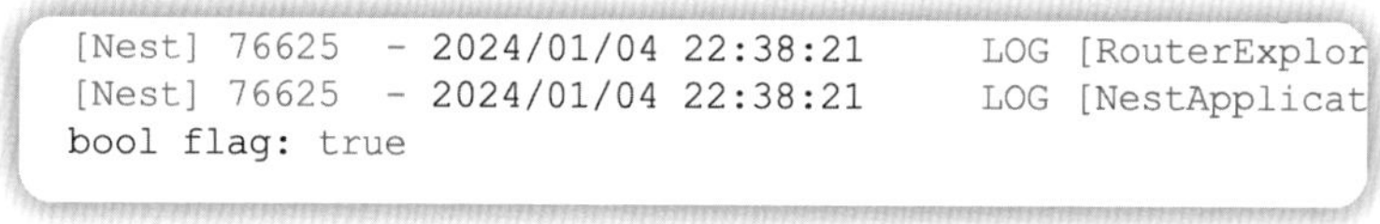

```
[Nest] 76625    - 2024/01/04 22:38:21      LOG [RouterExplor
[Nest] 76625    - 2024/01/04 22:38:21      LOG [NestApplicat
bool flag: true
```

▲ 圖 3-30 ParseBoolPipe 驗證參數

4. ParseUUIDPipe

uuid 是一串隨機生成且唯一的字元，廣泛應用於需要唯一性或保護隱私的場景，例如資料庫主鍵、訂單編號、使用者標識等。

uuid 有多個版本，包括 v1、v3、v4 和 v5。其中，v1 和 v4 是較為常用的版本。v1 版本透過結合時間戳記和節點資訊來保證時間順序和唯一性，而 v4 版本則生成完全隨機的 uuid。v3 和 v5 版本基於雜湊演算法，能夠確保對於相同的輸入，生成的 uuid 是一致的。

在接收 UUID 參數時，可以使用 ParseUUIDPipe 管道來驗證其格式正確性。範例程式如下：

```
@Get('uuid')
getHello4(@Query('id', ParseUUIDPipe) id: string): string {
  log('uuid id:', id)
  return this.appService.getHello()
}
```

在瀏覽器中輸入 localhost:3000/uuid?id=1262c798-aacf-4c41-ad73-7dd29a708a68，執行結果如圖 3-31 所示。

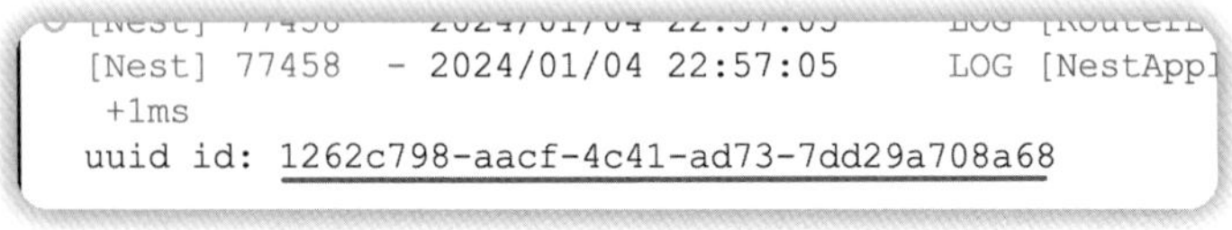

```
[Nest] 77458    - 2024/01/04 22:57:05    LOG [NestApp]
+1ms
uuid id: 1262c798-aacf-4c41-ad73-7dd29a708a68
```

▲ 圖 3-31 ParseUUIDPipe 驗證參數

5. ParseEnumPipe

ParseEnumPipe 是用於驗證列舉值的管道。如果接收到的參數不在預定的列舉值內，驗證將失敗。舉例來說，我們可以定義一組列舉值來描述某種狀態。範例程式如下：

```
enum StatusEnum {
  ACTIVE = 'active',
  INACTIVE = 'inactive',
  PENDING = 'pending',
}
```

同時，在路由方法中驗證參數是否合法：

```
@Get('enum')
getHello5(
  @Query('status', new ParseEnumPipe(StatusEnum)) status: string
): string {
  log('enum status:', status);
  return this.appService.getHello();
}
```

在瀏覽器中輸入 localhost:3000/enum?status=active，執行結果如圖 3-32 所示。

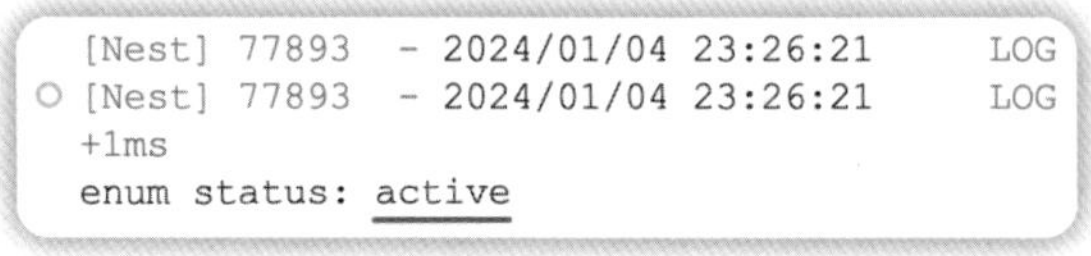

```
[Nest] 77893    - 2024/01/04 23:26:21    LOG
[Nest] 77893    - 2024/01/04 23:26:21    LOG
+1ms
enum status: active
```

▲ 圖 3-32 ParseEnumPipe 驗證參數

此時，如果傳遞一個不在列舉範圍內的參數，則會觸發異常，如圖 3-33 所示。

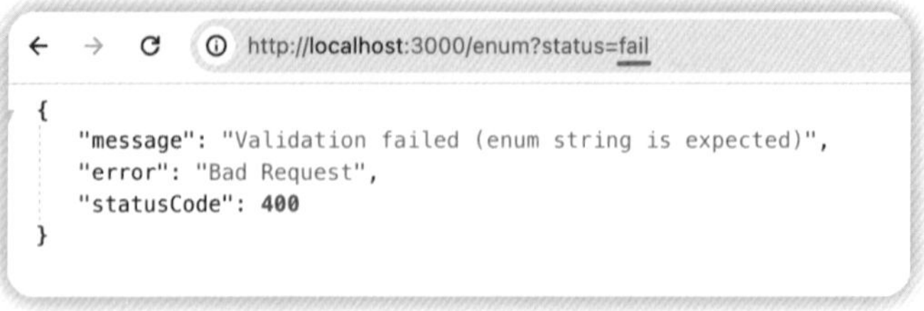

▲ 圖 3-33　enum 參數驗證異常

在實際開發中，驗證列舉參數的場景並不常見。然而，它能夠將前後端的列舉狀態統一起來維護，這對基於 MVC 架構系統的專案來說是一個不錯的選擇。

6. DefaultValuePipe

DefaultValuePipe 的作用非常直接：當請求中未傳遞參數時，它允許我們為該參數指定一個預設值。範例程式如下：

```
@Get('default')
getHello6(
  @Query('value', new DefaultValuePipe('jmin')) value: string
): string {
  log('default value:', value);
  return this.appService.getHello();
}
```

在瀏覽器中輸入 localhost:3000/default，執行結果如圖 3-34 所示。

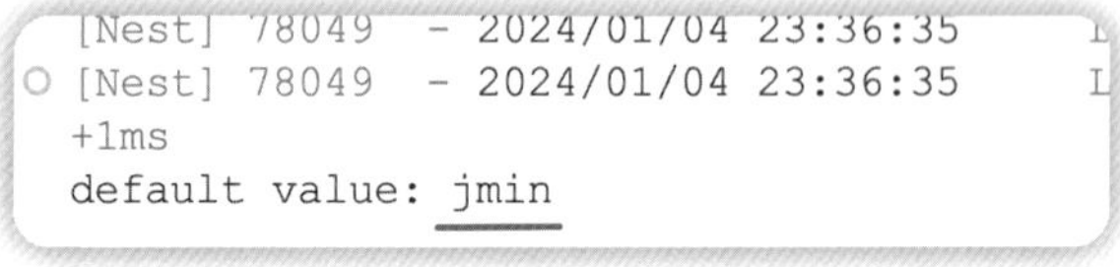

▲ 圖 3-34　DefaultValuePipe 驗證參數

需要注意的是，如果輸入位址為 localhost:3000/default?value=，則不會觸發預設賦值。Nest 會認為你傳遞的參數為空字串，執行結果如圖 3-35 所示。

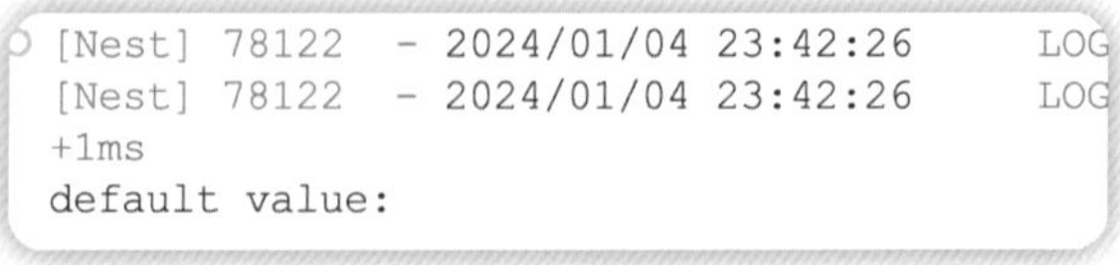

▲ 圖 3-35　傳遞空字串參數

7. ParseArrayPipe

ParseArrayPipe 用於將傳遞的字串參數轉為陣列類型。舉例來說，在需要傳遞多個使用者 ID 的場景中，用戶端可能會發送形如 /array/1,2,3 的請求。透過使用 ParseArrayPipe，這樣的請求可以被轉為 [1,2,3] 這樣的陣列形式。

由於 ParseArrayPipe 相依於 class-validator 和 class-transformer 這兩個 npm 套件，在開始使用它之前，需要先進行安裝。可以透過執行「pnpm add class-validator class-transformer –save」命令來增加這些相依。安裝完成後，可以繼續撰寫相關的程式：

```
@Get('array')
getHello7(@Query('ids', ParseArrayPipe) ids: Number[]): string {
  log('array ids:', ids);
  return this.appService.getHello();
}
```

在瀏覽器中輸入 localhost:3000/array?ids=1,2,3，執行結果如圖 3-36 所示。

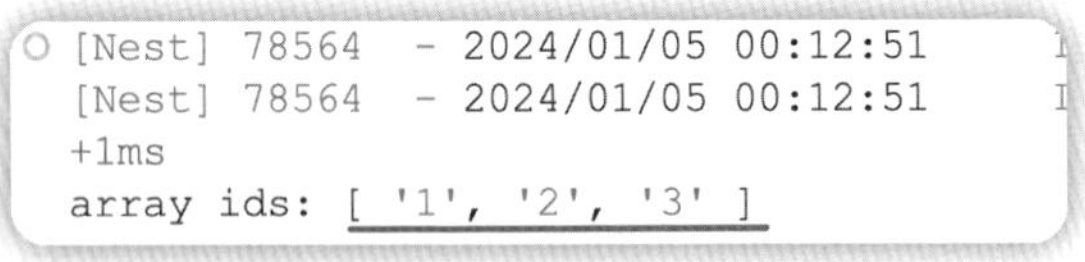

```
[Nest] 78564    - 2024/01/05 00:12:51
[Nest] 78564    - 2024/01/05 00:12:51
+1ms
array ids: [ '1', '2', '3' ]
```

▲ 圖 3-36　ParseArrayPipe 驗證參數

ParseArrayPipe 預設使用逗點（,）作為分隔符號來分隔字串參數。然而，在 Nest 框架中，我們可以靈活地自訂分隔符號以滿足不同的需求。例如：

```
@Get('array')
getHello7(
  @Query('ids', new ParseArrayPipe({
    separator: '-'
  })) ids: Number[]
): string {
  log('array ids:', ids);
  return this.appService.getHello();
}
```

這樣就可以透過 localhost:3000/array?ids=4-5-6 這種方式來傳參了，執行結果如圖 3-37 所示。

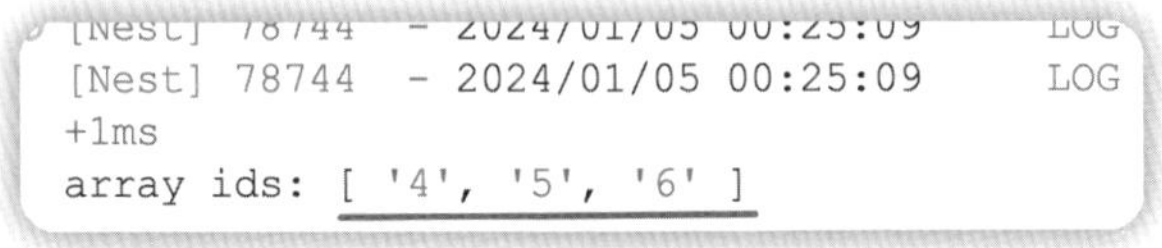

```
[Nest] 78744    - 2024/01/05 00:25:09     LOG
[Nest] 78744    - 2024/01/05 00:25:09     LOG
+1ms
array ids: [ '4', '5', '6' ]
```

▲ 圖 3-37　使用自訂參數分隔符號後的執行結果

8. ValidationPipe

前面介紹的都是 GET 請求的參數驗證，如果是 POST 請求發送的請求本體，又該如何驗證呢？這就要用到 ValidationPipe 了。POST 請求本體可以透過 @Body 裝飾器來接收，需要配合資料傳輸物件（Data Transfer Object，DTO）來使用。新建 src/dto/create-user.dto.ts 檔案。範例程式如下：

```
import { IsInt, IsNotEmpty, IsString } from "class-validator";
export class CreateUserDto {
    @IsNotEmpty()
    @IsString()
    name: string

    @IsInt({ message: "age 必須是數字類型 "})
    age: number

    @IsString({ message: "sex 必須是字串類型 "})
    sex: string

    phone: string

    email: string
}
```

ValidationPipe 的使用需要相依 class-validator 和 class-transformer 這兩個套件。class-validator 提供了一系列用於驗證欄位的裝飾器，例如 @IsNotEmpty、@IsString 和 @IsInt。這些裝飾器允許我們對欄位進行驗證，並可以自訂錯誤訊息資訊（message）。之後，我們可以在路由方法中綁定這些驗證器來執行資料驗證：

```
@Post('create-user')
createUser(@Body(ValidationPipe) createUserDto: CreateUserDto): Record<string, any> {
  return createUserDto
}
```

接下來，可以透過 Postman 或 AJAX 來呼叫介面。實際上，VS Code 也提供了一種更加快捷的方式來完成 mock 請求和回應的流程。具體步驟如下：

（1）安裝 VS Code 外掛程式 REST Client，如圖 3-38 所示。

▲ 圖 3-38　安裝外掛程式

（2）在專案根目錄下建立 .http 檔案，設置請求 URL、Content-type、請求參數 / 本體，多個介面用 ### 分隔。範例程式如下：

```
GET http://localhost:3000/array?ids=1-2-3
###

POST http://localhost:3000/create-user
Content-Type: application/json

{
  "name": "",
  "age": 22,
  "sex": "男",
  "phone": "",
  «email»: «mouse@example.com»,
}
```

（3）點擊 Send Request 發送請求，如圖 3-39 所示。

▲ 圖 3-39 發送請求

（4）發送請求後，在 VS Code 中可以看到回應本體，如圖 3-40 所示。

▲ 圖 3-40 查看回應結果

圖 3-40 中提示的 name 不能為空，這就是 ValidationPipe 的預設驗證結果。我們將 age 改為 String 類型，重新發送請求，可以看到傳回的自訂提示訊息，如圖 3-41 所示。

```
Send Request                                    utf-8
POST http://localhost:3000/create-user      4   Content-Length: 80
Content-Type: application/json              5   ETag: W/"50-9fU53baXMA/m0KbZbXht4QeIu
                                            6   Date: Fri, 05 Jan 2024 05:03:57 GMT
{                                           7   Connection: close
  "name": "mouse",                          8
  "age": "22",                              9 ∨ {
  "sex": "男",                              10 ∨   "message": [
  "phone": "",                              11        "age 必須是數字類型"
  "email": "mouse@example.com"              12     ],
}                                           13     "error": "Bad Request",
                                            14     "statusCode": 400
                                            15 }
```

▲ 圖 3-41 自訂驗證提示

在專案開發過程中，通常會有多個 POST 請求需要進行驗證。如果在每個路由方法中單獨綁定驗證器，可能會顯得相當煩瑣。為了簡化這一過程，我們可以將管道驗證器配置為全域生效。例如：

```
import { NestFactory } from '@nestjs/core';
import { AppModule } from './app.module';
import { ValidationPipe } from '@nestjs/common';

async function bootstrap() {
  const app = await NestFactory.create(AppModule);
  // 綁定全域驗證器
  app.useGlobalPipes(new ValidationPipe())
  await app.listen(3000);
}
bootstrap();
```

輸入正確的參數，重新點擊 Send Request，執行結果如圖 3-42 所示。

```
Send Request                                    5   ETag: W/"4c-ZkPM4vWihgQMYnXr16wurNcCY
POST http://localhost:3000/create-user      6   Date: Fri, 05 Jan 2024 05:07:57 GMT
Content-Type: application/json              7   Connection: close
                                            8
{                                           9 ∨ {
  "name": "mouse",                          10     "name": "mouse",
  "age": 22,                                11     "age": 22,
  "sex": "男",                              12     "sex": "男",
  "phone": "",                              13     "phone": "",
  "email": "mouse@example.com"              14     "email": "mouse@example.com"
}                                           15 }
```

▲ 圖 3-42 全域驗證器的執行結果

綜上所述，在 Nest 框架中，我們可以透過 DTO 來管理和驗證 POST 請求參數。DTO 作為一個強大的工具，能夠幫助我們確保資料的準確性和安全性。在實際應用中，為了保持程式的整潔和可維護性，通常會將 DTO 定義單獨取出到一個公共的模組或服務中進行管理。這樣做可以使得整個應用的分層結構更加清晰，同時也便於跨服務或跨模組重複使用 DTO 定義。

9. ParseFilePipe

ParseFilePipe 是專用於檔案上傳驗證的管道，能夠驗證上傳檔案的類型、大小等資訊。關於 ParseFilePipe 更詳細的使用方法和特性，將在下一節進行深入介紹。

3.6.3 自訂管道

至此，我們已經探討了 Nest 框架提供的資料驗證器，這些工具在大多數業務場景中已經足夠使用。然而，如果你面臨特定的訂製化需求，Nest 也支持透過自訂管道來滿足這些需求。

為了加深對前面討論的 ValidationPipe 的理解，我們將透過一個自訂實現的範例來演示其工作原理。範例程式如下：

```typescript
import { PipeTransform, Injectable, ArgumentMetadata,
    BadRequestException } from '@nestjs/common';
import { validate } from 'class-validator';
import { plainToInstance } from 'class-transformer';

@Injectable()
export class ValidationPipe implements PipeTransform<any> {
  async transform(value: any, { metatype }: ArgumentMetadata) {
    // 如果是原生的 JavaScript 類型，則跳過驗證
    if (!metatype || !this.toValidate(metatype)) {
      return value;
    }
    // 將普通物件轉為類型化物件才能夠驗證
    const object = plainToInstance(metatype, value);
    // 呼叫驗證方法
    const errors = await validate(object);
    if (errors.length > 0) {
      throw new BadRequestException('Validation failed');
    }
    return value;
  }
```

```
  private toValidate(metatype: Function): boolean {
    const types: Function[] = [String, Boolean, Number, Array, Object];
    return !types.includes(metatype);
  }
}
```

自訂管道需要實現 PipeTransform 介面的 transform 方法，其中 value 是我們需要處理的方法參數，而 metatype 是描述這個參數的中繼資料，用於標識當前處理的參數是原始 JavaScript 類型還是 DTO 類別。在這裡，我們需要驗證 DTO 類型的參數，因此需要透過 toValidate 方法把原生 JavaScript 資料型態過濾掉，這就是驗證器的原理。

在實際開發中，我們可以直接使用 Nest 中開箱即用的 ValidationPipe，它提供了豐富的選項配置，允許我們根據具體需求進行訂製化設置。

3.7　Nest 實現檔案上傳

檔案上傳幾乎是每個應用程式開發中必備的功能，包括單檔案上傳、多檔案上傳以及大檔案的分片上傳等多種場景。本節將透過 Nest 框架來學習如何實現最常見的單檔案和多檔案上傳。

首先，我們將介紹 Nest 框架中內建的上傳中介軟體 Multer。在前端部分，將使用 React 結合 Ant Design UI 元件庫來模擬一個接近實際開發環境的模式，建立一個簡單的上傳頁面。該頁面將允許使用者將檔案上傳到 Nest 後端專案中的 uploads 目錄。

3.7.1　初識 Multer

Multer 是一個 Node.js 中介軟體，主要用於處理檔案上傳，接收以 multipart/form-data 格式發送的資料，在 HTTP 中透過 POST 請求來實現。它提供了單檔案 .single(filename)、多檔案 .array(filename[, maxCount])、混合檔案 .fileds(fileds) 等多種形式的上傳介面，並且支援自訂檔案儲存引擎和自訂檔案儲存位置，可以更進一步地控制上傳內容。

Nest 在此基礎上做了一層封裝，透過裝飾器模式讓上傳檔案變得更加簡單和靈活。下面我們來建立一個前後端分離專案來進行演示。

先來建立 Nest 專案，執行「nest n nest-file-upload -p pnpm」命令，指定 pnpm 為專案套件管理器。專案建立成功後如圖 3-43 所示。

```
$ nest n nest-file-upload -p pnpm
⚡  We will scaffold your app in a few seconds..

CREATE nest-file-upload/.eslintrc.js (663 bytes)
CREATE nest-file-upload/.prettierrc (51 bytes)
CREATE nest-file-upload/README.md (3347 bytes)
CREATE nest-file-upload/nest-cli.json (171 bytes)
CREATE nest-file-upload/package.json (1957 bytes)
CREATE nest-file-upload/tsconfig.build.json (97 bytes)
CREATE nest-file-upload/tsconfig.json (546 bytes)
CREATE nest-file-upload/src/app.controller.spec.ts (617 bytes)
CREATE nest-file-upload/src/app.controller.ts (274 bytes)
CREATE nest-file-upload/src/app.module.ts (249 bytes)
CREATE nest-file-upload/src/app.service.ts (142 bytes)
CREATE nest-file-upload/src/main.ts (208 bytes)
CREATE nest-file-upload/test/app.e2e-spec.ts (630 bytes)
CREATE nest-file-upload/test/jest-e2e.json (183 bytes)

✔ Installation in progress... ☕

🚀  Successfully created project nest-file-upload
```

▲ 圖 3-43 建立後端專案

在 main.ts 主入口檔案中將通訊埠改為 8088，接下來前端透過代理這個通訊埠請求介面，程式如下：

```
async function bootstrap() {
  const app = await NestFactory.create(AppModule);
  await app.listen(8088);
}
```

執行「pnpm start:dev」命令啟動專案，如圖 3-44 所示。

```
[22:46:01] File change detected. Starting incremental compilation...

[22:46:01] Found 0 errors. Watching for file changes.

[Nest] 50152  - 2024/01/11 22:46:01     LOG [NestFactory] Starting Nest application...
[Nest] 50152  - 2024/01/11 22:46:01     LOG [InstanceLoader] MulterModule dependencies initial
[Nest] 50152  - 2024/01/11 22:46:01     LOG [InstanceLoader] AppModule dependencies initialize
[Nest] 50152  - 2024/01/11 22:46:01     LOG [RoutesResolver] AppController {/}: +6ms
[Nest] 50152  - 2024/01/11 22:46:01     LOG [RouterExplorer] Mapped {/, GET} route +1ms
[Nest] 50152  - 2024/01/11 22:46:01     LOG [RouterExplorer] Mapped {/upload, POST} route +0ms
```

▲ 圖 3-44 啟動後端專案

接下來建立前端 React 專案。首先，執行「create-react-app file-upload-front」命令初始化專案，一旦專案建立成功後，我們將繼續相依安裝，需要用到 antd、@antd-design/icons 和 axios 這三個套件。我們可以透過執行「pnpm add antd@4.x @ant-design/icons axios -S」命令來增加這些相依，結果如圖 3-45 所示。

```
Packages: +64
++++++++++++++++++++++++++++++++++++++++++++++++++++++++++++++
Progress: resolved 1329, reused 1329, downloaded 0, added 64, done

dependencies:
+ @ant-design/icons 5.2.6
+ antd 4.24.15 (5.12.8 is available)
+ axios 1.6.5
```

▲ 圖 3-45 安裝相依

最後，完善代理配置。在 React 專案中，反向代理透過 setupProxy.js 檔案實現。在 src 目錄下建立這個檔案，配置程式如下：

```
// 引入 http-proxy-middleware，react 鷹架已經安裝
const proxy = require("http-proxy-middleware");

module.exports = function (app) {
  app.use(
    // 遇見 /api 首碼的請求，就會觸發該代理配置
    proxy.createProxyMiddleware("/api", {
      // 請求轉發給誰
      target: "http://localhost:8088",
      // 控制伺服器收到的請求標頭中 Host 的值
      changeOrigin: true,
      // 重寫入請求路徑
      pathRewrite: { "^/api": "" },
    })
  );
};
```

配置完成後，執行「pnpm start」命令啟動專案，效果如圖 3-46 所示。

```
Compiled successfully!

You can now view file-upload-front in the browser.

  Local:            http://localhost:3000
  On Your Network:  http://192.168.0.100:3000

Note that the development build is not optimized.
To create a production build, use npm run build.

webpack compiled successfully
```

▲ 圖 3-46 啟動前端專案

3.7.2 單檔案上傳

在 Nest 框架中，上傳單個檔案的功能由 FileInterceptor() 攔截器和 @UploadedFile() 裝飾器共同實現。當 FileInterceptor 綁定到控制器方法時，它負責攔截請求中包含的檔案，並將這些檔案儲存到預設的位置。而 @UploadedFile 裝飾器用於從請求中提取已上傳的檔案，以便在控制器方法中進行進一步的處理。

為了演示這一功能，我們可以在 app.controller.ts 檔案中增加一個用於檔案上傳的介面。範例程式如下：

```
// 上傳單個檔案
@Post('/upload')
@UseInterceptors(FileInterceptor('file'))
uploadFile(@UploadedFile() file: Express.Multer.File, @Body() body) {
  console.log(file);
  return {
    message: '上傳成功',
    file: file.filename
  }
}
```

此時 TypeScript 會提示類型錯誤，如圖 3-47 所示。

▲ 圖 3-47 類型錯誤訊息

執行「pnpm add @types/multer -D」命令，為 Multer 增加類型安全。然後重新執行「pnpm start:dev」命令啟動服務。

我們還需要設置檔案上傳後的儲存位置。Multer 中介軟體提供了檔案儲存引擎，在 Nest 中的 app.module.ts 中可以這樣配置：

```
import { Module } from '@nestjs/common';
import { AppController } from './app.controller';
```

```
import { AppService } from './app.service';
import { MulterModule } from '@nestjs/platform-express';
import { diskStorage } from 'multer';
import { extname } from 'path';

@Module({
  imports: [
    MulterModule.register({
      // 定義儲存引擎
      storage: diskStorage({
        // 定義檔案儲存的目錄
        destination: './uploads',
        filename: (req, file, cb) => {
          // 建立隨機檔案名稱
          const randomName = Array(32)
            .fill(null)
            .map(() => Math.round(Math.random() * 16).toString(16))
            .join('');
          return cb(null, `${randomName}${extname(file.originalname)}`);
        },
      }),
    }),
  ],
  controllers: [AppController],
  providers: [AppService],
})
export class AppModule {}
```

在上述程式實現中，我們指定將上傳的檔案儲存在 uploads 目錄下。如果該目錄不存在，Nest 框架將自動建立它。為了確保檔案名稱的一致性和避免潛在的命名衝突，接下來將對上傳的檔案進行重新命名處理，然後將其儲存到指定的目錄中。

接下來完善前端程式，修改 App.js。範例程式如下：

```
import "./App.css";
import { UploadOutlined } from "@ant-design/icons";
import { Button, message, Upload } from "antd";
import { useState } from "react";
import axios from "axios";
// 上傳單個檔案
function fileUpload(data) {
  return axios.post("/api/upload", data, {
    headers: {
      "Content-type": "multipart/form-data",
    },
  });
```

```jsx
}
const App = () => {
  const [fileList, setFileList] = useState([]);
  // 移除檔案
  const handleRemove = (file) => {
    const files = fileList.filter((item) => item.uid !== file.uid);
    setFileList(files);
  };
  // beforeUpload 鉤子
  const handleBeforeUpload = async (file) => {
    const formData = new FormData();
    formData.append("file", file);
    await fileUpload(formData);
    setFileList([...fileList, file]);
    message.success("上傳成功");
  };

  return (
    <div className="App">
      <Upload
        fileList={fileList}
        onRemove={handleRemove}
        beforeUpload={handleBeforeUpload}
      >
        <Button icon={<UploadOutlined />}>選擇檔案</Button>
      </Upload>
    </div>
  );
};
export default App;
```

點擊「選擇檔案」按鈕上傳檔案，上傳成功後將傳回正確的結果，效果如圖 3-48 所示。

▲ 圖 3-48 單檔案上傳成功

上傳成功後，在 uploads 資料夾中也可以看到剛剛上傳的檔案，如圖 3-49 所示。

▲ 圖 3-49 上傳到指定資料夾

3.7.3 多檔案上傳

在成功實現單一檔案上傳的基礎上，接下來我們探討上傳多個檔案的場景。Nest 框架為此提供了兩個專用的 API：FilesInterceptor() 攔截器和 @UploadedFiles() 裝飾器。與處理單一檔案的 API 相比，它們在命名上僅多了一個 s，這表明它們用於處理多個檔案的上傳。為了演示這一功能，我們可以在 app.controller.ts 檔案中增加相應的介面方法。範例程式如下：

```
// 上傳多個檔案 (單表單欄位欄位)
@Post('/uploads')
@UseInterceptors(FilesInterceptor('files', 3)) // 限制檔案數量
uploadFiles(@UploadedFiles() files: Array<Express.Multer.File>, @Body() body) {
  console.log(files)
  return {
    files
  }
}
```

在以上程式中，files 用於接收表單欄位的欄位，並且限制了最多上傳 3 個檔案，@Body 用於接收除 files 之外的其他參數。

在之前的基礎上，修改前端程式如下：

```
import "./App.css";
import { UploadOutlined } from "@ant-design/icons";
import { Button, message, Upload } from "antd";
import { useState } from "react";
import axios from "axios";
// 上傳多個檔案 (單表單欄位欄位)
function filesUpload(data) {
  return axios.post("/api/uploads", data, {
    headers: {
      "Content-type": "multipart/form-data",
```

```
    },
  });
}

const App = () => {
  const [fileList, setFileList] = useState([]);
  // 移除檔案
  const handleRemove = (file) => {
    const files = fileList.filter((item) => item.uid !== file.uid);
    setFileList(files);
  };
  // 儲存選擇的檔案
  const handleBeforeUpload2 = async (file, files) => {
    setFileList([...fileList, ...files]);
  };
  // 上傳多個檔案
  const handleUploadAll = async () => {
    const formData = new FormData();
    fileList.forEach((item) => {
      formData.append("files", item);
    });
    await filesUpload(formData);
    message.success(" 上傳成功 ");
  };

  return (
    <div className="App">
      <Upload
        multiple
        fileList={fileList}
        onRemove={handleRemove}
        beforeUpload={handleBeforeUpload2}
      >
        <Button icon={<UploadOutlined />}> 選擇檔案 </Button>
      </Upload>
      <br />
      <Button onClick={handleUploadAll}> 點擊上傳多檔案 </Button>
    </div>
  );
};
export default App;
```

上面的程式應該很容易理解。相比單檔案上傳，這裡首先在 beforeUpload 鉤子中收集上傳的檔案，然後透過點擊事件發送請求將檔案上傳到 uploads 介面中。

接下來，選擇 3 個檔案後，點擊「點擊上傳多檔案」按鈕，執行結果如圖 3-50 所示。

▲ 圖 3-50 多檔案成功上傳

　　這是關於使用單表單欄位欄位 files 上傳的情況。另一種情景涉及多檔案上傳，需要使用不同的表單欄位欄位進行上傳。表單欄位的設置如下：

```
const formData = new FormData();
formData.append("files1", files[0]);
formData.append("files2", files[1]);
```

　　在後端，為了接收多個檔案欄位，我們可以使用 FileFieldsInterceptor 攔截器來定義各個檔案的欄位名稱，並設置上傳檔案的數量限制。這樣，我們可以在 app.controller.ts 中增加一個處理多檔案上傳的介面方法。範例程式如下：

```
// 上傳多個檔案（多表單欄位欄位）
@Post('/fieldUploads')
@UseInterceptors(
  FileFieldsInterceptor([
    { name: 'files1', maxCount: 3 },
    { name: 'files2', maxCount: 3 },
  ]),
) // maxCount 限制檔案數量
uploadFilesField(
  @UploadedFiles()
  files: { files1?: Express.Multer.File; files2?: Express.Multer.File },
  @Body() body,
) {
  console.log(files);
  return {
    files,
  };
}
```

對應的前端程式如下：

```
import "./App.css";
import { UploadOutlined } from "@ant-design/icons";
import { Button, message, Upload } from "antd";
import { useState } from "react";
import axios from "axios";
// 上傳多個檔案 (多表單欄位欄位)
function fieldFilesUploads(data) {
  return axios.post("/api/fieldUploads", data, {
    headers: {
      "Content-type": "multipart/form-data",
    },
  });
}

const App = () => {
  const [fileList, setFileList] = useState([]);
  // 移除檔案
  const handleRemove = (file) => {
    const files = fileList.filter((item) => item.uid !== file.uid);
    setFileList(files);
  };
  // 儲存選擇的檔案
  const handleBeforeUpload2 = async (file, files) => {
    setFileList([...fileList, ...files]);
  };
  // 上傳多個檔案
  const handleUploadAll = async () => {
    const formData = new FormData();
    fileList.forEach((item) => {
      formData.append("files1", item);
      formData.append("files2", item);
    });
    await fieldFilesUploads(formData);
    message.success(" 上傳成功 ");
  };

  return (
    <div className="App">
      <Upload
        multiple
        fileList={fileList}
        onRemove={handleRemove}
        beforeUpload={handleBeforeUpload2}
      >
        <Button icon={<UploadOutlined />}>選擇檔案 </Button>
      </Upload>
      <br />
```

```
      <Button onClick={handleUploadAll}>點擊上傳多檔案</Button>
    </div>
  );
};
export default App;
```

為了演示方便，上面的程式在 handleUploadAll 方法中設置了多個表單欄位 files1、files2 來模擬多個表單欄位的情況。上傳檔案的效果如圖 3-51 所示。

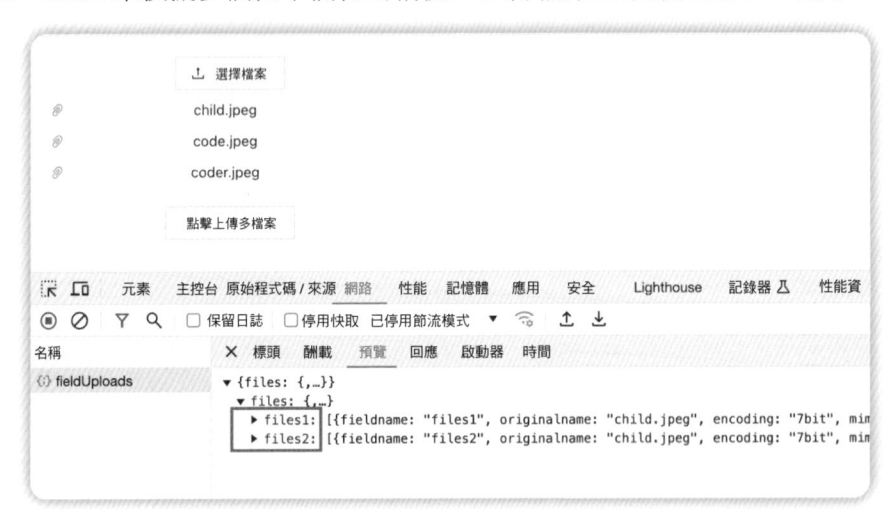

▲ 圖 3-51 多檔案上傳成功（多欄位）

3.7.4 上傳任意檔案

在某些情況下，如果不確定具體哪個欄位用於表示檔案資料，可以使用 AnyFilesInterceptor 攔截器來處理這種情況。此攔截器允許從請求的任何欄位中攔截檔案上傳。範例程式如下：

```
// 上傳任意檔案
@Post('/anyUploads')
@UseInterceptors(
  AnyFilesInterceptor({
    // 限制檔案數量
    limits: { files: 3 },
  }),
)
uploadAnyFiles(
  @UploadedFiles() files: Array<Express.Multer.File>,
  @Body() body,
) {
```

```
    console.log(files);
    return {
      files,
    };
}
```

對應的前端程式改動如下：

```
// 上傳多個檔案
const handleUploadAll = async () => {
  const formData = new FormData();
  fileList.forEach((item) => {
    formData.append("files1", item);
    formData.append("files2", item);
  });
  // 改為上傳任何檔案介面
  await anyFilesUpload(formData);
  message.success("上傳成功");
};
```

上傳效果如圖 3-52 所示。

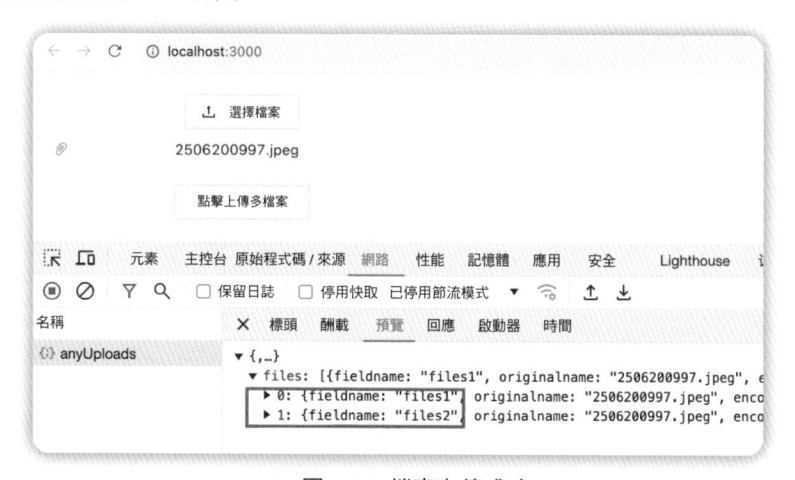

▲ 圖 3-52 檔案上傳成功

　　至此，我們介紹了 Nest 框架中幾種典型的檔案上傳方法。然而，為了建構一個健壯的檔案上傳功能，我們還需要進一步完善上傳邏輯。僅限制檔案數量是不夠的，我們還需要加入檔案驗證機制，這包括對檔案大小、類型等屬性進行驗證，以確保上傳的檔案符合特定的要求和標準。

3.7.5 檔案驗證

前面提到，檔案驗證通常透過管道來完成，Nest 中內建了用於檔案驗證的 ParseFilePipe 管道。接下來，在 anyUploads 介面中增加檔案大小和檔案類型驗證邏輯。範例程式如下：

```
// 上傳任何檔案
@Post('/anyUploads')
@UseInterceptors(
  AnyFilesInterceptor({
    // 限制檔案數量
    limits: { files: 3 },
  }),
)
uploadAnyFiles(
  @UploadedFiles(new ParseFilePipe({
    validators: [
      // 檔案大小限制
      new MaxFileSizeValidator({maxSize: 1024 * 1000}),
      // 檔案類型限制
      new FileTypeValidator({fileType: 'image/png'}),
    ]
  })) files: Array<Express.Multer.File>,
  @Body() body,
) {
  console.log(files);
  return {
    files,
  };
}
```

接下來上傳一幅大小超過 1024KB 的圖片，其執行結果如圖 3-53 所示。

▲ 圖 3-53 檔案大小限制

如果類型錯誤，應怎麼辦呢？可以嘗試上傳一個類型不符合要求的檔案，其執行結果如圖 3-54 所示。

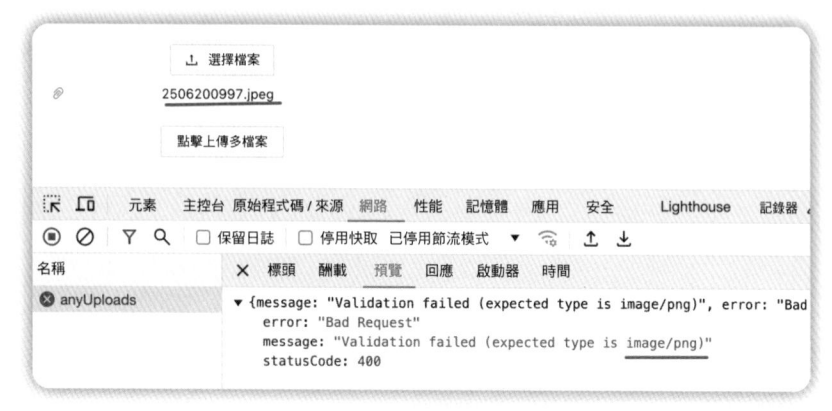

▲ 圖 3-54 檔案類型限制

在 Nest 框架中，檔案驗證主要依賴於這兩個核心組件：檔案大小驗證和檔案類型驗證。結合我們之前討論的自訂管道驗證器的實現方法，你應該已經掌握了足夠的知識來應用它們。現在，是時候將這些理論知識付諸實踐了。我鼓勵讀者立刻開始，實現自己的檔案驗證邏輯！

第 2 部分

進 階 篇

　　第 2 部分將深入探索 Nest 框架的進階部分，學習如何有效整合 MySQL 和 Redis 等資料庫技術，掌握後端服務中至關重要的身份驗證和授權機制，並將 Nest 應用部署到特定的伺服器環境。透過全面理解這些進階概念，我們將具備開發和部署自己的 Nest 應用服務的能力。

第 4 章
Nest 與資料庫

在前面的章節中，我們學習了 Nest 的基礎部分。從本章開始，我們將深入探討 Nest 的進階功能，包括如何使用 MySQL 資料庫進行資料持久化、使用 ORM（物件關係映射）來操作資料庫、以及如何透過 Redis 快取來提升性能。接著，我們將介紹應用中必不可少的身份驗證與授權，展示不同的實現方式來管理使用者許可權。最後，我們會介紹後端常用的運行維護工具 Docker，展示如何利用它快速建立和部署後端服務。

4.1 快速上手 MySQL

MySQL 是目前廣泛應用於各種 Web 應用的資料庫之一，扮演著關鍵的資料儲存和管理角色。對於任何資深的後端工程師而言，MySQL 都是一個不可或缺的工具。他們利用 MySQL 接收前端提交的資料，並高效率地將其儲存到資料庫中，同時能夠快速檢索資料以供前端著色使用。這些技能對我們後續章節的學習和專案實踐至關重要。

本節將引導讀者完成 MySQL 的安裝過程，確保它能夠在讀者的個人電腦上正常執行。此外，我們將透過命令列和視覺化工具這兩種方式，演示 MySQL 的常用命令，幫助讀者更進一步地理解和掌握這一強大的資料庫管理系統。

4.1.1 安裝和執行

MySQL 可以在各種作業系統上執行，如 Windows、Linux、macOS 等，也可以在容器平臺如 Docker 和 K8s 上執行。接下來，我們把 MySQL 安裝到個人主機上。在 MySQL 官網上，依次點擊 DOWNLOADS → MySQL Community Downloads，選擇適合自己主機的版本，然後點擊 Download 按鈕，如圖 4-1 所示。

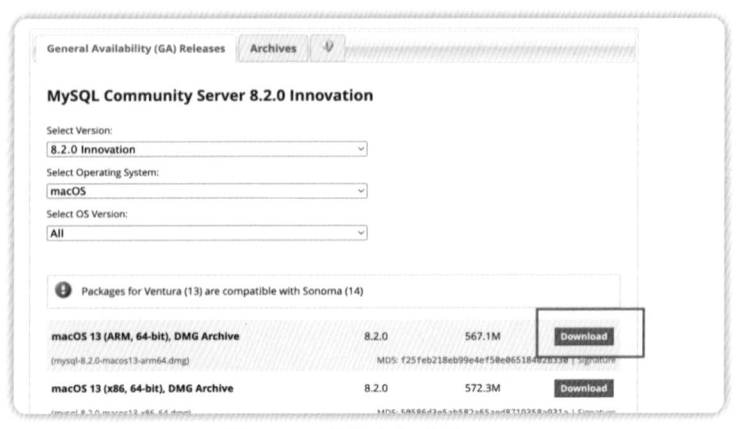

▲ 圖 4-1　下載指定版本的 MySQL

接著會提示登入或註冊，這裡可以直接選擇下載，如圖 4-2 所示。

▲ 圖 4-2　跳過登入註冊

下載完成後，開始安裝，依次點擊「繼續」按鈕，如圖 4-3 所示。

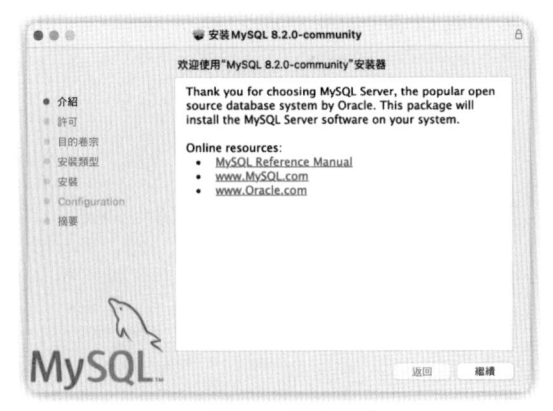

▲ 圖 4-3 開始安裝

到了這一步,需要輸入你的資料庫密碼,如圖 4-4 所示。連接資料庫時需要用到這個密碼,預設帳號是 root。

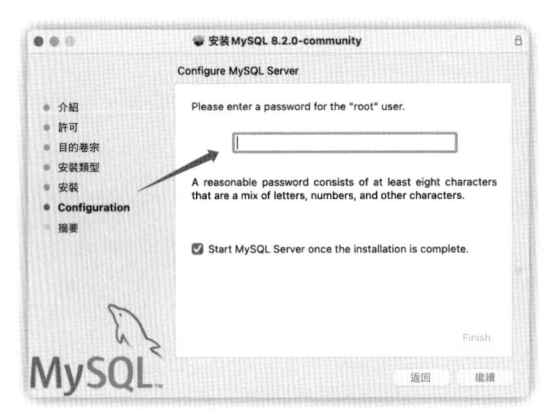

▲ 圖 4-4 輸入資料庫密碼

輸入完成後,點擊 Finish 按鈕完成安裝。MySQL 預設已經啟動,我們可以在「系統偏好設置」選項的底部找到它,如圖 4-5 所示。

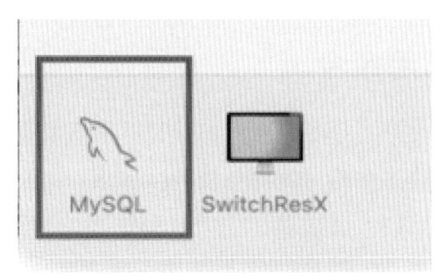

▲ 圖 4-5 MySQL 安裝位置

點擊 MySQL 圖示，可以看到 MySQL 已經成功執行了，如圖 4-6 所示。

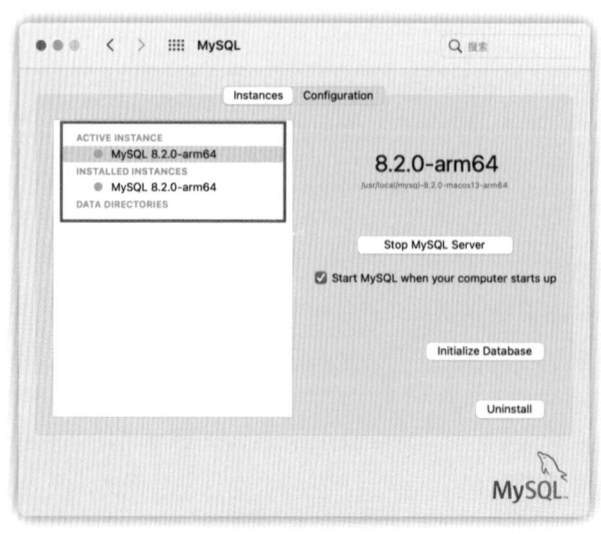

▲ 圖 4-6 執行成功

如果讀者在嘗試啟動 MySQL 時遇到問題，並且介面顯示如圖 4-7 所示，經過多次嘗試仍無法成功啟動，那麼可能需要採取進一步的解決措施。建議在這種情況下，卸載當前版本的 MySQL，並重新安裝一個與你的作業系統版本相容的 MySQL 版本。

▲ 圖 4-7 MySQL 版本與系統版本不匹配

接下來，我們分別透過命令列方式和視覺化用戶端來操作資料庫。

4.1.2 MySQL 的常用命令

操作 MySQL 的一種方式是透過命令列工具。要連接到 MySQL 伺服器，我們需要在系統的根目錄輸入命令「/usr/local/mysql/bin/mysql -u root –p」。

在此命令中，-u 參數後面跟著的是 MySQL 的帳號名稱，例如 root。執行該命令後，系統會提示你輸入該帳號的密碼，如圖 4-8 所示。請注意，具體的

MySQL 安裝路徑 /usr/local/mysql/bin/ 可能會因安裝方式和作業系統的不同而有所變化，因此讀者可能需要根據自己的實際情況調整該路徑。

```
~  17:10:16
$ /usr/local/mysql/bin/mysql -u root -p
Enter password:
```

▲ 圖 4-8 登入 MySQL

輸入密碼後，就可以進行命令列操作了。行「show databases;」命令，可以查看預設存在的資料庫，如圖 4-9 所示。

```
Copyright (c) 2000, 2022, Oracle and/or its affiliates.

Oracle is a registered trademark of Oracle Corporation and/or its
affiliates. Other names may be trademarks of their respective
owners.

Type 'help;' or '\h' for help. Type '\c' to clear the current input statement.

mysql> show databases;
+--------------------+
| Database           |
+--------------------+
| information_schema |
| mysql              |
| performance_schema |
| sys                |
+--------------------+
4 rows in set (0.04 sec)

mysql>
```

▲ 圖 4-9 查看資料庫

我們建立了一個名為 nest-mysql 的資料庫，執行「create database nest_mysql;」命令並展示，如圖 4-10 所示。

```
mysql> create database nest_mysql;
Query OK, 1 row affected (0.00 sec)

mysql> show databases;
+--------------------+
| Database           |
+--------------------+
| information_schema |
| mysql              |
| nest_mysql         |
| performance_schema |
| sys                |
+--------------------+
5 rows in set (0.01 sec)
```

▲ 圖 4-10 nest_mysql 資料庫

接下來，進入這個資料庫查看或建立表，執行「use nest_mysql;」命令使用資料庫，並透過「show tables;」命令查看有哪些表，如圖 4-11 所示。

```
mysql> use nest_mysql;
Database changed
mysql> show tables;
Empty set (0.01 sec)
```

▲ 圖 4-11 查看資料表

建立 user 表並插入一些基礎資料，執行以下命令：

```
create table user(id int(4) primary key, name char(20), age int(3));
```

其中，id、name、age 為表的列，int、char 為欄位類型，primary key 是主鍵，不能重複。

接下來執行「describe user;」或「desc user;」命令，查看表結構，如圖 4-12 所示。

```
mysql> create table user(id int(4) primary key, name char(20), age int(3));
Query OK, 0 rows affected, 2 warnings (0.01 sec)

mysql> desc user;
+-------+----------+------+-----+---------+-------+
| Field | Type     | Null | Key | Default | Extra |
+-------+----------+------+-----+---------+-------+
| id    | int      | NO   | PRI | NULL    |       |
| name  | char(20) | YES  |     | NULL    |       |
| age   | int      | YES  |     | NULL    |       |
+-------+----------+------+-----+---------+-------+
3 rows in set (0.00 sec)
```

▲ 圖 4-12 表結構

圖 4-12 中顯示的資料型態，MySQL 的資料型態還有很多，諸如：

- 整數類型：TINYINT、SMALLINT、MEDIUMINT、INIT、BIGINT。
- 浮點數字類型：FLOAT、DOUBLE。
- 定點數字類型：DECIMAL、MUMARIC。
- 字串類型：CHAR、VARCHAT、TEXT、BLOB。
- 日期類型：DATE、TIME、DATETIME、TIMESTAMP。

操作 MySQL 的命令也有很多，諸如：

- 建立資料庫：

```
create database <資料庫名稱>;
```

- 進入資料庫：
```
use <資料庫名稱>;
```
- 查看資料庫：
```
show databases;
```
- 刪除資料庫：
```
drop database <資料庫名稱>;
```
- 建立表：
```
create table 表名 (
  列名稱 1 資料型態 1 auto_increment,
  列名稱 2 資料型態 2,
  ...
  列名稱 n 資料型態 n,
  primary key (主鍵列名稱)
);
```
- 查看表結構：
```
desc <表名>;
```
- 修改表名稱：
```
alter table <表名> rename <新表名>;
```
- 刪除表：
```
drop table <表名>;
```
- 修改表欄位資訊：
```
alter table <表名> modify <列名稱> <資料型態>;
```
- 查詢表資料：
```
select * from <表名>;
```

你可能已經注意到，每次透過命令列執行資料庫操作可能會顯得有些繁瑣。尤其是當處理大量資料時，組織 SQL 敘述並逐一執行它們可能是一項相當耗時且工作量巨大的任務。為了簡化這一過程並提高效率，我們可以使用資料庫視覺化工具來操作資料庫。這些工具通常提供了使用者友善的介面，使得資料管理和 SQL 開發變得更加直觀和便捷。

4.1.3 視覺化操作 MySQL

MySQL 有許多 GUI 用戶端可供選擇，比如 MySQL 官方提供的 GUI 工具—MySQL Workbench、廣泛使用的 Navicat 以及適用於 VS Code 的 MySQL 外掛程式。在這裡，我們將使用最熟悉的 VS Code 來演示。

在 VS Code 應用商店中搜尋 mysql，從搜尋結果中選擇一個進行安裝即可，如圖 4-13 所示。

▲ 圖 4-13 安裝 MySQL 外掛程式

安裝完成後，從 VS Code 工具列中找到該 MySQL 外掛程式，並點擊「+」圖示以建立新的資料庫連接。此時將顯示如圖 4-14 所示的配置介面。

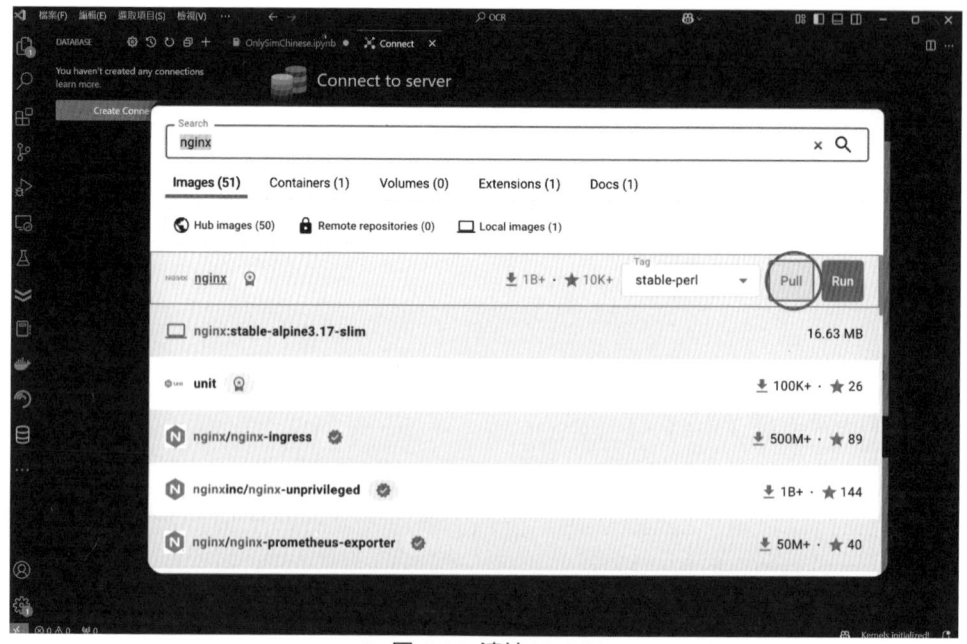

▲ 圖 4-14 連接 MySQL

從圖 4-14 可以看到，我們不僅可以連接 MySQL 資料庫，還可以連接 MySQL 之外的各種資料庫。在輸入連接名稱（例如案例中的 nest-test）、帳戶和密碼後，點擊下方的「連接」按鈕即可完成連接。

一旦連接成功，可以開啟之前建立的表，查看其表結構和新增的資料，如圖 4-15 所示。

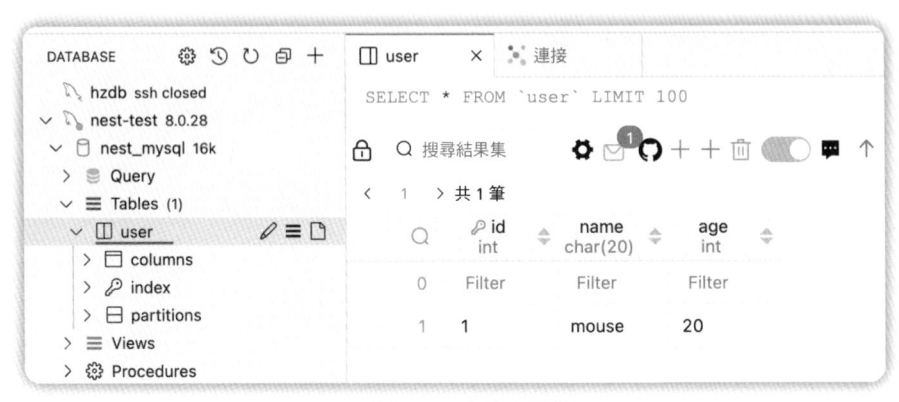

▲ 圖 4-15 查看表資料

如前所述，MySQL 提供了多種資料型態和操作命令。然而，在當前階段，我們並不需要記住這些細節。透過 GUI 面板，可以直觀地進行資料庫操作，因為外掛程式為我們提供了一個包含豐富操作元素的使用者介面，這簡化了許多常規任務，如圖 4-16 所示。

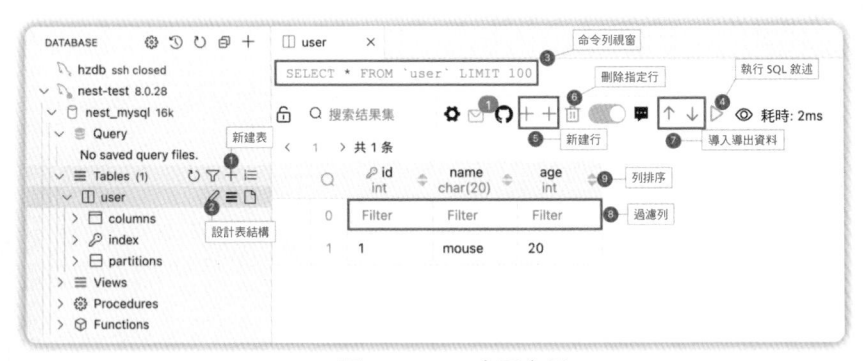

▲ 圖 4-16 GUI 介面介紹

比如新建一個 product 表，需要編輯預設的 SQL 敘述，如圖 4-17 所示。

▲ 圖 4-17 新建表

將其修改為以下的 SQL 敘述：

```
CREATE TABLE product(
    id int NOT NULL PRIMARY KEY AUTO_INCREMENT COMMENT 'Primary Key',
    create_time DATETIME COMMENT 'Create Time',
    name VARCHAR(255)
) COMMENT '產品表';
```

點擊 Execute 按鈕執行敘述，如圖 4-18 所示，可以看到成功建立 product 表。這種方式相比手動撰寫 SQL 敘述，確實方便得多，是吧？

▲ 圖 4-18 成功建立表

接下來設計 product 表，新增 updateTime（建立時間）和 price（價格）欄位，如圖 4-19 所示。

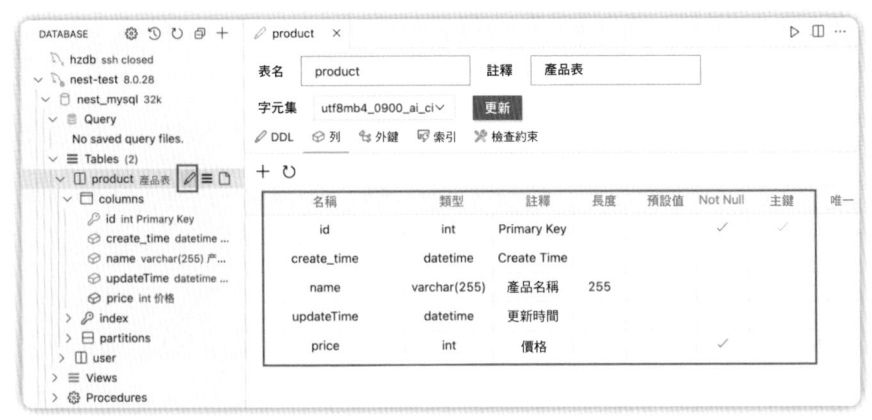

▲ 圖 4-19 表結構設計

在資料庫管理中，執行資料定義語言（Data Definition Language，DDL）敘述，用於設置和管理表結構，包括定義外鍵、建立索引以及設置檢查約束等。這些重要的概念將在後續章節中進行詳細介紹。

接下來，為了豐富我們的表結構，可以增加一個新的欄位，例如 description（產品描述），如圖 4-20 所示。

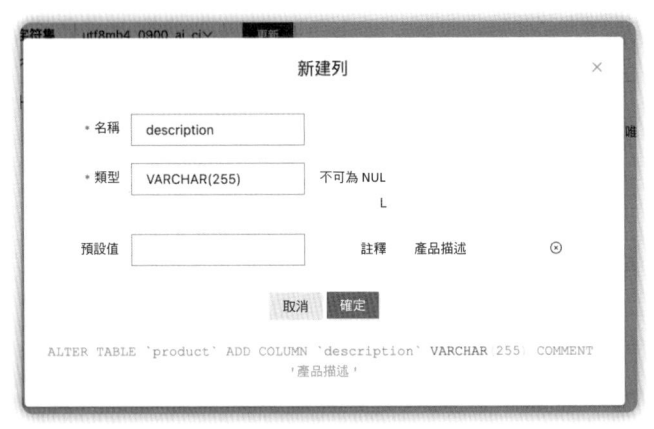

▲ 圖 4-20 新建列

設計完表結構之後，回到資料表中，就可以進行插入資料、匯入資料等操作了。插入資料的過程如圖 4-21 所示。

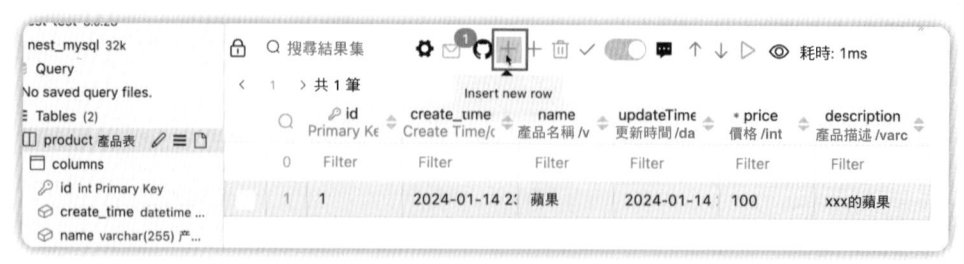

▲ 圖 4-21 插入資料

在圖 4-21 中，id 不用填，會自動自動增加；price 為不可為空欄位。我們也可以透過指定 SQL 敘述快速生成多筆資料，如圖 4-22 所示。

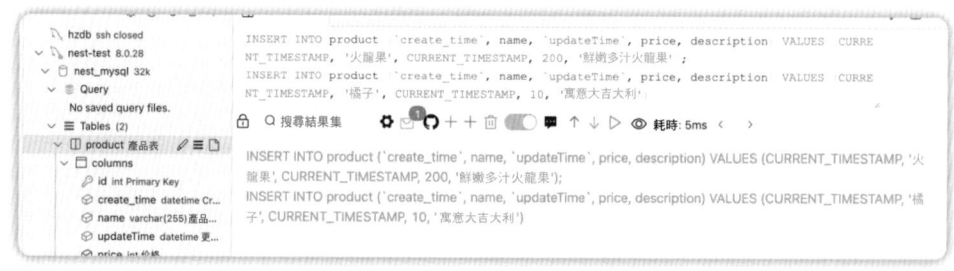

▲ 圖 4-22 執行 SQL 敘述生成資料

此外，還有更為高效的手段來輔助我們的開發和測試工作。我們可以執行「run mock」命令自動生成測試資料，這極大地簡化了我們的日常測試流程。具體操作如圖 4-23 所示。

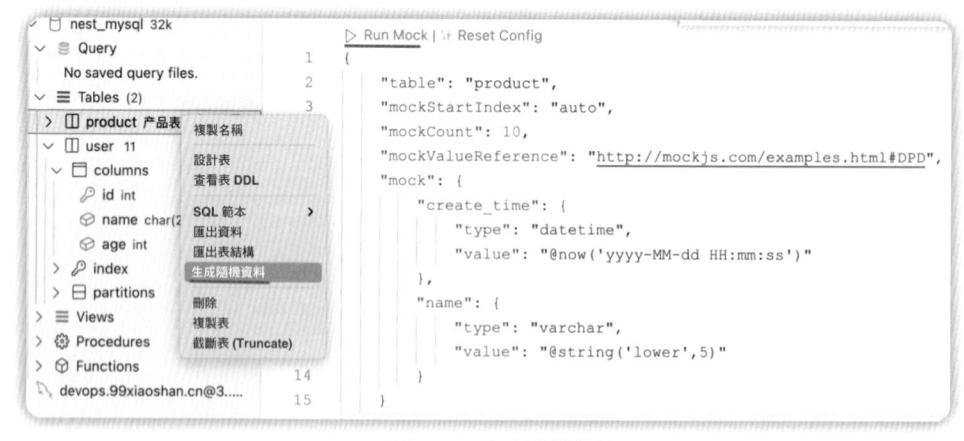

▲ 圖 4-23 生成隨機數據

接下來學習清空白資料表。按右鍵產品表，在彈出的快顯功能表中選擇「截斷表（Truncate）」，表中的資料就會被清空，如圖 4-24 所示。

▲ 圖 4-24 清空白資料表

最後學習如何刪除表。按右鍵產品表，在彈出的快顯功能表中選擇「刪除」命令來刪除表，如圖 4-25 所示。

▲ 圖 4-25 刪除表

至此，我們已經學習了常用的資料庫單表操作方法。在下一節中，我們將深入探討多表連結以及複雜 SQL 查詢的技巧，這些內容對於提升資料庫操作的效率和靈活性至關重要。

4.2 MySQL 表之間的關係

在上一節中,我們學習了如何透過命令列和視覺化工具這兩種方式對 MySQL 資料庫進行單表的增刪改查操作。在實際業務場景中,資料庫往往會包含多張表,這些表用於儲存不同類型的資訊,並在它們之間建立特定的聯繫。常見的關係類型包括一對一、一對多 / 多對一以及多對多關係。

本節內容將從實際應用出發,透過生活中的例子來具體說明這四種關係。我們會探討這些關係是如何透過外鍵在資料表之間建立聯繫的,以及如何對涉及這些關係的資料庫進行有效的操作和管理。

4.2.1 一對一關聯性

在資料庫設計中,有些資料表之間是一對一的關係。以下是一些典型的一對一關聯性範例:

- 一個人擁有一個獨一無二的身份證,而這個身份證也只屬於這個人。
- 一個學生分配有一個學號,而這個學號在系統中僅對應這一個學生。
- 每個人的指紋是獨一無二的,而每個指紋也只對應一個特定的人。

以身份證為例,我們可以設計兩個表來維護使用者資訊和身份證資訊,如圖 4-26 所示。

▲ 圖 4-26 使用者與身份證關係表

在此案例中,user 表擁有一個主鍵 id,該 id 是唯一的,用於標識單一使用者。為了實現使用者資訊與身份證資訊之間的一對一連結,我們可以在 id_card 表中引入一個外鍵欄位 user_id，該欄位用於儲存對應的 user 表的 id 值。這樣，透過 user 表的 id，我們可以快速地查詢到與之連結的身份證資訊，如圖 4-27 所示。

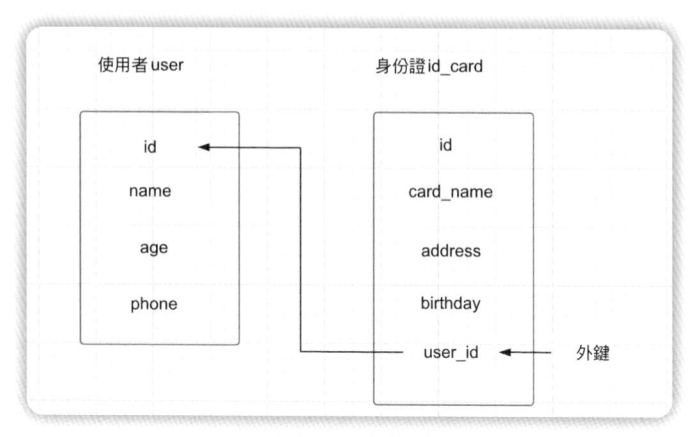

▲ 圖 4-27 外鍵連結

在圖 4-27 中，user_id 作為外鍵存在於 id_card 表中，它引用了 user 表的主鍵。在這種關係中，user 表被稱為主資料表，而 id_card 表則被稱為從表。這種設置定義了資料表之間的主從關係。

讀者可能會問，如何確定哪個表應該作為主資料表，哪個表作為從表呢？在設計主從關係時，我們需要綜合考慮多個因素，包括資料的唯一性、業務邏輯、資料表的存取頻率、資料完整性以及系統的擴充性等因素。在我們的案例中，user 表在業務流程中扮演核心角色，它不僅包含更豐富的使用者相關資訊，而且由於存取頻率較高，資料也更為完整。此外，user 表在業務發展中的擴充性也更強。

關於如何進行連結查詢，我們將透過實際建表演示來詳細說明。

1. JOIN 查詢

在上一節中，我們已經建立了 user 表。接下來，為了生成 id_card 表，需要在資料庫管理介面中點擊「新建表」按鈕。然後，根據你的需求修改系統預設的 SQL 敘述。完成這些步驟後，執行修改後的 SQL 敘述，即可建立新的 id_card 表，如圖 4-28 所示。

```
 8.0.28          1   -- Active: 1705225958726@@127.0.0.1@3306@nest_mysql   MySQL
mysql 16k            ▷ Execute | JSON | Copy
                 2   CREATE TABLE id_card(
ed query files. 3       id int NOT NULL PRIMARY KEY AUTO_INCREMENT COMMENT 'Primary Key',
s (1)            4       card_name VARCHAR(55) NOT NULL,
使用者表 11       5       address VARCHAR(255) NOT NULL,
lumns            6       user_id INT NOT NULL
d int            7   ) COMMENT '身份證表';
ame char(20)
```

▲ 圖 4-28 建立 id_card 表

在成功建立 id_card 表之後，接下來我們需要向該表增加一個外鍵 user_id，它將連結到 user 表的主鍵 id 上，以建立兩個表之間的一對一關聯性。這樣的設計允許 id_card 表透過 user_id 引用 user 表中的相應記錄。具體操作如圖 4-29 所示。

▲ 圖 4-29 連結外鍵

需要注意的是，On Update 和 On Delete 選項暫時保持預設的 NO ACTION，這涉及串聯操作，我們稍後會詳細討論。

接下來，為了向 id_card 表中增加隨機數據，我們需要對 SQL 敘述進行相應的修改，具體如圖 4-30 所示。修改完成後，點擊介面上的「Run Mock」按鈕來執行這些更改，並自動填充表資料。

在圖 4-30 中，@cname 用於表示生成的隨機名稱，而 @county 表示生成的隨機地址。這些特定的 mock 值允許自動為資料庫表填充範例資料。如果需要了解更多關於這些 mock 值的細節，可以參考 mockValueReference 提供的資訊。完成這些步驟後，生成的資料如圖 4-31 所示。

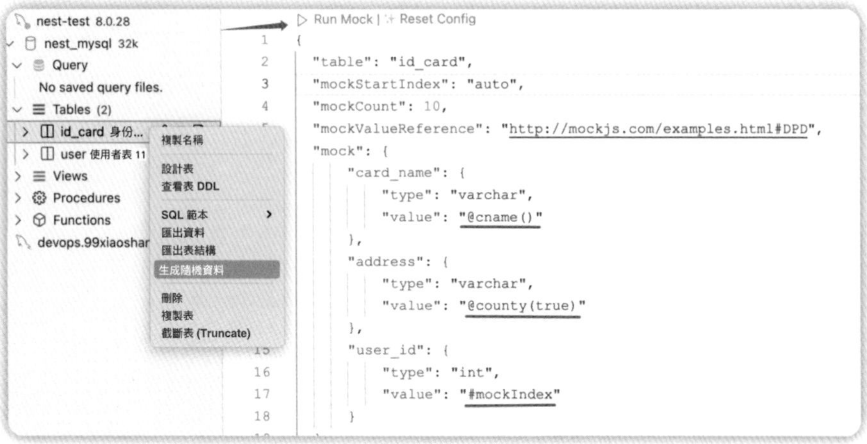

▲ 圖 4-30 生成 id_card 表資料

▲ 圖 4-31 id_card 表資料

以同樣的方式向 user 表中增加一些資料,具體如圖 4-32 所示。

▲ 圖 4-32 user 表資料

如何連結查詢呢?執行下面這筆 SQL 敘述:

```
SELECT * FROM user JOIN id_card ON user.id = id_card.user_id;
```

執行結果如圖 4-33 所示。

▲ 圖 4-33 連結查詢

可見 user 表和 id_card 表同時被查詢出來了，這得益於 JOIN ON 敘述，連結方式是 user.id = id_card.user_id，即主資料表中的 id 與從表中的 user_id 外鍵相連結。

2. 串聯

在資料庫設計中，我們經常會遇到需要定義串聯操作的情況，特別是涉及主資料表（也稱為父表）與從表（也稱為子表）之間的關係時。串聯操作定義了當父表中的記錄發生變化時，子表應如何回應。具體來說，有以下幾種可選項：

- RESTRICT：當父表中的某筆記錄被刪除或更新時，子表中存在與之相關的記錄，則會阻止父表的刪除或更新操作，即不允許對父表進行刪除或更新操作。
- CASCADE：當父表中的某筆記錄被刪除或更新時，子表中與之相關的記錄也會被自動刪除或更新。這樣可以確保資料的一致性，避免出現孤兒記錄。
- SET NULL：當父表中的某筆記錄被刪除或更新時，子表中與之相關的外鍵欄位被設置為空值（NULL）。這樣可以避免刪除或更新父表記錄時引發外鍵約束錯誤。
- NO ACTION：在 MySQL 中，NO ACTION 與 RESTRICT 具有相同的行為。

接下來，讓我們透過實際操作來演示如何應用這些串聯選項。首先，如果想要將串聯選項設置為 CASCADE，則需要在 VS Code 的 MySQL 外掛程式中進行一些調整。這通常涉及先刪除現有的外鍵約束，如圖 4-34 所示。

▲ 圖 4-34 刪除現有的外鍵約束

完成這一步驟後，我們可以重新定義外鍵，並設置其串聯行為為 CASCADE，以確保在父表記錄變更時，子表中的相關記錄也會自動更新或刪除，如圖 4-35 所示。

▲ 圖 4-35 設置 CASCADE 串聯方式

注　意

細心的讀者可能會注意到，在圖 4-35 中多了一個前文未提及的 SET DEFAULT 串聯選項。這是因為 MySQL 預設情況下不支援該串聯類型，而 VS Code 外掛程式支援連接多種類型的資料庫，如 Oracle、PostgreSQL 等，這些資料庫中支援 SET DEFAULT 選項。該選項代表當父表中的某筆記錄被刪除或更新時，子表中相關的外鍵欄位將被設置為預設值。

此時嘗試修改父表中的資料，切換到 user 表中，把 id=1 的記錄修改為 id=100，並按 Enter 鍵確認，如圖 4-36 所示。

▲ 圖 4-36 修改父表中的資料

再切換到 id_card 子表中，發現 id=1 的記錄中，user_id 已經被同步更新為 100，如圖 4-37 所示。

▲ 圖 4-37 子表資料同步更新

同樣地，CASCADE 串聯選項也適用於刪除操作。舉例來說，在 user 表中，如果我們選擇並刪除了 id 等於 100 的記錄，根據 CASCADE 規則，id_card 子表中與之連結的記錄也會自動被刪除，如圖 4-38 所示。因此，在執行刪除操作後，id_card 表中將不再存在 id_card 等於 100 的記錄。

▲ 圖 4-38 子表資料同步刪除

接下來，讓我們嘗試建立一個新的外鍵，並將其串聯方式設置為 SET NULL。然而，在點擊「確定」按鈕後，出現了一個錯誤，如圖 4-39 所示。錯誤訊息指出 user_id 欄位不能為 NULL，這與我們選擇的串聯選項相衝突。因為 SET NULL 串聯選項的意圖是，當父表中的記錄被更新或刪除時，子表中的相關外鍵欄位應被設置為 NULL。顯然，如果 user_id 欄位不允許為空，這就違反了外鍵約束的規則，導致操作無法進行。這就是所謂的外鍵約束衝突。

▲ 圖 4-39 外鍵不可為空約束

要解決這個外鍵約束衝突，我們需要調整 user_id 欄位的設置。首先，切換到資料庫表設計的「列」標籤頁。接下來，在「列」欄中找到 user_id 欄位，並修改其屬性以允許為空（NULL）。這樣，當父表記錄被更新或刪除時，子表中的 user_id 欄位就可以被設置為 NULL，而不會違反外鍵約束。更改後的介面如圖 4-40 所示。

▲ 圖 4-40 修改外鍵設置，允許設置為空

在成功刪除舊的外鍵並建立新的外鍵之後，我們可以繼續進行測試。現在，如果我們在主資料表中更新或刪除 id 等於 2 的記錄，應該可以觀察到 id_card 表中相應的 user_id 外鍵欄位被設置為 NULL。這一效果驗證了 SET NULL 串聯選項配置正確，並且外鍵約束獲得了適當的處理。操作結果如圖 4-41 所示。

▲ 圖 4-41 外鍵設置為 NULL

綜上所述，MySQL 提供了多種串聯選項，包括 RESTRICT、CASCADE、SET NULL 和 NO ACTION，以適應不同的業務需求和資料模型設計。在本案例中，考慮到當使用者記錄被刪除時，其對應的身份證資訊也隨之失去實際用途，因此選擇 CASCADE 作為串聯方式是恰當的。這樣，使用者記錄的刪除將自動觸發相關身份證資訊的刪除，確保資料的一致性和相關性得到維護。

4.2.2 一對多 / 多對一關係

除一對一關聯性外，生活中還普遍存在一對多或多對一的關係。以下是一些典型的例子：

- 一個公司有多個員工，公司通常只在一家公司工作。
- 一個學校有多名學生，每位學生通常只屬於一個學校。
- 一個作者可以寫多篇文章，文章也只屬於一個作者。

在資料庫設計中，一對多和多對一關係是相互連結的。為了在資料庫中準確描述這種關係，我們通常會使用外鍵來實現。以公司與員工的關係為例，員工表中會包含一個指向公司表的外鍵，以此表明一名員工只屬於一個公司，而公司可以有多個員工，如圖 4-42 所示。這種設計允許我們維護資料的完整性並清晰地表示實體間的連結。

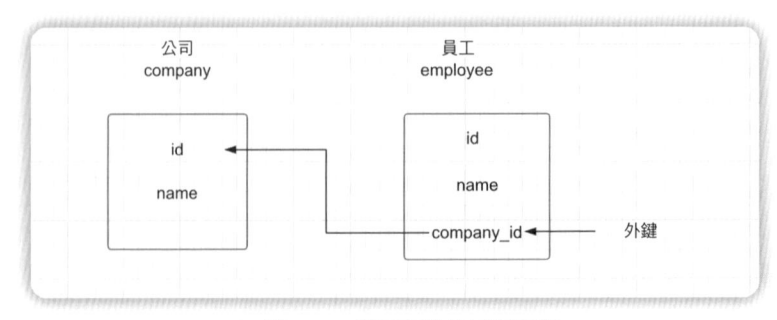

▲ 圖 4-42 公司表與員工表關係

4.2 MySQL 表之間的關係

在資料庫中，我們需要建立兩個表：company 表和 employee 表。對於 employee 表，我們將設置一個外鍵 company_id，該外鍵引用 company 表的主鍵。此外，我們將此外鍵的串聯方式設置為 CASCADE。這樣的設計表示，如果 company 表中的某筆記錄被刪除，那麼 employee 表中與之連結的所有員工記錄也會被自動刪除，以保證資料的一致性。同樣，如果 company 表中的記錄被更新（例如公司名稱變更），employee 表中相應的外鍵欄位也會自動更新，確保員工合約上的資訊保持最新。

完成表的建立和資料插入後，company 表和 employee 表的結構將如圖 4-43 所示。

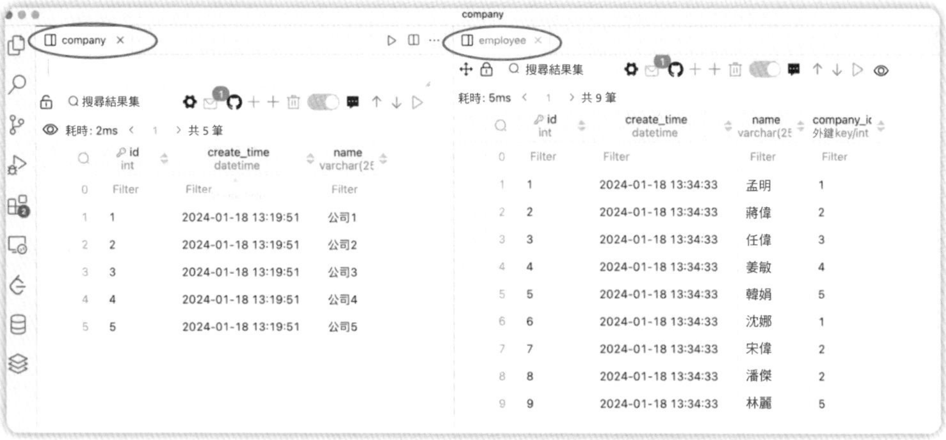

▲ 圖 4-43 company 表和 employee 表的結構

根據圖 4-43 的展示，我們可以看到 company 表和 employee 表都包含 id、create_time 和 name 列。除此之外，employee 表還增加了一個 company_id 列，作為外鍵連結到 company 表的主鍵 id 欄位。

現在，為了查詢特定公司（例如 id 為 2 的公司）的員工資訊，我們將執行一個 SQL 查詢，該查詢將傳回關鍵欄位 id、name 和 company_id。以下是執行此查詢的 SQL 敘述：

```
SELECT
company.id,
company.name,
employee.id as employee_id,
employee.name as employee_name,
company_id
FROM company
```

```
JOIN employee
ON company.id = employee.company_id
WHERE company.id = 2
LIMIT 100
```

　　在涉及多個表的查詢中，如果存在欄位名稱重複的情況，我們可以使用 AS 關鍵字對欄位進行重新命名，以確保查詢結果的清晰性。透過執行帶有 JOIN ... ON 條件的 SQL 敘述，我們可以檢索出 company 表中 id 等於 2 的公司所連結的員工資料。根據圖 4-44 的展示，該查詢結果顯示，該公司共有 3 名員工。

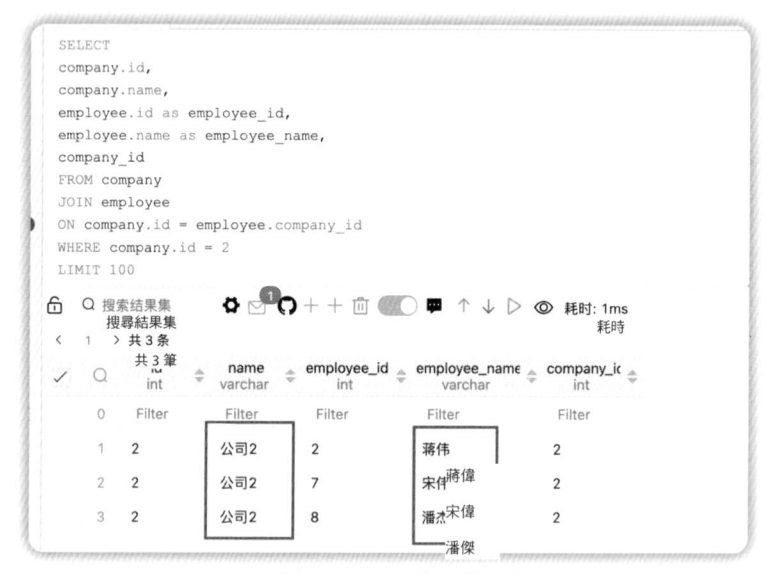

▲ 圖 4-44　一對多串聯查詢

　　接下來，我們將執行刪除操作。要刪除 company 表中 id 等於 2 的公司記錄，需要在資料庫管理介面中選中該記錄，然後點擊「刪除」按鈕來執行刪除操作。這一過程的結果將反映在資料庫中，如圖 4-45 所示。

▲ 圖 4-45　刪除指定公司

接下來，點擊介面上出現的「確認」按鈕，以執行之前的刪除查詢敘述。完成這一操作後，當我們再次查詢資料庫時，會發現先前與 id 等於 2 的公司記錄相連結的所有資料都已被清空，這反映了 CASCADE 串聯刪除的效果。如圖 4-46 所示，查詢結果現在顯示為空，這正是我們期望中 CASCADE 串聯操作的結果。

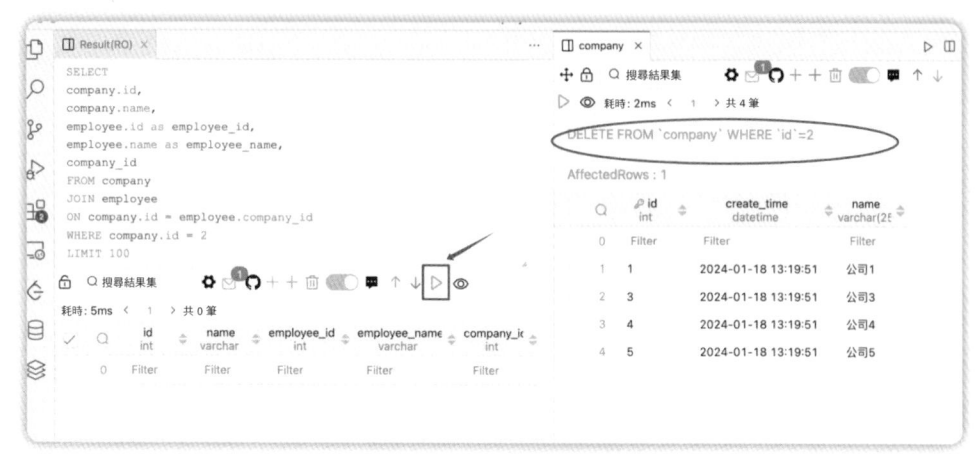

▲ 圖 4-46 一對多串聯刪除

透過上述操作，你可能已經注意到，一對多 / 多對一關係與一對一關聯性在表操作方面有許多相似之處。確實，這兩種關係在基本的資料庫操作上可能看起來差不多，但在實際的資料庫設計實踐中，它們各自扮演著不同的角色。

一對一關聯性通常用於標準化資料，它有助保持資料的一致性和減少容錯。而一對多關聯性則更常用於建立表之間的層次結構，這種結構可以清晰地表示實體間的從屬或連結關係。

在本案例中，公司與員工表之間的關係就是一個典型的父子關係範例。company 表可以被視為父表，因為它包含了公司的整體資訊；而 employee 表則作為子表，儲存了屬於各個公司的員工詳細資訊。透過在 employee 表中使用 company_id 外鍵，我們可以有效地表達和維護公司與員工之間的層次關係。

4.2.3 多對多關係

在資料庫設計中，多對多關係是一種更為複雜的實體關係。以下是一些典型的多對多關係範例：

- 一個訂單可以包含多種商品，而這些商品也可能出現在多個不同的訂單中。
- 一個學生可以註冊多門不同的課程，同時一門課程也可以有多名註冊學生。
- 一篇文章可以被標記為多個不同的標籤，而每個標籤也可能被用來分類多篇文章。
- 一個使用者可能擁有多個微信好友，同樣，一個好友也可能與多個使用者建立聯繫。

以訂單和商品的關係為例，它們之間的多對多關係可以透過一個連結表來實現。這個連結表通常包含兩個外鍵，分別指向訂單表和商品表的主鍵。圖 4-47 展示了這種關係是如何在資料庫中建模的。

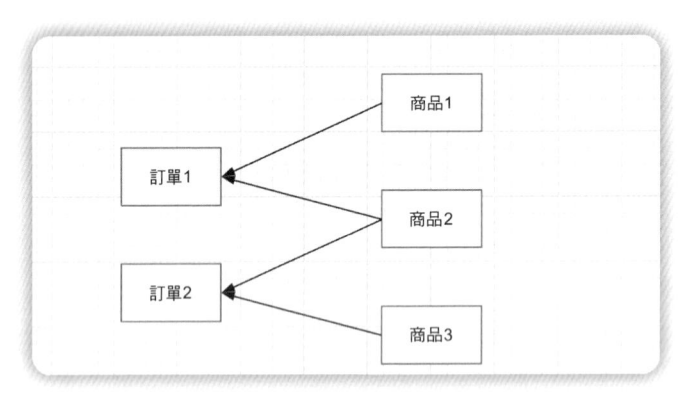

▲ 圖 4-47 訂單與商品的多對多關係

在現實世界中，多對多關係具有以下幾個顯著特點。

- 複雜性：多對多關係反映了兩個或多個實體之間複雜的相互作用，在許多實際應用場景中非常常見。
- 中間表（連結表）：為了在資料庫中實現多對多關係，通常需要建立一個中間表，也稱為連結表。該表儲存了參與關係的兩個實體的主鍵作為外鍵，並可能包含其他與這種關係直接相關的屬性資訊。

圖 4-48 將用於詳細描述這種多對多關係的結構和工作方式，它將幫助我們更清晰地理解涉及的實體如何透過中間表相互連結。

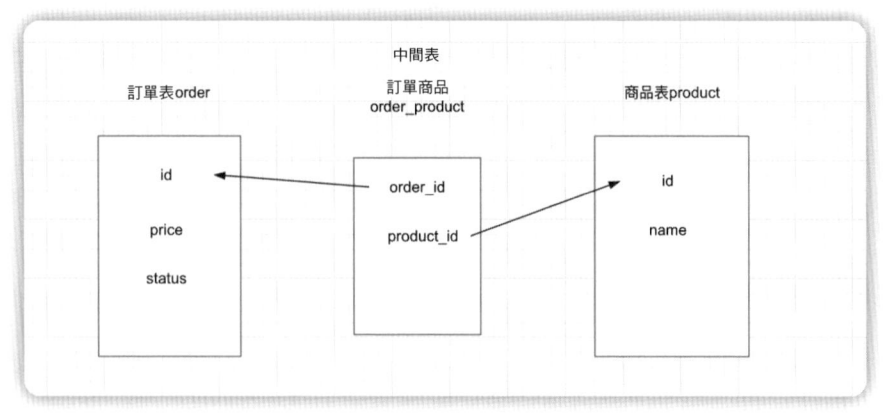

▲ 圖 4-48 中間表

　　圖 4-48 展示了 order 表和 product 表之間的關係。它們不是透過直接的外鍵關係來維護彼此之間的關係，而是透過一個名為 order_product 的中間表來維護對應的關係。這個中間表不僅維護了兩個表之間的外鍵關係，還可以儲存額外的資訊，例如訂單中商品的數量、價格和優惠折扣等。

　　為了演示多對多關係的實際應用，我們將建立這三個表：order、product 和 order_product。在建立 order_product 中間表時，我們將 order_id 和 product_id 都設置為表的主鍵，形成了一個複合主鍵。這樣的設計是因為 order_product 表中的每筆記錄都唯一地連結了 order 表和 product 表中的記錄，確保了這種組合的唯一性。

　　建立完成後的表結構和關係如圖 4-49 所示，清晰地展示了多對多關係如何在資料庫中得到實現和管理。

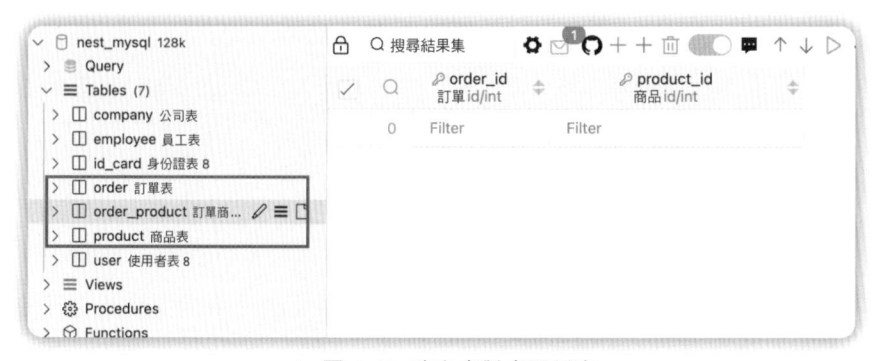

▲ 圖 4-49 建立多對多關係表

接下來，我們需要為 order_product 中間表分別增加指向 order 表和 product 表的外鍵。這些外鍵將確保中間表能夠維護與 order 表和 product 表的連結關係。此外，我們將這些外鍵的串聯選項設置為 CASCADE。這表示，如果 order 表或 product 表中的記錄被刪除，那麼 order_product 中間表中相應的連結記錄也將被自動刪除。這種設置保證了資料的一致性，並確保了資料庫中的資料關係保持有意義。

完成這些設置後，資料庫的結構如圖 4-50 所示。

▲ 圖 4-50 設置中間表外鍵

三個表都建立完畢後，接下來我們將使用 mock 方式生成隨機測試資料。在訂單表中，status 設定值範圍限制在 0 至 2 之間，價格限制在 0 至 100 之間。mock 的配置如下：

```
{
    "table": "order",
    "mockStartIndex": "auto",
    "mockCount": 10,
    "mockValueReference": "http://mockjs.com/examples.html#DPD",
    "mock": {
        "price": {
            "type": "int",
            "value": "@integer(1,100)"
        },
        "status": {
            "type": "int",
            "value": "@integer(0,2)"
        }
    }
}
```

在商品表中隨機生成「商品＋序號」格式的資料，配置如下：

```
{
    "table": "product",
    "mockStartIndex": "auto",
    "mockCount": 10,
    "mockValueReference": "http://mockjs.com/examples.html#DPD",
    "mock": {
        "name": {
            "type": "varchar",
            "value": "商品 @increment()"
        }
    }
}
```

中間表中兩個主鍵分別生成 0 至 10 範圍內的資料，配置如下：

```
{
    "table": "order_product",
    "mockStartIndex": "auto",
    "mockCount": 10,
    "mockValueReference": "http://mockjs.com/examples.html#DPD",
    "mock": {
        "order_id": {
            "type": "int",
            "value": "@integer(1,10)"
        },
        "product_id": {
            "type": "int",
            "value": "@integer(1,10)"
        }
    }
}
```

在成功生成資料之後，接下來的問題是，我們如何進行多對多關係的查詢操作呢？

正如前面所討論的，我們可以使用 JOIN 查詢來實現這一目標。不同於簡單的一對一或一對多查詢，多對多查詢需要連接多個表。以查詢特定訂單（例如 id 等於 5 的訂單）下所有連結的商品為例，我們需要執行一系列 JOIN 操作來連結 order 表、product 表以及 order_product 中間表。SQL 敘述如下：

```
SELECT * FROM `order`
JOIN order_product ON order.id = order_id
JOIN product ON product.id = product_id
WHERE order.id = 2
LIMIT 100
```

執行 SQL 查詢後，我們獲得了結果：訂單 id 為 2 的訂單中綁定了商品 id 為 103 和 104 的兩個商品，如圖 4-51 所示。

▲ 圖 4-51　多對多查詢結果

接下來選中 id=2 的這筆訂單記錄執行刪除操作，如圖 4-52 所示。

▲ 圖 4-52　刪除 order 表資料

此時，執行 SQL 敘述，可以看到中間表的查詢結果為空，如圖 4-53 所示，這說明 CASCADE 串聯方式已經生效。

▲ 圖 4-53　查詢結果為空

以上內容闡述了如何透過資料表來實現生活中的多對多關係模型。掌握本節知識後，我們能夠對生活中常見的實體關係進行有效的建模，例如朋友圈的互相點贊、社群網站中的好友關係以及許可權與角色之間的關係等。這些知識能幫助讀者更進一步地理解和分析業務場景。

4.3 快速上手 TypeORM

在 4.2 節中，我們學習了透過視覺化介面和命令列兩種方式來操作 MySQL 資料庫中的單表及多表的增刪改查。你可能會發現，命令列操作不僅需要記住各種 SQL 語法，而且在複雜場景下撰寫多表查詢、子查詢等敘述也相當煩瑣。ORM（Object-Relational Mapping，物件關係映射）框架可以有效地解決這類問題。本節將介紹主流 ORM 框架之一——TypeORM 的基本概念，並探討它如何實現資料庫操作。隨後，我們將透過程式演示來學習 TypeORM 的使用方法，將會有助我們在後續的 NestJS 專案中更有效地應用它。

4.3.1 基本概念

ORM 允許開發者透過建立一個「虛擬物件資料庫」來簡化資料庫操作。這種技術將關聯式資料庫的操作映射為對這些虛擬物件的操作，使得開發者可以直接在程式中進行資料庫操作。目前，TypeORM、Sequelize 和 Prisma 是一些主流的 ORM 框架。TypeORM 特別值得注意，因為它是一個成熟的 Node.js ORM 框架，使用 TypeScript 撰寫，並且能夠與 NestJS 框架友善整合。

4.3.2 專案準備

執行「npx typeorm init --name typeorm-test --database mysql2」命令來建立 TypeORM 專案，其中 --name 為專案名稱，--database 為資料庫類型。

> **提示**
>
> 如果你尚未安裝過 TypeORM 套件，使用 npx 命令時，系統會先檢測是否已安裝該套件。如果檢測到未安裝，npx 命令會詢問你是否要進行安裝。選擇「是」後，指令稿將自動安裝包括「reflect-metadata」在內的相關相依套件。

建立專案成功後，開啟終端並輸入「cd typeorm-test」命令以進入專案目錄。接下來，安裝 mysql2 套件，可以透過在終端執行「npm install mysql2 --save」命令來完成。將會安裝 mysql2 並將其作為相依項增加到你的 package.json 檔案中，如圖 4-54 所示。

```
$ npm install mysql2 -S
added 12 packages in 2m
```

▲ 圖 4-54 安裝 mysql2 套件

4.3.3 建立模型及實體

要處理資料庫，首先需要建立表。在 TypeORM 中，我們透過定義實體來告訴它如何建立一個資料表。以建立使用者表為例，其基本模型定義如下：

```
export class User {
    id: number
    name: string
    age: number
    nickname: string
    phone: string
}
```

然而，僅有模型還不夠，我們還需要將其定義為一個實體。這可以透過使用 @Entity 裝飾器來實現，程式如下：

```
import { Entity } from "typeorm"

@Entity()
export class User {
    id: number
    name: string
    age: number
    nickname: string
    phone: string
}
```

4.3.4 定義資料列及類型

建立資料表後，下一步是在表中增加列。在 TypeORM 中，我們使用裝飾器來修飾實體類別的屬性，以定義列的特性。用於裝飾列的裝飾器包括 @Column、@PrimaryColumn 和 @PrimaryGeneratedColumn。其中，@PrimaryColumn 用於定義普通主鍵列，而 @PrimaryGeneratedColumn 用於定義自動增加主鍵列。裝飾器還允許我們指定列的資料型態。範例程式如下：

```
import { Entity, Column, PrimaryGeneratedColumn } from "typeorm"
@Entity()
export class User {
    @PrimaryGeneratedColumn()
    id: number

    @Column({
        length: 100,
    })
    name: string

    @Column()
    age: number

    @Column({
        length: 100
    })
    nickname: string

    @Column()
    phone: string

    @Column("text")
    desc: string

    @Column("double")
    other: number
}
```

請注意，string 類型預設映射為資料庫中的 VARCHAR(255) 類型（以 MySQL 為例），而 number 類型預設映射為整數類型。如果我們希望某個欄位是文字（TEXT）類型或浮點（DOUBLE）類型，可以透過 @Column 裝飾器指定。舉例來說，將 description 欄位設置為 TEXT 類型，將 other 欄位設置為 DOUBLE 類型。完成這些設置後，將上述程式替換到 src/entity/User.ts 檔案中，實體的定義就完成了。

4.3.5 連接資料庫

為了與資料庫進行互動，我們需要設置一個資料來源（DataSource）。資料來源包含了所有與資料庫連接相關的配置資訊。透過這些配置，我們可以建立與資料庫的初始連接或連接池。範例程式如下：

```
import "reflect-metadata"
import { DataSource } from "typeorm"
```

```
import { User } from "./entity/User"

export const AppDataSource = new DataSource({
    type: "mysql",
    host: "localhost",
    port: 3306,
    username: "root",
    password: "你的資料庫密碼",
    database: "typeorm_mysql",
    entities: [User],
    synchronize: true,
    logging: true,
})
```

在以上程式中，資料庫配置屬性說明如下。

- type：表示資料庫類型，例如 mysql、oracle、sqlite、mongodb 等。
- database：表示你需要連接的資料庫名稱。
- entities：用於定義需要載入的實體，它接收實體類別或目錄路徑，如 ./**/entity/*.ts。
- synchronize：表示在應用啟動時是否自動建立和更新資料庫結構。
- logging：用於開機記錄記錄，記錄實際執行了哪些 SQL 敘述。
- host、port、username、password：為資料庫的基礎配置，包括資料庫伺服器地址、通訊埠、使用者名稱和密碼。

建立 DataSource 實例後，需要呼叫該實例的 initialize 方法來建立資料庫連接。一旦資料庫連接成功，我們可以使用物理管理器（manager）來執行資料的插入和查詢操作。修改 src/index.ts 的程式如下：

```
import { AppDataSource } from "./data-source"
import { User } from "./entity/User"

async function bootstrap() {
    await AppDataSource.initialize()
    const user = new User()
    user.name = "mouse"
    user.nickname = "乳酪"
    user.age = 25
    user.phone = '13000000000'
    user.desc = '預設描述'
    user.other = 1.0
    // 插入資料
    await AppDataSource.manager.save(user)
```

```
    // 查詢資料
    const users = await AppDataSource.manager.find(User)
    console.log(" 查詢使用者： ", users)
}
bootstrap()
```

執行「npm run start」命令啟動服務，可以看到成功插入資料並列印查詢結果，如圖 4-55 所示。

```
                  `rc`.`CONSTRAINT_NAME` = `kcu`.`CONSTRAINT_NAME`
query: SELECT VERSION() AS `version`
query: SELECT * FROM `INFORMATION_SCHEMA`.`COLUMNS` WHERE `TABLE_SCHEMA` = 'typeorm_mysql' AND `TABLE_NAME` = 'typeorm_metadata
`
query: COMMIT
query: START TRANSACTION
query: INSERT INTO `user`(`id`, `name`, `age`, `nickname`, `phone`, `desc`, `other`) VALUES (DEFAULT, ?, ?, ?, ?, ?, ?) -- PARA
METERS: ["mouse",25,"乳酪","13000000000","預設描述",1]
query: COMMIT
query: SELECT `User`.`id` AS `User_id`, `User`.`name` AS `User_name`, `User`.`age` AS `User_age`, `User`.`nickname` AS `User_ni
ckname`, `User`.`phone` AS `User_phone`, `User`.`desc` AS `User_desc`, `User`.`other` AS `User_other` FROM `user` `User`
查詢使用者： [
  User {
    id: 2,
    name: 'mouse',
    age: 25,
    nickname: '乳酪',
    phone: '13000000000',
    desc: '預設描述',
    other: 1
  }
]
```

▲ 圖 4-55 TypeORM 插入、查詢資料

4.3.6 使用 Repository 操作 CRUD

除使用物理管理器（EntityManager）來管理實體外，我們還可以使用儲存庫（Repository）來管理每個實體。下面透過程式演示如何進行資料庫的增刪改查操作。

首先是使用 Repository 進行新增和查詢操作，程式如下：

```
import { AppDataSource } from "./data-source"
import { User } from "./entity/User"

async function bootstrap() {
    // 使用儲存庫方式
    await AppDataSource.initialize()
    const user = new User()
    user.name = "mouse2"
    user.nickname = " 乳酪 2"
    user.age = 25
    user.phone = '13000000000'
    user.desc = ' 預設描述 2'
    user.other = 1.0

    const userRepository = AppDataSource.getRepository(User)
```

```
    // 新增
    await userRepository.save(user)
    console.log('新增使用者成功');
    // 查詢
    const saveUsers = await userRepository.find()
    console.log('查詢使用者：', saveUsers);
}
bootstrap()
```

執行「npm run start」命令，主控台列印結果如圖 4-56 所示。

▲ 圖 4-56 新增並查詢資料

使用 VS Code 的 MySQL 視覺化外掛程式也可以查詢到這筆資料，如圖 4-57 所示。

▲ 圖 4-57 視覺化資料結果

接著嘗試更新 id 為 1 的記錄中的 name 和 nickname 欄位。在 Repository 中，可以這樣操作：

```
const userUpdate = await userRepository.findOneBy({
    id: 1,
})
userUpdate.name = 'mouse3'
userUpdate.nickname = '乳酪3'
```

```
await userRepository.save(userUpdate)
console.log('更新成功');
```

再次執行「npm run start」命令，執行結果如圖 4-58 所示，可以看到成功更新了 name 和 nickname 欄位。

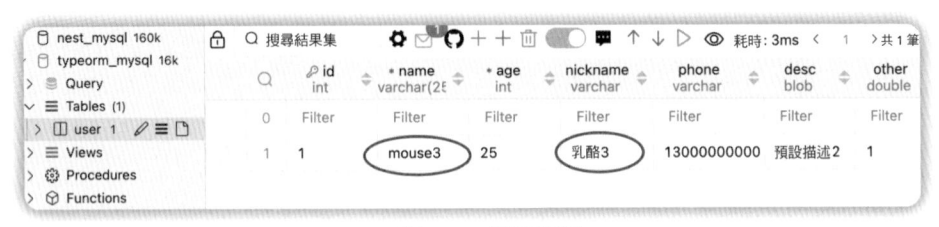

▲ 圖 4-58 更新資料

最後，在 Repository 中，刪除 id 為 1 的記錄，程式如下：

```
// 刪除
const userDelete = await userRepository.findOneBy({
    id: 1,
})
await userRepository.delete(userDelete)
console.log('< 刪除成功');
```

成功刪除資料，結果如圖 4-59 所示。

▲ 圖 4-59 刪除資料

至此，我們完成了在 Repository 中進行資料庫的增刪改查操作，它就像操作物件一樣簡單。

4.3.7 使用 QueryBuilder 操作 CRUD

QueryBuilder 是 TypeOrm 中最強大的功能之一，它提供了更加底層的資料庫查詢操作，靈活度高，支援組合更複雜的 SQL 查詢敘述，如多條件查詢或多表查詢等，在本書的專案實戰中會頻繁使用它。

接下來介紹建立 QueryBuilder 的 3 種方式。

第一種是使用 DataSource 來建立，例如：

```
const user = await dataSource
    .createQueryBuilder()
    .select("user")
    .from(User, "user")
    .where("user.id = :id", { id: 1 })
    .getOne();
```

這種方式提供了較大的靈活性，適合跨實體的複雜查詢操作和資料庫等級的操作。

第二種是使用 Repository 來建立，例如：

```
const user = await dataSource
    .getRepository(User)
    .createQueryBuilder("user")
    .where("user.id = :id", { id: 1 })
    .getOne();
```

透過 Repository 建立的 QueryBuilder 比較常用，它直接與實體相連結，提供了比直接使用 Repository 方法更加靈活的資料庫查詢能力。

第三種是使用 Manager 來建立，例如：

```
const user = await dataSource.manager
    .createQueryBuilder(User, "user")
    .where("user.id = :id", { id: 1 })
    .getOne();
```

這種方式並不常用，因此這裡不做過多介紹。

接下來透過程式演示 QueryBuilder 的基本用法，首先使用 DataSource 方式新增一個使用者，程式如下：

```
await AppDataSource.initialize()
const user = new User()
user.name = "mouse3"
user.nickname = " 乳酪 3"
user.age = 25
user.phone = '13000000000'
user.desc = ' 預設描述 3'
user.other = 1.0

const queryBuilder = AppDataSource.createQueryBuilder()
// 新增
await queryBuilder.insert().into(User).values(user).execute()
console.log(' 新增使用者成功 ');
```

以上程式使用 insert 和 into 方法進行插入操作，語法接近原始 SQL 敘述。在 values 方法中可以傳入物件或陣列，如果要插入多筆記錄，則選擇傳入一個陣列。最後呼叫 execute 方法執行敘述。

執行「npm run start」命令後，結果如圖 4-60 所示，可以看到成功插入了一筆資料。

```
query: SELECT VERSION() AS `version`
query: SELECT * FROM `INFORMATION_SCHEMA`.`COLUMNS` WHERE `TABLE_SCHEMA` = 'typeorm_mysql' AND `TABLE_NAME` = 'ty
query: COMMIT
query: INSERT INTO `user`(`id`, `name`, `age`, `nickname`, `phone`, `desc`, `other`) VALUES (DEFAULT, ?, ?, ?, ?,
ouse3",25,"乳酪 3","13000000000","預設描述 3",1]
新增使用者成功
```

▲ 圖 4-60 新增資料

接下來查詢插入的資料，相關程式如下：

```
// 查詢
const saveUsers = await queryBuilder.select('u').from(User, 'u').getMany()
console.log('查詢使用者:', saveUsers);
```

同樣地，查詢操作使用類似於「select from」的 SQL 敘述，u 是 User 實體的別名，getMany 方法可以把多筆記錄查詢出來，查詢結果的資料欄位名稱與實體表中定義的欄位名稱相同，如圖 4-61 所示。

```
query: SELECT `u`.`id` AS `u_id`, `u`.`name` AS `u_name`, `u`.`age` AS `u_age`, `u`.`nickname` AS `u_nickname`, `u`.`phone`
`u`.`desc` AS `u_desc`, `u`.`other` AS `u_other` FROM `user` `u`
查詢使用者: [
  User {
    id: 1,
    name: 'update name',
    age: 25,
    nickname: '乳酪1',
    phone: '13000000000',
    desc: '預設描述1',
    other: 1.25
  },
  User {
    id: 2,
    name: 'mouse1',
    age: 25,
    nickname: '乳酪1',
    phone: '13000000000',
    desc: '預設描述1',
    other: 1.25
```

▲ 圖 4-61 getMany 查詢結果

TypeORM 還提供了 getRawMany 方法，允許我們執行更底層的資料庫查詢並獲取原始資料，查詢結果如圖 4-62 所示。

```
query: SELECT `u`.`id` AS `u_id`, `u`.`name` AS `u_name`, `u`.`age` AS `u_age`, `u`.`nickname` AS `u_nickname`, `u`.`pho
`u`.`desc` AS `u_desc`, `u`.`other` AS `u_other` FROM `user` `u`
查詢使用者: [
  {
    u_id: 1,
    u_name: 'update name',
    u_age: 25,
    u_nickname: '乳酪1',
    u_phone: '13000000000',
    u_desc: '預設描述1',
    u_other: 1.25
  },
  {
    u_id: 2,
    u_name: 'mouse1',
    u_age: 25,
    u_nickname: '乳酪1',
    u_phone: '13000000000',
    u_desc: '預設描述1',
    u_other: 1.25
  },
```

▲ 圖 4-62 getRawMany 查詢結果

使用此方法時，傳回的資料通常包含資料庫查詢的實際結果，其中欄位名稱可能會與自訂的表別名拼接在一起。

接下來修改使用者資訊，先查詢出指定 id 為 4 的使用者，修改資訊後，再進行更新操作，程式如下：

```
// 修改
const userUpdate = await queryBuilder.select('u').from(User, 'u').where('u.id = :id',
                   { id: 4 }).getOne()
userUpdate.name = 'mouse4'
userUpdate.nickname = '乳酪 4'
queryBuilder.update(User).set(userUpdate).where('u.id = :id', { id: 4 }).execute()
console.log('更新成功');
```

在上述程式中，首先使用 where 子句增加查詢準則，並透過 getOne 方法獲取滿足條件的單一記錄。如果找到了使用者，修改其 name 和 nickname 屬性，然後使用 update 和 set 方法來更新資料庫中的記錄，如圖 4-63 所示。注意，set 方法需要一個物件，包含要更新的欄位和新值。

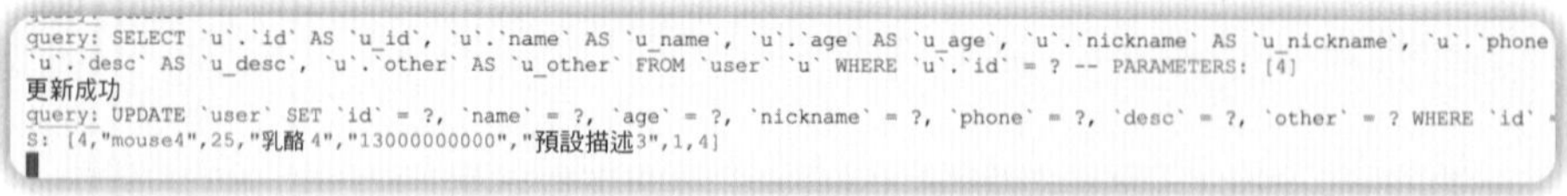

▲ 圖 4-63 修改使用者資訊

最後刪除使用者，把剛剛測試的 id 為 4 的使用者刪除，程式如下：

```
// 刪除
await queryBuilder.delete().from(User).where('id = :id', { id: 4 }).execute()
console.log('刪除成功')
```

執行結果如圖 4-64 所示。

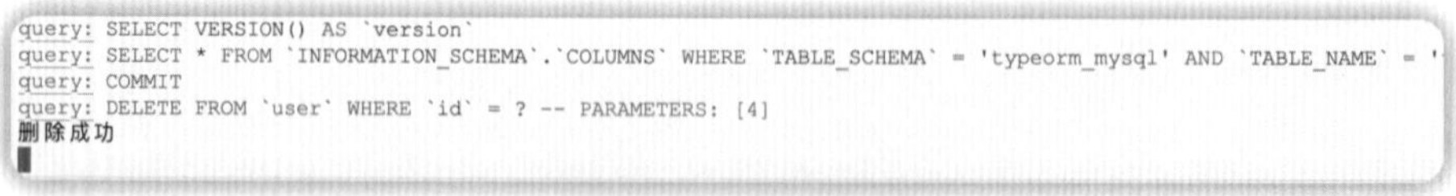

▲ 圖 4-64 刪除使用者

至此，我們完成了用 TypeORM 來建立資料模型和實體，並演示了如何連接資料庫服務。最後，透過使用 TypeORM 提供的物理管理器（EntityManager）、儲存庫（Repository）及查詢建構器（QueryBuilder）三種方式對資料庫進行 CRUD 操作。這些操作不僅幫助我們進一步理解 ORM 框架的使用，也大大簡化了操作資料庫的過程。

4.4 使用 TypeORM 處理多表關係

MySQL 中表之間存在多種關係,如一對一、一對多/多對一以及多對多關係。本節將基於第 4.2 節中提到的實體關係範例,演示如何在 TypeORM 中透過表來描述這些關係。

4.4.1 一對一關聯性

本小節介紹資料庫中表和表的一對一關聯性,以使用者表和身份證表為例。在 src/entity 資料夾下建立一個 IdCard 實體,程式如下:

```
import {
  Entity,
  Column,
  PrimaryGeneratedColumn,
  OneToOne,
  JoinColumn,
} from "typeorm";
import { User } from "./User";

@Entity()
export class IdCard {
  @PrimaryGeneratedColumn()
  id: number;

  @Column()
  name: string;

  @Column()
  address: string;

  @Column()
  birthday: Date;

  @Column()
  email: string;

  @OneToOne(() => User)
  @JoinColumn()
  user: User;
}
```

在上述程式中,我們使用 @OneToOne 裝飾器來建立與 User 實體之間的一對一關聯性,同時用 @JoinColumn 裝飾器來定義 user 列,以維護一個外鍵。

TypeORM 會自動生成外鍵的 id。

注意：請不要忘記把 IdCard 實體增加到 DataSource 配置選項 entities 中，並建議配置為 ./**/entity/*.ts。

接下來，在 index.ts 中增加以下的一段邏輯，分別建立使用者表和身份證表，並插入資料，並確保把它們連結起來。範例程式如下：

```
// 一對一關聯性
// 建立一個使用者
await AppDataSource.initialize()
const user = new User()
user.name = "mouse1"
user.nickname = " 乳酪 1"
user.age = 25
user.phone = '13000000000'
user.desc = ' 預設描述 1'
user.other = 1.25
// 建立一個身份證資訊
const idCard = new IdCard()
idCard.name = "my name is mouse"
idCard.address = " 廣東省廣州市 "
idCard.birthday = new Date()
idCard.email = 'jmin95@163.com'
// 連結兩個實體
idCard.user = user

// 獲取實體的儲存庫
const userRepository = AppDataSource.getRepository(User)
const idCardRepository = AppDataSource.getRepository(IdCard)
// 首先儲存使用者
await userRepository.save(user)

// 使用者已儲存。現在我們需要儲存使用者的身份證資訊
await idCardRepository.save(idCard)

// 完成
console.log(" 資料已儲存，並且在資料庫中建立了使用者與身份證之間的連結關係 ")
```

在這裡，我們分別建立了一個名為 user 的使用者和一個身份證資訊，隨後將它們連結起來，再呼叫儲存庫的 save 方法來儲存這些資訊。

執行「npm run start」命令，我們可以看到成功建立了兩個表，如圖 4-65 所示。

4.4 使用 TypeORM 處理多表關係

▲ 圖 4-65 建立一對一表

由圖 4-65 可見，id_card 表中已生成了 userId 欄位（外鍵），它連結的是 user 表中的 id 欄位。在前面的學習中我們已了解 JOIN 查詢，接下來執行下面的 SQL 敘述來查詢資料：

```
SELECT * FROM user JOIN id_card ON user.id = id_card.`userId`;
```

執行結果如圖 4-66 所示。

▲ 圖 4-66 SQL 命令一對一查詢

當然，我們也可以用 Repository 提供的 find 方法來查詢。由於兩個實體的關係是單向的，即 IdCard 實體持有 User 實體的外鍵列，使得它有權存取 User，但反過來 User 不知道 IdCard 的存在。因此，在這種情況下應該使用 idCardRepository 來查詢，並且設置連結（relations），在 index.ts 中增加以下程式：

```
const idCardRes = await idCardRepository.find({
    relations: {
        user: true
    }
})
console.log(idCardRes);
```

再次執行「npm run start」命令，成功查詢出兩筆連結了 user 欄位的記錄，如圖 4-67 所示。

▲ 圖 4-67 Repository 一對一查詢

為了讓 User 實體能夠存取 IdCard 實體，我們需要修改實體類別以建立雙向連結，程式如下：

```
@Entity()
export class IdCard {
    /* 其他列 */

    @OneToOne(() => User, (user) => user.card)
    @JoinColumn()
    user: User
}

@Entity()
export class User {
    /* 其他列 */

    @OneToOne(() => IdCard, (card) => card.user)
    card: IdCard
}
```

此時，用 userRepository 載入連結表的資料，程式如下：

```
// 雙向連結
const userRes = await userRepository.find({
    relations: {
        card: true
    }
})
console.log(userRes);
```

執行結果如圖 4-68 所示。

```
query: COMMIT
資料已儲存，並且在資料庫中建立了使用者與身份證之間的連結關係
query: SELECT `User`.`id` AS `User_id`, `User`.`name` AS `User_name`, `User`.`age` AS `User_age`, `User`.`nickname` AS `User_nickname`, `User`.`phone` AS `User_phone`, `User`.`desc` AS `User_desc`, `User`.`other` AS `User_other`, `User__User_card`.`id` AS `User__User_card_id`, `User__User_card`.`name` AS `User__User_card_name`, `User__User_card`.`address` AS `User__User_card_address`, `User__User_card`.`birthday` AS `User__User_card_birthday`, `User__User_card`.`email` AS `User__User_card_email`, `User__User_card`.`userId` AS `User__User_card_userId` FROM `user` `User` LEFT JOIN `id_card` `User__User_card` ON `User__User_card`.`userId`=`User`.`id`
[
  User {
    id: 1,
    name: 'mouse1',
    age: 25,
    nickname: '乳酪1',
    phone: '13000000000',
    desc: '預設描述1',
    other: 1.25,
    card: IdCard {
      id: 1,
      name: 'my name is mouse',
      address: '廣東省廣州市',
      birthday: 2024-01-28T07:02:38.000Z,
      email: 'jmin95@163.com'
    }
  }
]
```

▲ 圖 4-68 雙向關係查詢

前面我們學習了如何設置串聯自動更新和刪除連結資料，那麼在 TypeORM 中如何實現呢？可以在 User 實體中定義串聯選項為 cascade: true，程式如下：

```
@Entity()
export class User {
    /* 其他列 */

    @OneToOne(() => IdCard, (card) => card.user, {
        cascade: true
    })
    card: IdCard;
}
```

此時儲存或刪除 user 物件會自動更新 idCard 物件。我們來看下面這段程式：

```
// 串聯儲存、更新資料
// 建立一個使用者
await AppDataSource.initialize()
const user = new User()
user.name = "mouse1"
user.nickname = " 乳酪 1"
user.age = 25
user.phone = '13000000000'
user.desc = ' 預設描述 1'
user.other = 1.25
// 建立一個身份證資訊
const idCard = new IdCard()
idCard.name = "my name is mouse"
idCard.address = " 廣東省廣州市 "
```

```
idCard.birthday = new Date()
idCard.email = 'jmin95@163.com'
// 注意這裡需要透過 user 來連結 idCard
user.card = idCard

// 獲取實體的儲存庫
const userRepository = AppDataSource.getRepository(User)
// 儲存使用者
await userRepository.save(user
```

在這裡,我們分別建立了一個使用者和一個身份證資訊,並使用 user 實體中的 card 欄位連結 idCard 物件。接著透過 user 實體儲存 user 實例物件。刪除之前的表後,重新執行「npm run start」命令,執行結果如圖 4-69 所示。

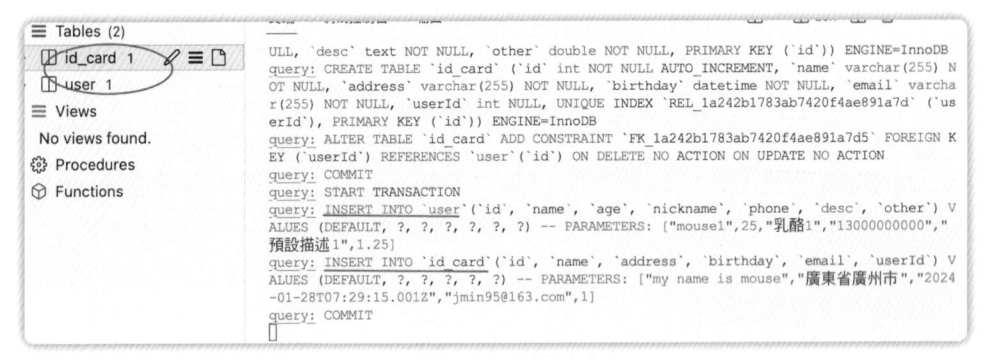

▲ 圖 4-69 串聯建立

從圖 4-69 中可以看出,TypeORM 成功地執行了 SQL 敘述並建立了兩個表。有了資料後,我們接下來演示如何更新資料,程式如下:

```
// 查詢 id 為 1 的資料
const loadedUser = await userRepository.findOne({
    where: {
        id: 1
    },
    relations: {
        card: true
    }
})
console.log(loadedUser);
// 更新欄位
loadedUser.name = 'update name'
loadedUser.card.name = 'new idCard name'

await userRepository.save(loadedUser)
```

執行結果如圖4-70所示，id_card表和user表中的name欄位已經被成功更新。

▲ 圖 4-70 串聯更新

4.4.2 一對多 / 多對一關係

在 TypeORM 中，要表示公司與員工之間的一對多關聯性（即一個公司可以有多個員工）和多對一關係（即每個員工屬於一個公司），我們需要建立兩個實體類別：Company 和 Employee。範例程式如下：

```
import {
  Entity,
  Column,
  PrimaryGeneratedColumn,
  OneToMany,
  JoinColumn,
} from «typeorm»
import { Employee } from «./Employee»

@Entity()
export class Company {
  @PrimaryGeneratedColumn()
  id: number

  @Column()
  name: string

  @OneToMany(() => Employee, (employee) => employee.company, {
    // 設置串聯自動更新連結表
    cascade: true
  })
  employees: Employee[]
}
```

在 Company 類別中，我們使用 @OneToMany 裝飾器來定義與員工表的連結，並設置串聯（cascade）選項以自動更新相連結的 Employee 記錄。這表示，當儲存 Company 物件時，相連結的 Employee 記錄也會自動更新。此外，我們在 Company 類別中定義了一個 employees 屬性，用於儲存與該公司相連結的員工列表。

@OneToMany 裝飾器不能單獨使用，它必須與 @ManyToOne 裝飾器配對使用，以確保關係的雙向性。現在讓我們來看看 Employee 類別是如何定義這種關係的：

```
import {
  Entity,
  Column,
  PrimaryGeneratedColumn,
  JoinColumn,
  ManyToOne,
} from «typeorm»
import { Company } from «./Company»

@Entity()
export class Employee {
  @PrimaryGeneratedColumn()
  id: number

  @Column()
  name: string

  @ManyToOne(() => Company, (company) => company.employees)
  @JoinColumn()
  company: Company
}
```

在上述程式中，我們同樣連結了與之對應的 Company 類別，並使用 @JoinColumn 裝飾器定義了外鍵列。在一對一 / 多對一關係中，外鍵列始終出現在「多」的一方。

雙方物理定義完成後，接下來插入資料，在 index.ts 中增加以下程式：

```
// 一對多 / 多對一關係
// 建立一個公司
await AppDataSource.initialize()
const company = new Company()
company.name = " 甜筒公司 "
// 建立員工 1
const employee1 = new Employee()
employee1.name = " 員工 1 號 "
// 建立員工 2
```

```
const employee2 = new Employee()
employee2.name = "員工 2 號"
// 注意，這裡需要透過 Company 來連結 Employee
company.employees = [employee1, employee2]

// 獲取實體的儲存庫
const companyRepository = AppDataSource.getRepository(Company)
// 儲存公司
await companyRepository.save(company)
```

執行「npm run start」命令，執行結果如圖 4-71 所示，id 為 1 的公司下有兩筆員工記錄，並透過 companyId 進行連結。

▲ 圖 4-71 建立一對多 / 多對一資料

4.4.3 多對多關係

接下來，我們將建立訂單（Order）和商品（Product）之間的多對多關係。在多對多關係中，一個訂單可以包含多個商品，同時一個商品也可以出現在多個訂單中。首先，我們需要定義 Order 實體類別，它將代表資料庫中的訂單表。範例程式如下：

```
import {
  Entity,
  PrimaryGeneratedColumn,
  Column,
  ManyToMany,
  JoinTable,
} from 'typeorm'
import { Product } from './Product'

@Entity()
```

```
export class Order {
  @PrimaryGeneratedColumn()
  id: number

  @Column()
  name: string

  @ManyToMany(() => Product, (product) => product.orders, {
    cascade: true
  })
  products: Product[]
}
```

在上述程式中,多對多關係透過 @ManyToMany 裝飾器來連結,設置串聯 cascade 選項會在儲存訂單時自動儲存商品資訊。而 Product 實體類別的定義如下:

```
import {
  Entity,
  PrimaryGeneratedColumn,
  Column,
  ManyToMany,
  JoinTable,
} from 'typeorm'
import { Order } from './Order'

@Entity()
export class Product {
  @PrimaryGeneratedColumn()
  id: number

  @Column()
  name: string

  @ManyToMany(() => Order, (order) => order.products)
  @JoinTable()
  orders: Order[]
}
```

在這裡用 @JoinTable() 來生成連接表(中間表),TypeORM 會建立一個中間表來管理多對多關係。我們在 index.ts 中增加以下程式,用於生成多對多關聯資料。

```
// 多對多關係
// 建立訂單 1
await AppDataSource.initialize()
const order = new Order()
order.name = "訂單 xxx"
// 建立商品 1
```

```
const product1 = new Product()
product1.name = "商品 1 號"
// 建立商品 2
const product2 = new Product()
product2.name = "商品 2 號"
// 注意這裡訂單 1 綁定多個商品
order.products = [product1, product2]

// 獲取實體的儲存庫
const orderRepository = AppDataSource.getRepository(Order)
// 儲存訂單
await orderRepository.save(order)
```

上述程式建立了一個訂單，並同時綁定了兩個商品。由於使用了串聯，因此在儲存訂單時會自動建立訂單表。執行結果如圖 4-72 所示。

▲ 圖 4-72 建立多對多資料

由圖 4-72 可見，ORM 自動建立了一個名為 product_orders_order 的中間表，生成 productId 和 orderId 列作為複合主鍵。同時，orderId=1 中綁定了兩個商品。

除此之外，我們還可以自訂中間表的名稱和列名稱。範例程式如下：

```
@ManyToMany(() => Order, (order) => order.products)
@JoinTable({
  name: 'order_products',
  joinColumn: {
    name: "order",
    referencedColumnName: "id"
  },
  inverseJoinColumn: {
    name: "product",
    referencedColumnName: "id"
  }
})
orders: Order[]
```

刪除舊表後重新執行，結果如圖 4-73 所示。

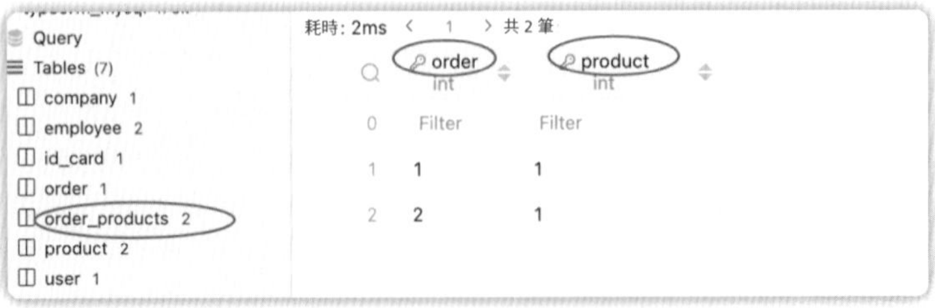

▲ 圖 4-73 自訂中間表資訊

至此，我們已經學習了 TypeORM 中的幾種基本關係：一對一、一對多 / 多對一以及多對多。掌握 @OneToOne、@OneToMany/@ManyToOne 和 @ManyToMany 這幾個裝飾器的用法至關重要。理解不同表之間的關係有助我們更進一步地抽象和表示現實生活中的各種複雜關係。

4.5 在 Nest 中使用 TypeORM 操作 MySQL

在掌握了 MySQL 和 TypeORM 的基礎知識之後，本節將學習如何在 Nest 框架中整合 TypeORM 函數庫來操作 MySQL 資料庫。我們將從建立一個基礎的 Nest 專案開始，並介紹如何在 Nest 中配置資料庫連接。此外，我們還將學習 TypeORM 提供的三種操作資料庫的方法：使用 EntityManager、Repository 和 QueryBuilder。

4.5.1 專案準備

執行「nest n nest-typeorm-mysql -p pnpm」命令，建立名為 nest-typeorm-mysql 的專案，指定套件管理為 pnpm，效果如圖 4-74 所示。

```
$ nest n nest-typeorm-mysql -p pnpm
⚡  We will scaffold your app in a few seconds..

CREATE nest-typeorm-mysql/.eslintrc.js (663 bytes)
CREATE nest-typeorm-mysql/.prettierrc (51 bytes)
CREATE nest-typeorm-mysql/README.md (3347 bytes)
CREATE nest-typeorm-mysql/nest-cli.json (171 bytes)
CREATE nest-typeorm-mysql/package.json (1959 bytes)
CREATE nest-typeorm-mysql/tsconfig.build.json (97 bytes)
CREATE nest-typeorm-mysql/tsconfig.json (546 bytes)
CREATE nest-typeorm-mysql/src/app.controller.spec.ts (617 bytes)
CREATE nest-typeorm-mysql/src/app.controller.ts (274 bytes)
CREATE nest-typeorm-mysql/src/app.module.ts (249 bytes)
CREATE nest-typeorm-mysql/src/app.service.ts (142 bytes)
CREATE nest-typeorm-mysql/src/main.ts (208 bytes)
CREATE nest-typeorm-mysql/test/app.e2e-spec.ts (630 bytes)
CREATE nest-typeorm-mysql/test/jest-e2e.json (183 bytes)

✔ Installation in progress... ☕

🚀  Successfully created project nest-typeorm-mysql
```

▲ 圖 4-74 建立專案

4.5 在 Nest 中使用 TypeORM 操作 MySQL

接著執行「nest g resource user」命令,生成 user 模組,裡面包含 REST 風格的 CURD 介面和 entities 實體類別,我們稍後會用到它。

接下來,安裝 typeorm、mysql2 和 @nestjs/typeorm 套件,執行以下命令:

```
pnpm add typeorm mysql2 @nestjs/typeorm -S
```

其中,typeorm 和 mysql2 這兩個套件我們之前已經使用過了,而 @nestjs/typeorm 是 Nest 對 TypeORM 的整合。它進一步封裝了 TypeORM API 的用法,支援模組動態匯入和相依注入等功能。

TypeORM 提供了初始化資料庫連接的方式,在入口檔案 app.modules.ts 中引入:

```
import { Module } from '@nestjs/common';
import { AppController } from './app.controller';
import { AppService } from './app.service';
import { UserModule } from './user/user.module';
import { TypeOrmModule} from '@nestjs/typeorm';
import { User } from './user/entities/user.entity';

@Module({
  imports: [
    // 初始化 MySQL 連接
    TypeOrmModule.forRoot({
      type: 'mysql',
      host: 'localhost',
      port: 3306,
      username: 'root',
      password: 'jminjmin',
      database: 'nest_typeorm',
      entities: [User],
      synchronize: true
    }),
    UserModule
  ],
  controllers: [AppController],
  providers: [AppService],
})
export class AppModule {}
```

上述程式透過 forRoot 方法在根模組(AppModule)中註冊了 TypeORM 模組。這樣做可以使得 TypeORM 服務在應用程式的各個模組中都可以共用使用。需要注意的是,在開發過程中,可以在 TypeORM 配置中設置 synchronize: true,這樣程式啟動時會自動載入或更新資料庫結構以匹配物理定義。然而,在生產環境中，我們應該禁用自動同步（synchronize: false），並採用手動遷移或其他安全的方式來管理資料庫結構的變更。

接下來，在 user.entity.ts 中完善 User 實體，新增以下幾列欄位：

```typescript
import { Entity, PrimaryGeneratedColumn, Column } from "typeorm";

@Entity()
export class User{

    @PrimaryGeneratedColumn()
    id: number;

    @Column()
    name: string;

    @Column()
    sex: string;

    @Column()
    createTime: Date;

    @Column()
    updateTime: Date;

}
```

執行「pnpm run start:dev」命令啟動服務。TypeORM 會根據資料庫配置，在內部呼叫 initialize 方法連接資料庫，並根據實體映射自動建立 user 表，如圖 4-75 所示。

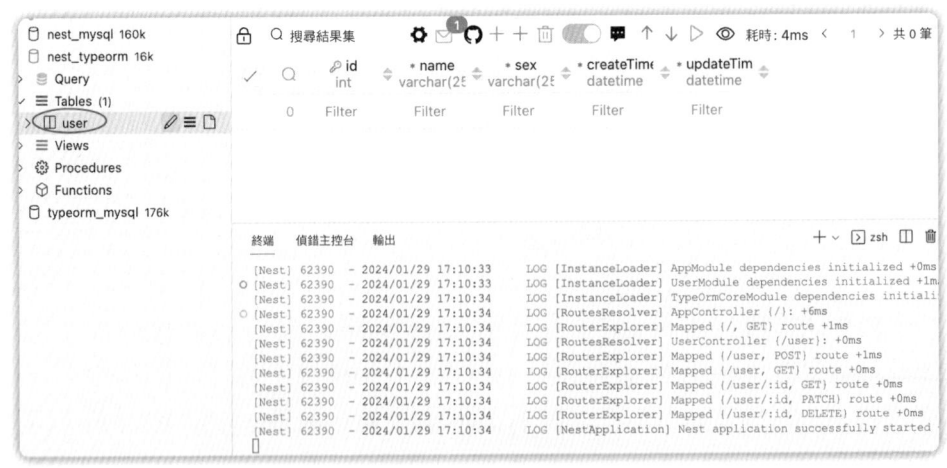

▲ 圖 4-75 啟動服務

4.5.2 使用 EntityManager 操作實體

一旦資料庫表建立完成，我們就可以利用 EntityManager 對實體執行增刪改查的操作。在 Nest 中，這些資料操作邏輯通常位於服務層（Service）。為了使 UserService 類別具備操作資料庫的能力，我們需要對其進行相應的修改，程式如下：

```
import { Injectable } from '@nestjs/common';
import { CreateUserDto } from './dto/create-user.dto';
import { UpdateUserDto } from './dto/update-user.dto';
import { InjectEntityManager } from '@nestjs/typeorm';
import { EntityManager } from 'typeorm';
import { User } from './entities/user.entity';

@Injectable()
export class UserService {
  @InjectEntityManager()
  private manage: EntityManager
  async create(createUserDto: CreateUserDto) {
    this.manage.save(User, createUserDto)
  }

  async findAll() {
    return await this.manage.find(User);
  }

  async findOne(id: number) {
    return this.manage.findBy(User, {id})
  }

  async update(id: number, updateUserDto: UpdateUserDto) {
    return await this.manage.save(User, {
      id,
      ...updateUserDto,
    })
  }

  async remove(id: number) {
    return await this.manage.delete(User, {id})
  }
}
```

在上述程式中，我們透過 @InjectEntityManager 裝飾器注入了 EntityManager（物理管理器），並呼叫它的 API 進行增刪改查操作。每次操作需要指定要處理的實體類別，本例中是 User。

首先，我們建立一個使用者。為了模擬前端發送請求，在根目錄下建立 .http 檔案（使用 VS Code 的 REST Client 外掛程式），配置請求介面如下：

```
@createdAt = {{$datetime iso8601}}

// 建立使用者
POST http://localhost:3300/user
Content-Type: application/json

{
  "name": "mouse",
  "sex": "男",
  "createTime": "{{createdAt}}",
  "updateTime": "{{createdAt}}"
}

###

// 查詢使用者
GET http://localhost:3300/user
```

在這裡透過 DTO 來接收 POST 請求資料，同樣在 create-user.dto.ts 檔案中增加以下欄位：

```
export class CreateUserDto {
  name: string;
  sex: string;
  createTime: Date;
  updateTime: Date;
}
```

點擊 Send Request，在資料庫中可以看到新增了一筆使用者記錄，如圖 4-76 所示。

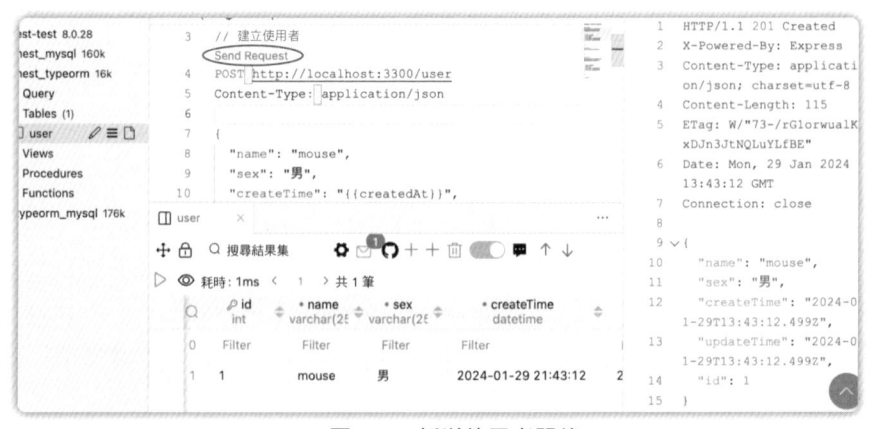

▲ 圖 4-76 新增使用者記錄

同樣，點擊 Send Request，如圖 4-77 所示，在右側可以看到剛剛新建的使用者資料。

```
###

// 查詢使用者
Send Request
GET http://localhost:3300/user
```

```
  7   Connection: close
  8
  9 ∨ [
 10 ∨   {
 11       "id": 1,
 12       "name": "mouse",
 13       "sex": "男",
 14       "createTime": "2024-01-29T10:15:0
          3.000Z",
 15       "updateTime": "2024-01-29T10:15:0
          3.000Z"
 16     }
 17   ]
```

▲ 圖 4-77 查詢使用者資料

4.5.3 使用 Repository 操作實體

除了使用 EntityManager 操作實體外，我們還可以使用 TypeORM 提供的 Repository 物件來進行增刪改查操作。Repository 物件擁有與 EntityManager 同樣的功能，區別在於使用 Repository 時不需要每次操作都傳入實體物件。以查詢為例，使用 EntityManager 的程式如下：

```
async findAll() {
  return await this.manage.find(User);
}
```

而使用 Repository 時，可以直接獲取指定的實體儲存庫並操作，具體程式如下：

```
async findAll() {
  return await this.userRepository.find();
}
```

下面透過 Repository 演示更新和刪除操作，我們把 user.service.ts 的程式修改如下：

```
import { Injectable } from '@nestjs/common';
import { CreateUserDto } from './dto/create-user.dto';
import { UpdateUserDto } from './dto/update-user.dto';
import { InjectRepository } from '@nestjs/typeorm';
import { Repository } from 'typeorm';
import { User } from './entities/user.entity';
```

```
@Injectable()
export class UserService {
  @InjectRepository(User) private userRepository: Repository<User>
  async create(createUserDto: CreateUserDto) {
    createUserDto.createTime = createUserDto.updateTime = new Date()
    return await this.userRepository.save(createUserDto);
  }

  async findAll() {
    return await this.userRepository.find();
  }

  async findOne(id: number) {
    return this.userRepository.findBy({
      id
    });
  }

  async update(id: number, updateUserDto: UpdateUserDto) {
    updateUserDto.updateTime = new Date()
    return await this.userRepository.update(id, updateUserDto);
  }

  async remove(id: number) {
    return await this.userRepository.delete(id);
  }
}
```

在上述程式中，透過 @InjectRepository 裝飾器實現相依注入，以獲取指定實體類別的 Repository。然後，可以在方法中使用 userRepository 實例提供的方法進行增刪改查操作。由於 Nest 的相依注入作用域限定在模組層級別，因此還需要在 Module 中透過 forFeature 方法匯入並註冊這些儲存庫。程式用法如下：

```
import { Module } from '@nestjs/common';
import { UserService } from './user.service';
import { UserController } from './user.controller';
import { TypeOrmModule } from '@nestjs/typeorm';
import { User } from './entities/user.entity';

@Module({
  imports: [
    TypeOrmModule.forFeature([User])
  ],
  controllers: [UserController],
  providers: [UserService]
})
export class UserModule {}
```

接下來修改和刪除 id 為 1 的使用者記錄，在 .http 檔案中增加請求配置：

```
###

// 更新使用者
PATCH http://localhost:3300/user/1
Content-Type: application/json

{
  "name": "Minnie",
  "sex": "女",
  "updateTime": "{{createdAt}}"
}

###

// 刪除使用者
DELETE http://localhost:3300/user/1
```

點擊 Send Request，如圖 4-78 所示，從右側的回應結果可以看出已成功更新了一組資料。下方的使用者記錄中的 name、sex 和 updateTime 已更新。

▲ 圖 4-78 更新使用者資料

最後點擊 Send Request，發送刪除 id 為 1 的請求，此時資料庫記錄已經被刪除，如圖 4-79 所示。

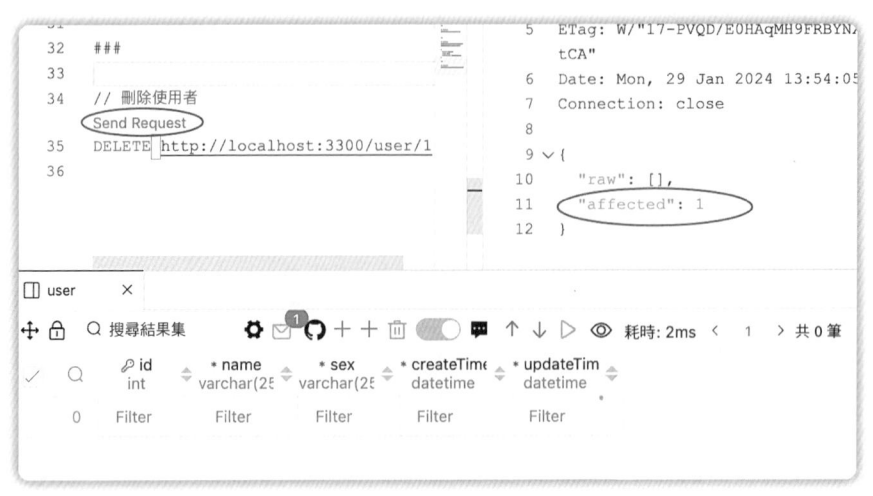

▲ 圖 4-79 刪除使用者資料

　　至此，我們已經掌握了在 TypeORM 中使用 EntityManager 和 Repository 進行資料的增刪改查操作。

4.5.4 使用 QueryBuilder 操作實體

　　QueryBuilder 是 TypeOrm 中最強大的功能之一，它提供了更加底層的資料庫查詢操作，具有高靈活度高，支援組合更複雜的 SQL 查詢敘述，如多條件查詢或多表查詢等。在本書的專案實戰中，我們會頻繁使用它。

　　依然以使用者資訊模組為例，新增使用者資訊。範例程式如下：

```
async create(createUserDto: CreateUserDto) {
  createUserDto.createTime = createUserDto.updateTime = new Date()
  return await this.dataSource
  .createQueryBuilder()
  .insert()
  .into(User)
  .values(createUserDto)
  .execute()
}
```

　　上述程式使用 TypeOrm 提供的 dataSource 物件建立查詢器，將使用者資訊插入 user 表中，測試效果如圖 4-80 所示。

▲ 圖 4-80 新增資料

重點來看查詢操作，如果需要查詢使用者名稱為 mouse 或性別為「男」的記錄，使用 Repository 時可以增加 where 條件。先來看下面的程式：

```
async findAll(name: string, sex: string) {
  return this.userRepository.find({
    where: {
      name,
      sex
    }
  });
}
```

事實上，這樣不能查詢出滿足上面條件的記錄，因為 where 中的條件是透過 And 連接的，而非 Or。如果要實現上述功能，則需要使用 QueryBuilder，程式如下：

```
async findAll(name: string, sex: string) {
  return await this.dataSource
  .createQueryBuilder(User, 'u')
  .where('u.name = :name OR u.sex = :sex', { name, sex })
  .getMany()
}
```

從測試結果可以看出，所有符合條件的記錄都被成功查詢出來了，如圖 4-81 所示。

```
0 ∨   {
1       "id": 2,
2       "name": "mouse",
3       "sex": "男",
4       "createTime": "2024-04-28T08:21:28.000Z",
5       "updateTime": "2024-04-28T08:21:28.000Z"
6     },
7 ∨   {
8       "id": 3,
9       "name": "mouse",
20      "sex": "女",
21      "createTime": "2024-04-28T09:28:14.000Z",
22      "updateTime": "2024-04-28T09:28:14.000Z"
23    },
24 ∨  {
25      "id": 4,
26      "name": "mouse2",
27      "sex": "男",
28      "createTime": "2024-04-28T09:33:45.000Z",
```

▲ 圖 4-81 查詢記錄

QueryBuilder 的查詢功能遠不止如此，它在多表查詢的場景中被頻繁使用，這部分將在專案實戰中表現。

更新和刪除比較簡單，使用 Repository 可以滿足大部分場景的需求。範例程式如下：

```
async update(id: number, updateUserDto: UpdateUserDto) {
  updateUserDto.updateTime = new Date()
  return await this.dataSource
  .createQueryBuilder()
  .update(User)
  .set(updateUserDto)
  .where("id = :id", { id })
  .execute()
}

async remove(id: number) {
  return await this.dataSource
  .createQueryBuilder()
  .delete()
  .from(User)
  .where("id = :id", { id })
  .execute()
}
```

在上述程式中，使用 update 和 set 方法來更新具有特定 id 的使用者資訊，效果如圖 4-82 所示。

4.5 在 Nest 中使用 TypeORM 操作 MySQL

▲ 圖 4-82 更新使用者資訊

刪除使用者透過 delete 方法實現，效果如圖 4-83 所示。

▲ 圖 4-83 刪除使用者資訊

此時，user 表中少了 id 為 2 的記錄，如圖 4-84 所示。

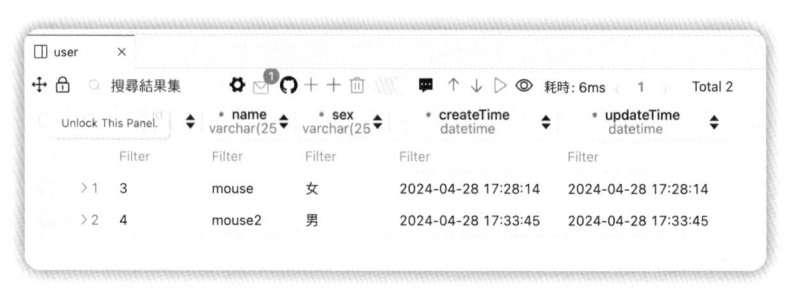

▲ 圖 4-84 使用者列表

以上就是在 Nest 框架中運用 TypeORM 函數庫進行資料庫操作的用法，介紹了 3 種不同的操作實體的方法，包括 EntityManager、Repository 和 QueryBuilder。讀者可根據具體的查詢需求和場景複雜度來選擇最合適的方法進行資料庫操作。

第5章
性能最佳化之資料快取

在第 4 章中，我們學習了 MySQL 資料持久化技術，該技術透過將資料儲存在硬碟上來實現資料的持久化。然而，後端服務往往要求能夠快速回應，而 MySQL 由於其儲存機制和查詢效率等因素，在需要快速回應的情況下可能表現不盡如人意。因此，本章我們將介紹另一種儲存技術—Redis。

我們將從 Redis 的基本使用開始學習，包括透過命令列和視覺化工具兩種方式來操作 Redis。隨後，我們會學習如何在 NestJS 應用程式中引入 Redis，以提升資料處理的效率。隨著課程內容的不斷深入，本書的專案實戰部分還將展示如何在真實場景中應用 Redis，幫助讀者更進一步地理解和掌握這門技術。

5.1 快速上手 Redis

Redis 是一個開放原始碼的記憶體資料結構儲存系統，採用鍵 - 值（Key-Value）對的形式儲存資料。它支援多種資料結構，如 String（字串）、Hash（雜湊）、List（清單）、Set（集合）、Sorted Set（有序集合）、Geospatial（地理資訊）、Bitmap（點陣圖）和 JSON 等。Redis 將資料儲存在記憶體中，以實現快速查詢，因此被廣泛用於服務端開發中的中介軟體。在提升應用性能和處理高併發場景方面，起著至關重要的作用。

5.1.1 安裝和執行

Redis 支援兩種操作方式：視覺化介面和命令列介面。本節將首先指導如何在本地機器上安裝 Redis，並分別介紹這兩種使用方法。

對於 macOS 作業系統的使用者，可以使用軟體套件管理器 Homebrew 來安裝 Redis。Windows 使用者可以直接從 Redis 官網下載安裝套件並解壓使用，或參考第 7 章內容了解如何使用 Docker 安裝 Redis 鏡像。安裝過程和執行結果如圖 5-1 所示。

```
~ 18:51:51
$ brew install redis
Updating Homebrew...
==> Downloading https://mirrors.ustc.edu.cn/homebrew-bottles/bottles/redis-6.2.4.arm64_big_sur.bottle.tar.
#=#=#
curl: (22) The requested URL returned error: 404
Warning: Bottle missing, falling back to the default domain...
==> Downloading https://ghcr.io/v2/homebrew/core/redis/manifests/6.2.4
######################################################################### 100.0%
==> Downloading https://ghcr.io/v2/homebrew/core/redis/blobs/sha256:097941b5d4c9845b3006820798b23a6599de86
==> Downloading from https://pkg-containers.githubusercontent.com/ghcr1/blobs/sha256:097941b5d4c9845b30068
######################################################################### 100.0%
==> Pouring redis--6.2.4.arm64_big_sur.bottle.tar.gz
==> Caveats
To start redis:
  brew services start redis
Or, if you don't want/need a background service you can just run:
  /opt/homebrew/opt/redis/bin/redis-server /opt/homebrew/etc/redis.conf
==> Summary
  /opt/homebrew/Cellar/redis/6.2.4: 14 files, 2MB
==> `brew cleanup` has not been run in 30 days, running now...
Removing: /Users/jmin/Library/Logs/Homebrew/pandoc... (64B)
```

▲ 圖 5-1 安裝過程和執行結果

安裝成功後，執行「brew services start redis」命令啟動 Redis，如圖 5-2 所示。

```
~ 18:54:36
$ brew services start redis
==> Successfully started `redis` (label: homebrew.mxcl.redis)
```

▲ 圖 5-2 啟動 Redis

Redis 內建了 redis-cli 命令列工具，執行 redis-cli 命令即可進入互動模式。接下來，將介紹如何使用這個工具來操作 Redis 支援的幾種資料結構。

5.1.2 Redis 的常用命令

Redis 提供了一系列操作資料的命令，接下來我們將逐一演示並說明這些命令的使用方法。

1. 字串操作

字串操作命令如下：

- SET key value：設置鍵 - 值對。
- GET key：獲取鍵對應的值。
- MGET：獲取多個鍵的值。
- DEL key：刪除鍵 - 值對。

- INCR key：將鍵的值增加 1。
- INCRBY key increment：將鍵的值增加指定的增量。

儲存字串是很常見的操作，可以在 CLI 中使用 SET、GET、MGET、DEL 命令操作字串，如圖 5-3 所示。

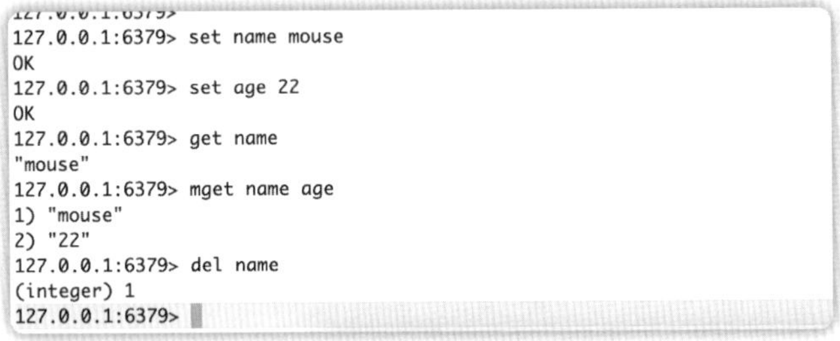

```
127.0.0.1:6379> set name mouse
OK
127.0.0.1:6379> set age 22
OK
127.0.0.1:6379> get name
"mouse"
127.0.0.1:6379> mget name age
1) "mouse"
2) "22"
127.0.0.1:6379> del name
(integer) 1
127.0.0.1:6379>
```

▲ 圖 5-3 操作字串

而 INCR 和 INCRBY 是用於計數的命令（計數器），我們常見的點贊數和瀏覽量就是透過它們來實現的。在 CLI 中執行這兩個命令的結果如圖 5-4 所示。

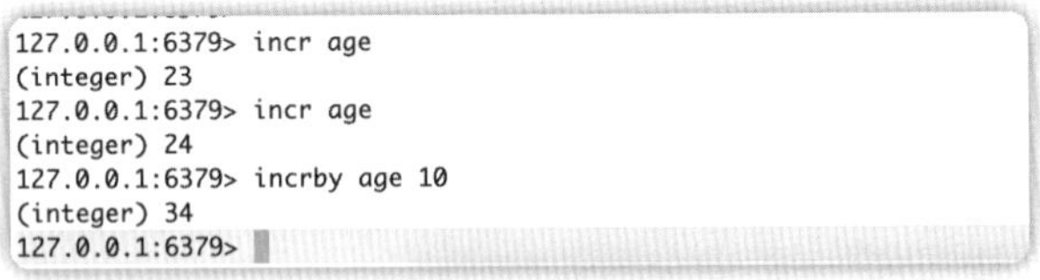

```
127.0.0.1:6379> incr age
(integer) 23
127.0.0.1:6379> incr age
(integer) 24
127.0.0.1:6379> incrby age 10
(integer) 34
127.0.0.1:6379>
```

▲ 圖 5-4 計數器

2. 列表操作

列表操作命令如下：

- LPUSH key value：將值推入列表左側。
- RPUSH key value：將值推入列表右側。
- LPOP key value：從列表左側刪除值。
- RPOP key value：從列表右側刪除值。
- LLEN key：獲取列表長度。
- LRANGE key start stop：獲取列表指定範圍的值。

在 Redis 中，列表通常應用於訊息佇列、任務佇列、時間線等場景，它的操作類似於 JavaScript 中的陣列操作，讀者應該很容易理解。範例如圖 5-5 所示。

```
127.0.0.1:6379> lpush list 1
(integer) 1
127.0.0.1:6379> lpush list 2
(integer) 2
127.0.0.1:6379> lpush list 3
(integer) 3
127.0.0.1:6379> rpush list 4
(integer) 4
127.0.0.1:6379> rpush list 5 6
(integer) 6
127.0.0.1:6379> llen list
(integer) 6
127.0.0.1:6379> lrange list 0 -1
1) "3"
2) "2"
3) "1"
4) "4"
5) "5"
6) "6"
127.0.0.1:6379> lpop list
"3"
127.0.0.1:6379> rpop list
"6"
127.0.0.1:6379>
```

▲ 圖 5-5 操作列表

值得注意的是，lrange 用於獲取列表資料，其中 0 表示索引的開始位置，而 -1 表示結尾，即查詢全部清單資料。

3. 集合操作

集合操作命令如下：

- SADD key member：向集合增加成員。
- SREM key member：移除集合中的成員。
- SMEMBERS key：獲取集合的所有成員。
- SINTER key1 key2：獲取多個集合的交集。

在 Redis 中，集合由一組無序但唯一的成員組成。使用集合可以對資料執行交集、並集、差集等操作。集合常用於標籤系統，如進行文章標籤和商品標籤管理，從而輕鬆地實現標籤的組合和篩選功能。

集合操作如圖 5-6 所示。

顯然，使用 Redis-CLI 工具不夠直觀和方便，我們可以改用官方的 RedisInsight 視覺化工具。下載並安裝完成後，開啟該工具，我們可以看到它已經自動連接上本地的 Redis 服務，如圖 5-7 所示。

```
127.0.0.1:6379> sadd myset nestJS nodeJS
(integer) 2
127.0.0.1:6379> sadd myset javascript
(integer) 1
127.0.0.1:6379> sadd myset 'typescript'
(integer) 1
127.0.0.1:6379> smembers myset
1) "javascript"
2) "nodeJS"
3) "nestJS"
4) "typescript"
127.0.0.1:6379> sadd myset2 javascript typescript
(integer) 2
127.0.0.1:6379> sinter myset myset2
1) "javascript"
2) "typescript"
127.0.0.1:6379> srem myset nodeJS
(integer) 1
127.0.0.1:6379> smembers myset
1) "javascript"
2) "nestJS"
3) "typescript"
127.0.0.1:6379>
```

▲ 圖 5-6 操作集合

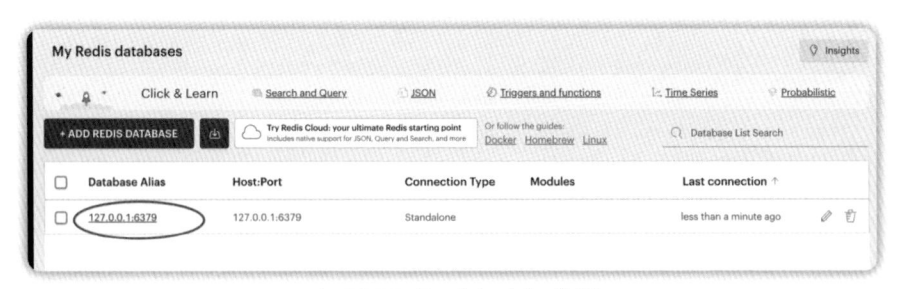

▲ 圖 5-7 RedisInsight 介面

點擊這個資料庫，展示前面建立的各種類型的 key 和 value，如圖 5-8 所示。

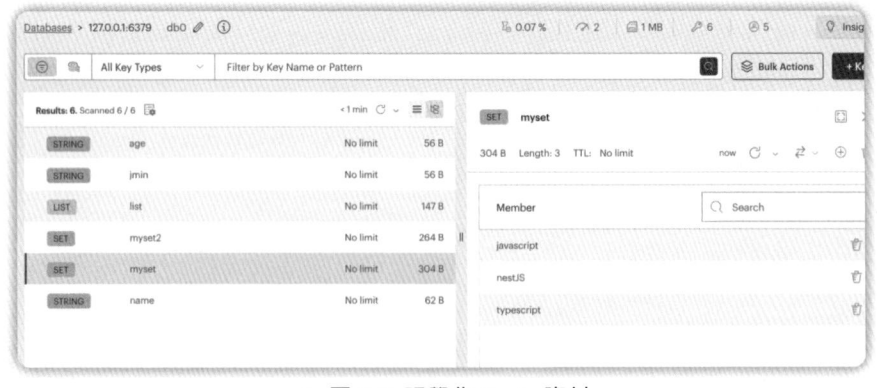

▲ 圖 5-8 視覺化 Redis 資料

接下來，透過這個 GUI 工具演示其他資料結構。

4. 有序集合操作

有序集合操作命令如下：

- ZADD key score member：向有序集合增加成員及其分數。
- ZRANGE key start stop：按分數範圍獲取有序集合的成員。

與集合不同，Sorted Set（有序集合，也被稱為 ZSet）是由一組按照分數排序並且唯一的資料組成的，通常應用在遊戲的排行榜中。在 GUI 工具中增加一個 ZSet 資料結構的 key，如圖 5-9 所示。

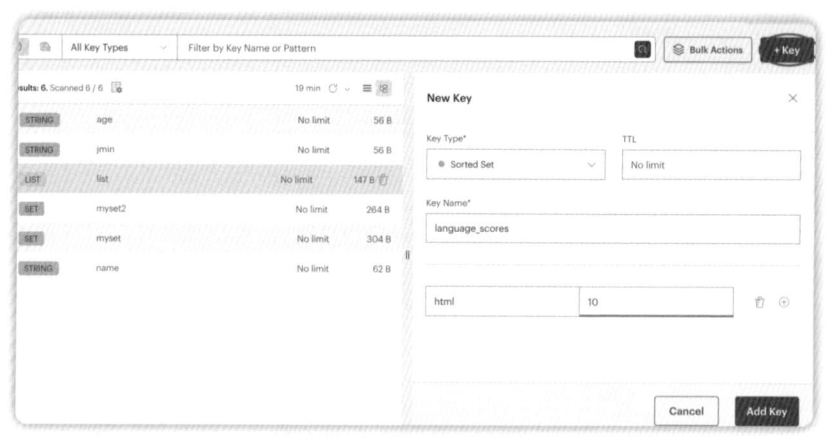

▲ 圖 5-9 視覺化新增 key

接下來，點擊　按鈕增加有序集合成員，並為它設置分數，如圖 5-10 所示。

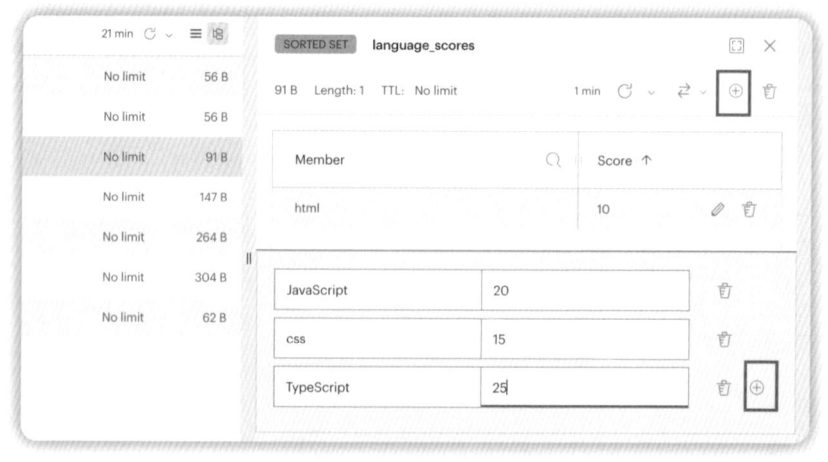

▲ 圖 5-10 視覺化新增成員

當然，如果我們需要執行一段 Redis 命令，依然可以使用命令列工具，如圖 5-11 所示。

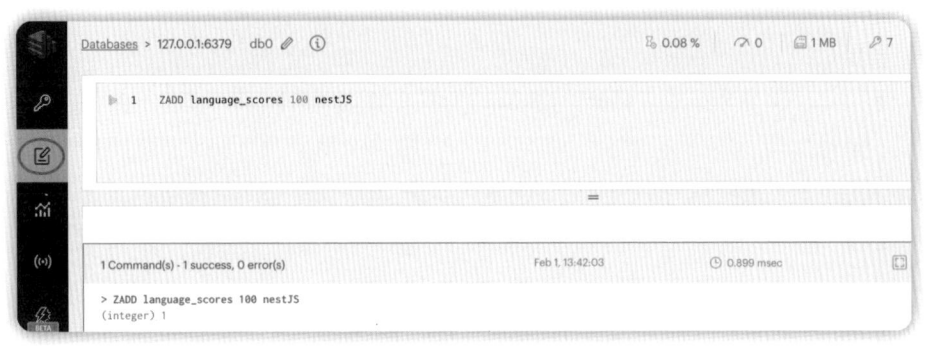

▲ 圖 5-11 在 GUI 中執行命令

5. 雜湊操作

雜湊操作命令如下：

- HSET key field value：設置雜湊欄位的值。
- HGET key field：獲取雜湊欄位的值。
- HGETALL key：獲取雜湊的所有欄位和值。

雜湊結構適用於儲存複雜物件結構的資料，例如使用者資訊、訂單資訊等，每個 key 代表不同的物件，field value 代表物件的屬性和值。以儲存 userInfo 為例，執行以下命令設置欄位：

```
HSET userInfo name mouse
HSET userInfo age 22
HSET userInfo userId 123321
```

生成的資料結構如圖 5-12 所示。

▲ 圖 5-12 儲存雜湊資料

執行 HGET userInfo name 和 HGETALL userInfo 命令查詢資料，如圖 5-13 所示。

▲ 圖 5-13 查詢雜湊資料

6. 地理空間操作

地理空間操作命令如下：

- GEOADD key longitude latitude member：根據經緯度增加座標成員。
- GEOPOS key member [member⋯]：獲取一個或多個成員的地理位置座標。
- GEOSEARCH key <FROMMEMBER | FROMLONLAT > <BYRADIUS | BYBOX> <⋯>：根據不同條件獲取成員座標。
- GEODIST key member1 member2 [unit]：計算座標成員之間的距離。

Redis 可以用於儲存地理空間的座標，並支持在替定半徑和範圍邊界內進行搜尋。這種功能常應用於共用汽車、自行車、充電寶等場景中。

執行以下命令給 share_cars（共用汽車）欄位增加幾個座標位置：

```
GEOADD share_cars -122.27652 37.805186 car1
GEOADD share_cars -122.2674626 37.8062344 car2 -122.2469854 37.8104049 car3
```

執行後看到 Redis 地理空間是透過有序集合來儲存的，如圖 5-14 所示。

5.1 快速上手 Redis

▲ 圖 5-14 新增地理座標

接下來獲取 car1 與 car2 的座標資訊，查詢結果如圖 5-15 所示。

▲ 圖 5-15 查詢指定成員的座標

接下來增加一些條件，給定一個經緯度座標，查詢這個座標 5km 範圍內的汽車，結果按照由近到遠的順序輸出，如圖 5-16 所示。

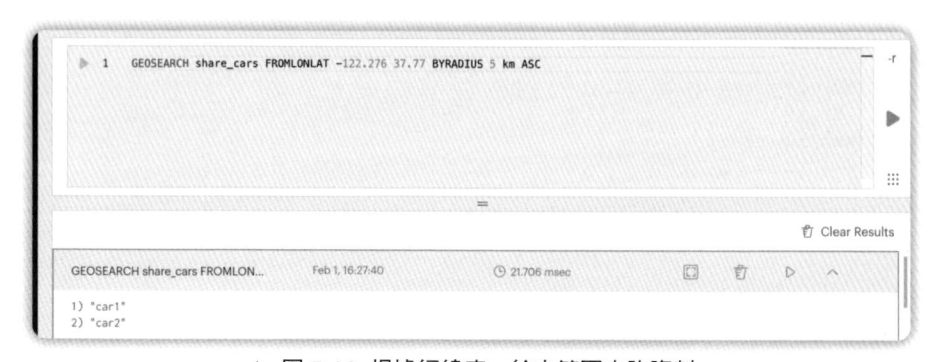

▲ 圖 5-16 根據經緯度、給定範圍查詢資料

至此，讀者應該已經理解了如何實現查詢附近共用充電寶和自行車的功能原理。

7. 點陣圖操作

點陣圖操作命令如下：

- SETBIT key offset value：給指定偏移量的位設置 0 或 1。
- GETBIT key offset：獲取指定偏移量的位。
- BITCOUNT key：獲取指定位為 1 的總計數。

在 Redis 中，點陣圖的應用也很廣泛，如記錄使用者的活躍狀態、線上狀態、存取頻率等。我們以統計使用者 1001 在 30 天內的存取頻率為例，假設他在第 10、20、25 天時有存取記錄，程式如下：

```
SETBIT user:1001:visit 10 1
SETBIT user:1001:visit 20 1
SETBIT user:1001:visit 25 1
```

此時，要判斷該使用者在第 15 天時是否有存取記錄，如圖 5-17 所示，我們可以使用 GETBIT 進行查詢。

▲ 圖 5-17 查詢使用者是否有存取記錄

接下來統計使用者在 30 天內的存取頻率，如圖 5-18 所示，可以使用 BITCOUNT 進行查詢。

▲ 圖 5-18 查詢使用者的存取頻率

至此，Redis 中常用的資料結構與命令介紹完畢。結合實際生活中的應用場景，相信讀者可以快速掌握並有效地運用 Redis。

5.2 在 Nest 中使用 Redis 快取

在 5.1 節中，我們學習了 Redis 中各種資料結構的使用方法，並探討了它們的多種應用場景，包括排行榜、附近的共用裝置、計數器和使用者存取頻率等。實際上，Redis 的用途遠不止這些。它還可以廣泛應用於快取實現、訊息佇列、單點登入和分散式鎖等多種場景。Redis 的靈活性和高性能等特點使其成為一個強大而多用途的工具，為我們提供了廣泛的解決方案，以滿足多樣化的應用需求。

接下來，我們將在 Nest 應用程式中實現 Redis 的資料快取功能，以便更深入地了解 Redis 的實際應用。

5.2.1 專案準備

本小節實現這樣一個需求：透過 Redis 快取使用者的購物車資訊。當使用者查詢購物車資訊時，首先從 Redis 中查詢，如果快取為空，再去 MySQL 中查詢。當使用者在購物車中增加商品數量時，需要將更新儲存到 MySQL 中，並同步更新到 Redis，以確保快取資料的一致性。

先建立一個名為 nest-redis 的專案，執行「nest n nest-redis -p pnpm」命令，指定套件管理器為 pnpm。專案建立成功後，如圖 5-19 所示。

```
$ nest new nest-redis -p pnpm
⚡ We will scaffold your app in a few seconds..
CREATE nest-redis/.eslintrc.js (663 bytes)
CREATE nest-redis/.prettierrc (51 bytes)
CREATE nest-redis/README.md (3347 bytes)
CREATE nest-redis/nest-cli.json (171 bytes)
CREATE nest-redis/package.json (1951 bytes)
CREATE nest-redis/tsconfig.build.json (97 bytes)
CREATE nest-redis/tsconfig.json (546 bytes)
CREATE nest-redis/src/app.controller.spec.ts (617 bytes)
CREATE nest-redis/src/app.controller.ts (274 bytes)
CREATE nest-redis/src/app.module.ts (249 bytes)
CREATE nest-redis/src/app.service.ts (142 bytes)
CREATE nest-redis/src/main.ts (208 bytes)
CREATE nest-redis/test/app.e2e-spec.ts (630 bytes)
CREATE nest-redis/test/jest-e2e.json (183 bytes)

✔ Installation in progress... 🍭

🚀 Successfully created project nest-redis
```

▲ 圖 5-19 建立專案

接下來，執行「pnpm install typeorm mysql2 redis @nestjs/typeorm -S」命令，安裝我們需要的相依套件。

同時，執行「nest g resource shopping-cart --no-spec」命令，生成一個名為 shopping-cart 的購物車模組，這裡不生成單元測試檔案，執行結果如圖 5-20 所示。

```
$ nest g resource shopping-cart --no-spec
? What transport layer do you use? REST API
? Would you like to generate CRUD entry points? Yes
CREATE src/shopping-cart/shopping-cart.controller.ts (1055 bytes)
CREATE src/shopping-cart/shopping-cart.module.ts (298 bytes)
CREATE src/shopping-cart/shopping-cart.service.ts (721 bytes)
CREATE src/shopping-cart/dto/create-shopping-cart.dto.ts (38 bytes)
CREATE src/shopping-cart/dto/update-shopping-cart.dto.ts (202 bytes)
CREATE src/shopping-cart/entities/shopping-cart.entity.ts (29 bytes)
UPDATE src/app.module.ts (417 bytes)
```

▲ 圖 5-20 生成購物車模組

在 VS Code 中執行「pnpm start:dev」命令啟動專案，如圖 5-21 所示。

```
[21:24:08] Starting compilation in watch mode...

[21:24:09] Found 0 errors. Watching for file changes.

[Nest] 29761  - 2024/02/01 21:24:09     LOG [NestFactory] Starting Nest application...
[Nest] 29761  - 2024/02/01 21:24:09     LOG [InstanceLoader] AppModule dependencies initialized +5ms
[Nest] 29761  - 2024/02/01 21:24:09     LOG [RoutesResolver] AppController {/}: +8ms
[Nest] 29761  - 2024/02/01 21:24:09     LOG [RouterExplorer] Mapped {/, GET} route +1ms
[Nest] 29761  - 2024/02/01 21:24:09     LOG [NestApplication] Nest application successfully started +0ms
```

▲ 圖 5-21 啟動專案

5.2.2 Redis 初始化

Redis 通常會在多個模組中使用。為了更進一步地管理它，我們新建一個 redis.module.ts 檔案，專門用來配置和匯出 Redis 模組，其他模組可以透過相依注入的方式使用它。根據使用頻率和需求，使用者甚至可以把 Redis 定義為全域模組。RedisModule 的程式如下：

```typescript
import { Module } from '@nestjs/common';
import { createClient } from 'redis';

const createRedisClient = async () => {
  return await createClient({
    socket: {
      host: 'localhost',
      port: 6379,
    }
  }).connect();
};

@Module({
  providers: [
    {
      provide: 'NEST_REDIS',
      useFactory: createRedisClient,
    },
  ],
  exports: ['NEST_REDIS'],
})
export class RedisModule {}
```

在程式實現中，createClient 方法負責根據提供的 Redis 配置資訊（如主機地址、通訊埠編號等）來註冊 Redis 用戶端，並透過 connect 方法與 Redis 服務建立連接。在 @Module 裝飾器中，我們使用 providers 屬性定義一個相依項，其 token 為 'NEST_REDIS'。useFactory 屬性用於定義一個工廠方法，該方法將建立並傳回 Redis 用戶端實例。最後，透過 exports 屬性將該用戶端匯出，使其成為可在其他模組中使用的共用物件。

接下來，在 shopping-cart.service.ts 中透過 @Inject 裝飾器來匯入並使用 Redis 用戶端，核心程式如下：

```typescript
@Injectable()
export class ShoppingCartService {
  @Inject('NEST_REDIS')
  private redisClient: RedisClientType
```

```
async create(createShoppingCartDto: CreateShoppingCartDto) {
  await this.redisClient.set('xxx', JSON.stringify(createShoppingCartDto));
}
```

注入 Redis 相依後,我們可以在服務方法中透過 this.redisClient 呼叫它上面的所有方法。

5.2.3 建表並建構快取

完善購物車服務邏輯之前,在 app.module.ts 中初始化 MySQL 連接,程式如下:

```
import { Module } from '@nestjs/common';
import { AppController } from './app.controller';
import { AppService } from './app.service';
import { ShoppingCartModule } from './shopping-cart/shopping-cart.module';
import { TypeOrmModule } from '@nestjs/typeorm';

@Module({
  imports: [
    // 初始化 MySQL 連接
    TypeOrmModule.forRoot({
      type: 'mysql',
      host: 'localhost',
      port: 3306,
      username: 'root',
      password: 'jminjmin',
      database: 'nest_redis',
      entities: [__dirname + '/**/*.entity{.ts,.js}'],
      autoLoadEntities: true,
      synchronize: true
    }),
    ShoppingCartModule
  ],
  controllers: [AppController],
  providers: [AppService],
})
export class AppModule {}
```

這裡我們用 TypeORM 的儲存庫模式來操作實體,完善 shopping-cart.entity.ts 檔案,程式如下:

```
import { Column, Entity, PrimaryGeneratedColumn } from "typeorm"
@Entity()
export class ShoppingCart {
  @PrimaryGeneratedColumn()
  id: number
```

```typescript
  @Column()
  userId: number

  @Column({type: 'json'})
  cartData: Record<string, number>
}
```

定義 create、findOne 和 update 方法用於增加、查詢、更新購物車資訊，並保持 Redis 與 MySQL 資料的一致性。即在更新 MySQL 資料時，Redis 快取必須同時更新。範例程式如下：

```typescript
import { Inject, Injectable } from '@nestjs/common';
import { CreateShoppingCartDto } from './dto/create-shopping-cart.dto';
import { UpdateShoppingCartDto } from './dto/update-shopping-cart.dto';
import { RedisClientType } from 'redis';
import { InjectRepository } from '@nestjs/typeorm';
import { ShoppingCart } from './entities/shopping-cart.entity';
import { Repository } from 'typeorm';

@Injectable()
export class ShoppingCartService {
  @Inject('NEST_REDIS')
  private redisClient: RedisClientType
  @InjectRepository(ShoppingCart)
  private shoppingCartRepository: Repository<ShoppingCart>

  async create(createShoppingCartDto: CreateShoppingCartDto) {
    // 儲存到 db 中
    await this.shoppingCartRepository.save(createShoppingCartDto);
    // 更新 Redis 快取
    await this.redisClient.set(`cart:${createShoppingCartDto.userId}`, JSON.stringify(createShoppingCartDto));
    return {
      msg: '增加成功',
      success: true
    }
  }

  async findOne(id: number): Promise<ShoppingCart> {
    // 先從 Redis 中查詢快取，沒有再查 db
    const data = await this.redisClient.get(`cart:${id}`)
    const cartEntity = data ? JSON.parse(data) : null
    if (cartEntity) return cartEntity
    return await this.shoppingCartRepository.findOne({
      where: {
        userId: id
```

```
      }
    });
  }

  async update(updateShoppingCartDto: UpdateShoppingCartDto) {
    const { userId, cartData: { count = 1 } } = updateShoppingCartDto
    // 查詢資料
    const cartEntity = await this.findOne(userId)
    const cart = cartEntity ? cartEntity.cartData : {}
    let quality = (cart.count || 0) + count
    // 更新 count
    cart.count = quality

    // 更新 db 資料
    await this.shoppingCartRepository.update({userId}, cartEntity)
    // 更新 Redis 快取
    await this.redisClient.set(`cart:${userId}`, JSON.stringify(cartEntity))
    return {
      msg: '更新成功',
      success: true
    }

  }
}
```

首先來看程式中用於增加購物車的 create 方法，在 createShoppingCartDto 中定義以下介面欄位：

```
export class CreateShoppingCartDto {
  userId: number

  cartData: Record<string, number>
}
```

userId 將作為 Redis 中儲存的鍵（key），而 cartData 表示購物車資訊。在 create 方法中，我們首先使用 TypeORM 儲存庫將資料儲存到 MySQL 資料庫中，然後將資料設置（set）到 Redis 快取中。userId 作為快取鍵的一部分，用於儲存購物車資訊，這樣每個使用者都可以透過其 userId 獲取到對應的快取資料。

findOne 查詢方法的工作原理很直觀。當 NestJS 接收到請求時，它會首先嘗試根據 userId 從 Redis 快取中獲取資料。如果快取命中，則將快取中的資料傳回給前端；如果沒有命中，則從資料庫中獲取查詢結果。在處理那些使用者資料更新不頻繁但讀取頻繁的場景時，使用快取可以顯著減輕資料庫的負擔。

至於 update 方法，它首先會查詢出指定的記錄進行修改，然後再次將更新後的記錄儲存到資料庫中，並刷新 Redis 快取，以此完成整個更新操作。

5.2.4 執行程式

一切準備就緒後，下一步是發送請求以測試應用程式的效果。為此，我們可以使用 VS Code 的 REST Client 外掛程式。首先，在專案的根目錄下建立一個以 .http 為副檔名的檔案，然後在該檔案中配置所需的介面。具體配置程式如下：

```
// 增加購物車
POST http://localhost:3000/shopping-cart
Content-Type: application/json

{
  "userId": 111,
  "cartData": {
    "count": 1
  }
}

###
// 更新購物車數量
PATCH  http://localhost:3000/shopping-cart
Content-Type: application/json

{
  "userId": 111,
  "cartData": {
    "count": 1
  }
}
```

重新執行「pnpm start:dev」命令以啟動服務。服務啟動後，TypeORM 會自動建立所需的表結構，點擊 Send Request 增加幾筆資料，如圖 5-22 所示。

▲ 圖 5-22 建立資料

在 VS Code 中查看資料記錄，如圖 5-23 所示。

▲ 圖 5-23　查看資料記錄

同時，Redis 快取中也儲存了相應的資料記錄。要查看這些快取資料，可以使用 RedisInsight 工具，並利用 cart* 作為篩選條件來檢索特定的購物車資料，如圖 5-24 所示。

▲ 圖 5-24　查看 Redis 快取資料

接下來，更新資料庫中的 count 欄位，將 userId 為 111 的使用者的購物車商品數增加為 10，點擊 Send Request，結果如圖 5-25 所示。

▲ 圖 5-25　更新購物車商品數量

根據我們的設計方案，程式首先會在快取中查詢 userId 為 111 的使用者資料，並傳回這些資料。一旦資料被成功修改，程式接下來會將更新後的資料同步到資料庫和快取中。在完成這些步驟後，你可以查詢資料庫和快取中的記錄，以驗證資料是否已經更新。圖 5-26 展示了查詢結果，從中可以看到資料庫和快取中的資料都已更新。

▲ 圖 5-26　更新資料庫及快取

5.2.5　設置快取有效期

在實際業務中，Redis 通常會設置快取過期時間，以避免資料不一致或快取長時間未存取（更新）導致記憶體空間浪費等問題。我們為快取設置了 30 秒的過期時間，程式如下：

```
// 更新 Redis 快取
await this.redisClient.set(
  `cart:${createShoppingCartDto.userId}`,
  JSON.stringify(createShoppingCartDto),
  { EX: 30 }
);
```

對於 userId 為 111 的記錄，執行完更新操作後，如圖 5-27 所示，其有效期（TTL）已經從 30 秒開始倒計時。

▲ 圖 5-27 設置 Redis 快取有效期

有讀者可能會問：為什麼要設置快取有效期呢？筆者舉出了以下幾個理由：

- 釋放記憶體空間：如果快取長時間不被存取或更新，這部分快取可能會持續佔用大量的空間，從而不被釋放，這在一定程度上會導致 Redis 頻繁擴充。設置過期時間可以自動釋放記憶體供其他快取使用。
- 保證資料的即時性：當快取對應的業務邏輯發生變更時，失效的快取一直在記憶體中可能會導致業務邏輯錯誤，即我們常說的「無效資料」，這會影響系統的穩定性。設置自動過期可以保證快取在一定時間內是有效的，有效避免這種問題。
- 保證資料的安全性：過期時間是一種容錯機制，快取長時間存活在記憶體中，如果遇到記憶體洩露或惡意軟體攻擊，快取中的隱私資料可能會洩露。設置有效期可以定時清理記憶體中的隱私資料，減少資料洩露風險。
- 保證資料的一致性：在併發場景或快取服務異常時，最新快取並未更新到記憶體中，此時獲取到的舊快取資料可能會因為資料不一致問題導致系統異常。設置一定的有效期讓舊快取過期，可以保證快取的資料與資料庫中的資料一致。

5.2.6 選擇合理的有效期

設置 Redis 快取的有效期並沒有統一的最佳時長，這完全取決於具體的業務場景需求。一般來說可以根據以下三種策略來選擇：

- 短期快取：在資料即時性要求高且頻繁變動的情況下，可以設置較短的快取時間，如幾分鐘或幾小時，以確保快取資料及時與資料庫同步。這常見於新聞資訊推送、熱點頭條及天氣預報等。

- 中期快取：對於一些變動不頻繁，但要求具有一定即時性的資料，可以設置較長的快取有效期，如幾小時或幾天，以盡可能減輕資料庫的存取壓力。這常見於電子商務購物車資料、使用者登入資料等。
- 長期快取：對於相對穩定且變動少的資料，可以設置較長的有效期，如幾天或幾周。常見於靜態資源快取、地理位置資訊更新等。

透過在 Nest 中整合 Redis 快取，能夠有效減少對 MySQL 資料庫的存取。同時，合理設置有效期，不僅可以提升系統性能，還能有效降低資料庫的負擔。

第 6 章
身份驗證與授權

本章將深入探討身份驗證和授權這兩個安全概念。在應用程式中，身份驗證是驗證使用者身份的過程，而授權是確定使用者是否有許可權執行特定操作的過程。本章首先介紹在身份驗證和授權過程中涉及的常見概念，例如 Cookie、Session、Token、JWT（JSON Web Token）和 SSO（Single Sign-On，單點登入）等，以便讀者能夠清晰地理解它們之間的區別和適用場景。

接下來，我們將在 Nest 中演示如何使用 JWT 和 Passport.js 進行身份驗證，並使用主流的 RBAC（Role-Based Access Control，基於角色的存取控制）模型實現許可權控制，最後基於 Redis 實現單點登入。

透過本章的學習，相信讀者將更加深入地了解身份驗證和授權的原理和實現方式。

6.1 Cookie、Session、Token、JWT、SSO 詳解

身份驗證和授權是設計安全、可靠系統的核心功能之一。在開始對身份驗證和許可權控制進行開發之前，有必要先回答以下幾個問題：

- 什麼是身份驗證？
- 什麼是授權？
- 實現身份驗證或授權的方式有哪幾種？
- Cookie、Session、Token、JWT、SSO、OAuth 分別是什麼，它們之間有什麼區別？

接下來，我們將一一解答以上問題。

6.1.1 什麼是身份驗證

在網際網路安全領域，身份驗證的通俗解釋是確認當前使用者確實是其自稱的那個人。這個過程類似於手機上的指紋解鎖功能：只有當你的指紋與手機中儲存的指紋資料匹配時，手機才能被成功解鎖。

常見的身份驗證方式有以下幾種：

- 帳號密碼登入驗證
- 手機簡訊驗證
- 電子郵件驗證碼驗證
- 人臉辨識驗證
- 指紋登入驗證

6.1.2 什麼是授權

使用者授予第三方應用存取某些資源的許可權，稱為授權。舉例來說，當你登入微信或小程式時，它會詢問你是否允許授權存取昵稱、微信名稱、手機號碼等個人資訊；當你首次安裝 App 時，App 會詢問你是否允許授權存取相機、相簿、麥克風、通訊錄或地理位置等個人資訊。

6.1.3 什麼是憑證

實現身份驗證和授權的前提是需要一種媒介來標識存取者的身份。在實際生活中，身份證或護照是最為常見的證明公民身份的憑證，透過它我們可以乘坐各種交通工具或辦理各種銀行業務。而在網際網路中，我們需要註冊或登入帳號才能進行下單購物、評論點贊等操作。當使用者登入某個網站時，伺服器會傳回一個 SessionID（階段 ID）或 Token（權杖）到使用者瀏覽器中，用於標識當前使用者的身份。再次請求存取網站時帶上這個憑證，伺服器才能辨識出該使用者的身份，並授予相應的操作許可權。

6.1.4 什麼是 Cookie

HTTP 協定是一種無狀態協定，這表示服務端在每次請求中無法辨識請求是由哪個使用者發起的。為了解決這個問題，需要一種機制來儲存使用者的身份資訊，並在使用者進行後續請求時提供給伺服器，以便辨識是否為同一使用者。這就是 Cookie 的用武之地。

Cookie 儲存在用戶端瀏覽器中，由伺服器發送並以文字檔形式存在，其大小通常限制在 4KB 以內。當使用者再次向同一伺服器發起請求時，Cookie 會被自動包含在請求標頭中並發送至伺服器。需要注意的是，Cookie 不支持跨域存取，即每個 Cookie 都與特定域名綁定，無法被其他域名下的頁面存取，但在相同域名下的頁面之間可以共用使用。有關 Cookie 工作流程如圖 6-1 所示。

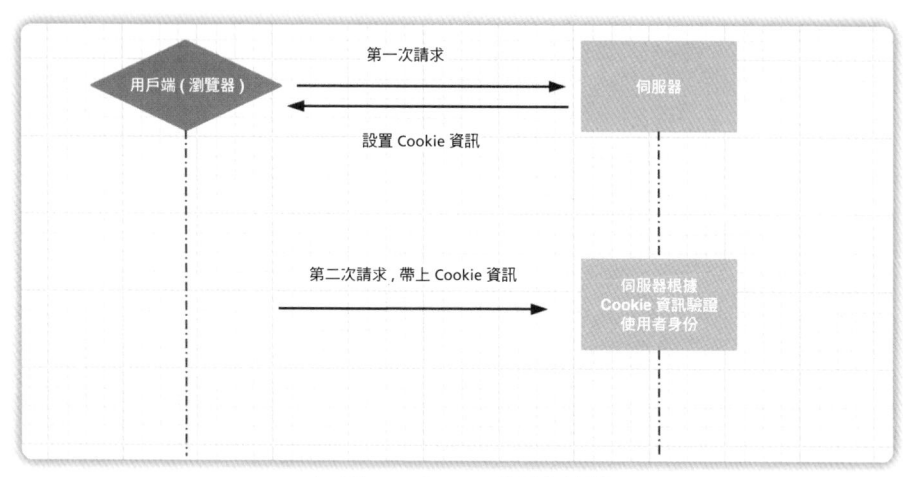

▲ 圖 6-1 Cookie 的工作流程

由於 Cookie 儲存在用戶端，使用者可以輕易地查看和修改這些資訊，這可能導致使用者資訊洩露，帶來安全隱憂。為了解決這一問題，通常會採用結合使用 Session 和 Cookie 的方法。在這種方法中，使用者的敏感資訊儲存在服務端的 Session 中，而 Cookie 僅儲存 Session 的識別字。這樣，即使用戶端的 Cookie 被存取，攻擊者也無法直接獲取到敏感的使用者資訊。

6.1.5 什麼是 Session

Session 與 Cookie 之間有著奇妙的因緣，Session 是另一種維持服務端與用戶端階段狀態的機制，通常基於 Cookie 實現。Session 儲存在服務端中，而 SessionID 儲存在用戶端的 Cookie 欄位中。

使用者首次向伺服器發起請求時，伺服器根據使用者資訊生成對應的 Session，並將該 Session 的唯一標識 SessionID 傳回到瀏覽器中，並儲存在 Cookie

內。當使用者再次向伺服器發起請求時，瀏覽器會自動判斷 Cookie 中是否存在與當前域名匹配的資訊，並將其發送到服務端。服務端根據 Cookie 中的 SessionID 來查詢是否存在對應的 Session 資訊。如果沒有找到，說明使用者尚未登入或登入失敗；不然執行使用者登入後的操作。Session 工作流程如圖 6-2 所示。

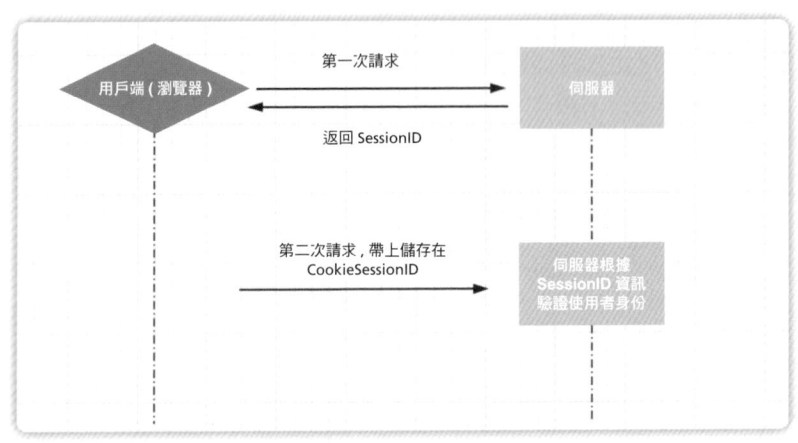

▲ 圖 6-2 Session 工作流程

6.1.6 Session 與 Cookie 的區別

Session 與 Cookie 在 Web 程式中的不同用途和實現方式決定了它們之間存在以下區別：

- 儲存位置：Cookie 通常以文字形式儲存在用戶端瀏覽器中，而 Session 資料則儲存在服務端，可能是資料庫或記憶體中，可以以任意資料型態存在。

- 安全性：由於 Cookie 儲存在用戶端，因此可以被使用者查看和修改，存在安全風險。相比之下，Session 資料儲存在服務端，使用者無法直接存取或修改，因此相對更安全。不過，Session 仍然可能面臨 CSRF（Cross-Site Request Forgery，跨站請求偽造）攻擊的風險，這種攻擊經常透過釣魚網站進行。

- 生命週期：Cookie 可以設置一個較長的有效期，實現如預設登入等功能。而 Session 的生命週期通常較短，會在用戶端關閉或 Session 逾時後失效（預設情況下）。

- 儲存大小：單一 Cookie 的資料儲存大小通常限制在 4KB 以內，而 Session 所能儲存的資料量遠大於 Cookie，沒有 4KB 的限制。

結合使用 Session 和 Cookie 的方案相較於僅使用 Cookie 具有明顯優勢，但它並非沒有缺陷。除了容易受到 CSRF（跨站請求偽造）攻擊之外，Session 由於儲存在服務端，還面臨著一些挑戰。特別是在多伺服器環境中，當出現高併發時，使用者的請求可能會被負載平衡器分配到不同的伺服器上。如果使用者的 Session 資訊僅儲存在某一台伺服器上，那麼其他伺服器將無法存取該 Session 資訊，從而導致階段管理問題。

為了解決這一問題，通常需要採用特殊的方案在多台伺服器之間同步 Session 資訊。一種常見的做法是使用像 Redis 這樣的分散式快取系統來儲存 Session 資料，以確保所有伺服器都能存取到最新的使用者階段狀態。

除了 Session 和 Cookie 的結合使用，我們還可以考慮使用 Token（如 JWT）來實現服務端的無狀態化。無狀態化表示使用者的階段資訊被編碼在 Token 中，每次請求時由用戶端發送，服務端透過解析 Token 來驗證使用者身份，無須在服務端儲存 Session 資訊。

6.1.7 什麼是 Token

Token 是存取受限資源時需要攜帶的一種憑證，通常是一個加密字串，常見的 Token 有以下幾種。

1. 存取權杖

存取權杖（Access Token）是一種用於授權使用者存取特定資源或執行特定操作的憑證。使用者登入成功後，伺服器會頒發一個存取權杖，使用者在存取受保護的資源時需要攜帶該權杖進行驗證。存取權杖的工作流程如圖 6-3 所示。

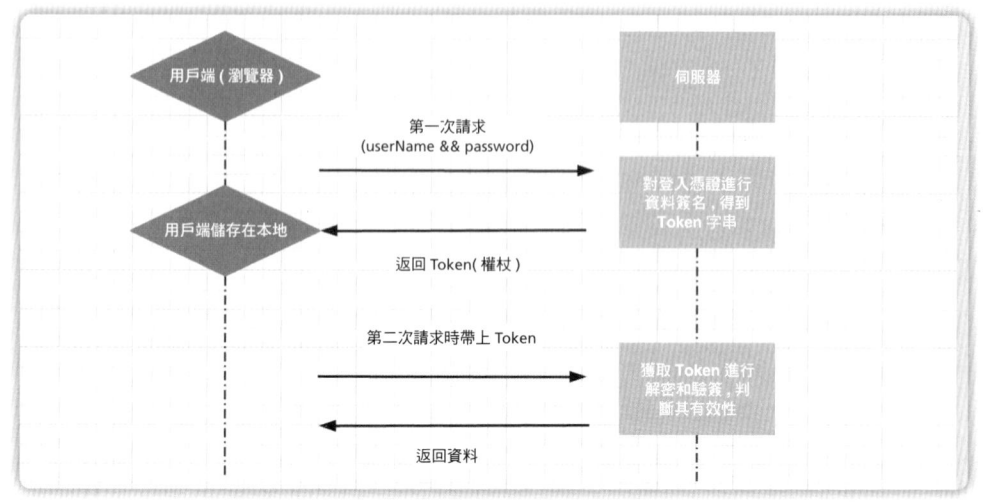

▲ 圖 6-3 存取權杖的工作流程

圖 6-3 中的流程詳細描述如下：

（1）用戶端發送登入請求：用戶端發送包含使用者名稱和密碼的登入請求。

（2）服務端驗證使用者名稱和密碼：服務端接收到登入請求後，對使用者名稱和密碼進行驗證。

（3）服務端簽發 Token：帳號和密碼驗證成功後，服務端會生成一個包含使用者身份資訊的存取權杖，並將其發送給用戶端。

（4）用戶端儲存 Token：用戶端收到 Token 後，通常會將其儲存在 Cookie、localStorage 或 sessionStorage 等地方，以便在後續的請求中使用。

（5）用戶端請求介面：用戶端在發送請求時，會將 Token 作為身份憑證附加在請求的頭部（通常是 Authorization 標頭部）或請求參數中。

（6）服務端驗證 Token：服務端在接收到用戶端的請求後，會從請求中提取 Token 並進行驗證。如果 Token 有效且包含足夠的許可權，服務端會處理請求並傳回相應的資料。

2. 刷新權杖

刷新權杖（Refresh Token）專門用於刷新存取權杖的 Token，它的有效期通常比存取權杖長，以保證使用者登入狀態的持久性。如果沒有刷新權杖，使用者每次存取時都需要輸入帳號和密碼以重新驗證，這顯然很麻煩。因此，用戶端通常會維護一個刷新權杖，定時刷新造訪權杖，從而提升網站的使用者體驗。刷新權杖的工作流程如圖 6-4 所示。

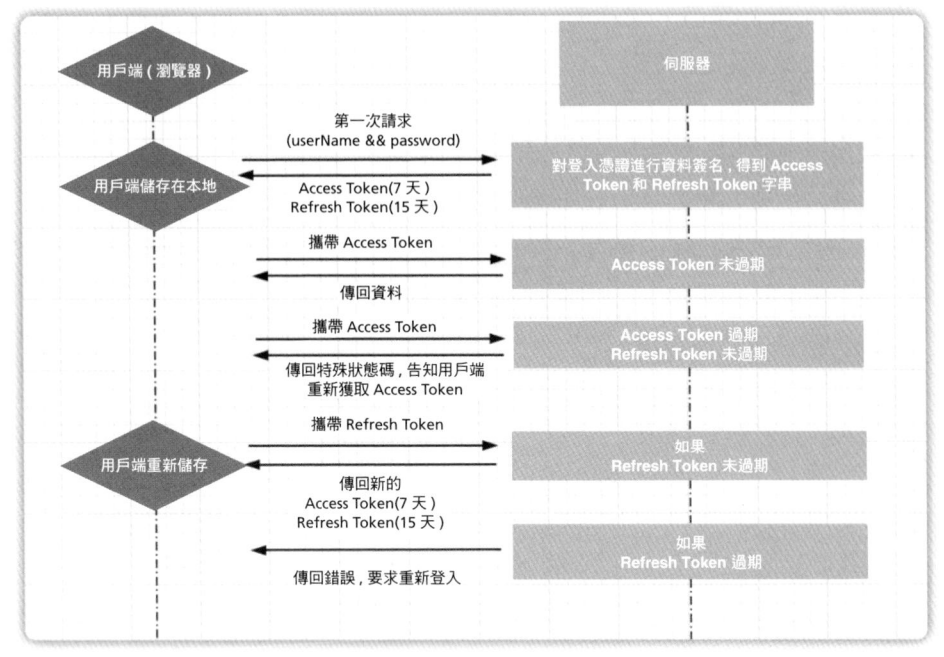

▲ 圖 6-4 雙 Token 工作流程

圖 6-4 中的流程詳細描述如下：

（1）用戶端發送登入請求：用戶端發送包含使用者名稱和密碼的登入請求。

（2）服務端驗證使用者名稱和密碼：服務端接收到登入請求後，對使用者名稱和密碼進行驗證。

（3）服務端簽發存取權杖和刷新權杖：如果帳號和密碼驗證成功，服務端會生成一個包含使用者身份資訊的存取權杖和一個用於刷新存取權杖的刷新權杖，並將它們一起發送給用戶端。

（4）用戶端儲存 Token：用戶端收到 Token 後，通常會將存取權杖儲存在 Cookie、localStorage 或 sessionStorage 等地方，刷新權杖則儲存在安全的地方，比如 HTTP Only 的 Cookie 中，以免被 XSS（Cross-Site Scripting，跨站指令稿）攻擊。

（5）用戶端請求介面：用戶端在發送請求時，會將存取權杖作為身份憑證附加在請求的頭部（通常是 Authorization 標頭部）或請求參數中。

（6）服務端驗證存取權杖：服務端在接收到用戶端的請求後，從請求中提取存取權杖並進行驗證。如果存取權杖有效，服務端處理請求並傳回相應的資料。如果存取權杖過期或無效，則傳回指定狀態提示或錯誤，告知用戶端用刷新權杖獲取新的存取權杖。

（7）用戶端用刷新權杖獲取新的存取權杖：用戶端發送一個特殊的請求，攜帶刷新權杖向服務端請求新的存取權杖。

（8）服務端驗證刷新權杖：服務端接收到用戶端的請求後，驗證刷新權杖的有效性。如果刷新權杖有效且未過期，服務端會生成一個新的存取權杖和刷新權杖，並一同發送給用戶端。

（9）用戶端更新 Token：用戶端收到新的 Token 後，更新舊的 Token，並使用新的 Token 進行後續請求。

（10）重新登入：如果刷新權杖失效或過期，則傳回錯誤資訊，要求使用者重新登入。

6.1.8 什麼是 JWT

JWT 是一種特殊的 Token，遵循基於 JSON 物件結構的開放標準（RFC 7519），用於安全地在網路應用中傳遞認證資訊。JWT 允許在使用者和服務端之間傳遞經認證的使用者資訊，而無需暴露使用者的認證憑據。

使用者登入成功後，服務端會認證使用者的身份，並把一個 JWT 傳回給用戶端。此後，用戶端在請求受保護資源時需要將 JWT 作為認證資訊包含在請求中。服務端會驗證 JWT 的有效性，以確認使用者是否有權存取所請求的資源。JWT 的工作流程如圖 6-5 所示。

圖 6-5 中的流程詳細描述如下：

（1）用戶端發送登入請求：用戶端發送包含使用者名稱和密碼的登入請求。

（2）服務端驗證使用者名稱和密碼：服務端接收到登入請求後，對使用者名稱和密碼進行驗證。

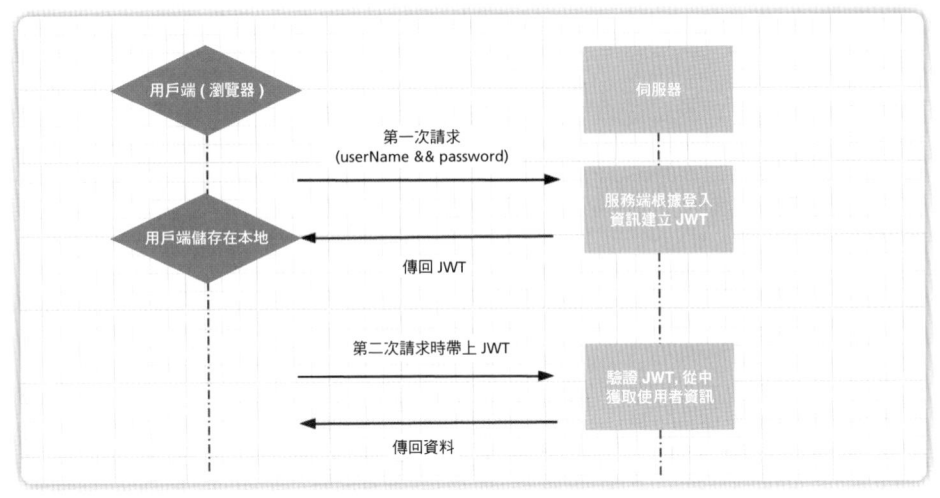

▲ 圖 6-5 JWT 的工作流程

（3）服務端簽發 JWT：帳號和密碼驗證成功後，服務端會建立一個包含使用者身份資訊的 JWT，並將其發送給用戶端。

（4）用戶端儲存 JWT：用戶端收到 Token 後，通常會將其儲存在 localStorage 或 Cookie 等地方，以便在後續的請求中使用。

（5）用戶端請求介面：用戶端在發送請求時，會在請求標頭部的 Authorization 欄位中使用 Bearer 模式增加 JWT，如 Authorization: Bearer <token>。

（6）服務端驗證 JWT：服務端在接收到用戶端的請求後，會從請求中提取 JWT 並進行驗證。如果 JWT 有效且包含足夠的許可權，那麼服務端會處理請求並傳回相應的資料。

細心的讀者可能會發現，JWT 與傳統的 Token 認證方式在工作流程上有相似之處：在首次認證回應時，伺服器都會將認證資訊傳回給用戶端，用戶端隨後將其儲存起來；當用戶端再次請求伺服器資源時，都需要攜帶這些認證資訊以便進行驗證。儘管工作流程相似，但它們之間還是存在一些關鍵的區別，下一小節將詳細介紹。

6.1.9 JWT 與 Token 的區別

JWT 與傳統 Token 在身份驗證和資源存取控制中都扮演著權杖的角色，它們有一些共同點。

- 存取資源的權杖：無論是 JWT 還是傳統 Token，它們都被用作存取資源的權杖。
- 記錄使用者認證資訊：它們都用於記錄使用者的認證資訊，以便服務端進行驗證。
- 服務端無狀態化：兩者都支援服務端的無狀態化，即服務端不需要儲存階段資訊。
- 跨域資源分享：JWT 和 Token 都能夠支持跨域資源分享（Cross-Origin Resource Sharing，CORS）。

然而，JWT 與傳統 Token 之間也存在一些關鍵的區別。

- 鑑權機制：傳統 Token 鑑權通常需要服務端解析用戶端發送的 Token，並根據該 Token 去資料庫查詢使用者資訊，以此驗證 Token 的有效性。
- 多樣性：Token 可以採取多種形式，例如隨機字串、OAuth 權杖、Session Token 等。而 JWT 是 Token 的一種具體實現形式。
- 資訊儲存與加密：JWT 在發行時會將使用者資訊加密並儲存在 Token 自身中，用戶端攜帶 JWT 進行請求，服務端透過金鑰進行解密驗證，無須再次查詢資料庫。

6.1.10 什麼是 SSO

SSO（Single Sign On，單點登入）是在多個應用系統中，只需要登入一次，就可以存取其他相互信任的應用系統。

在企業中，多個應用系統可能在同一域名下，子系統透過二級域名來區分。當然，多個應用系統也可能分佈在不同域名下，下面分別來說明這兩種場景中單點登入的工作流程。

1. 同域名下的單點登入

企業通常擁有一個主域名，並利用二級域名來區分其不同的子系統。舉例來說，如果企業的頂層網域名是 xxx.com，那麼業務系統 A 和 B 可能分別位於 a.xxx.com 和 b.xxx.com。為了實現單點登入，企業需要一個集中的登入系統，比如位於 sso.xxx.com。

在單點登入機制下，使用者只需在 sso.xxx.com 進行一次登入，隨後便能無縫存取系統 A 和 B，無須重複登入。這種登入認證機制允許使用者在多個相連結的系統間自由切換，同時保持階段的連續性。具體的工作流程如圖 6-6 所示。

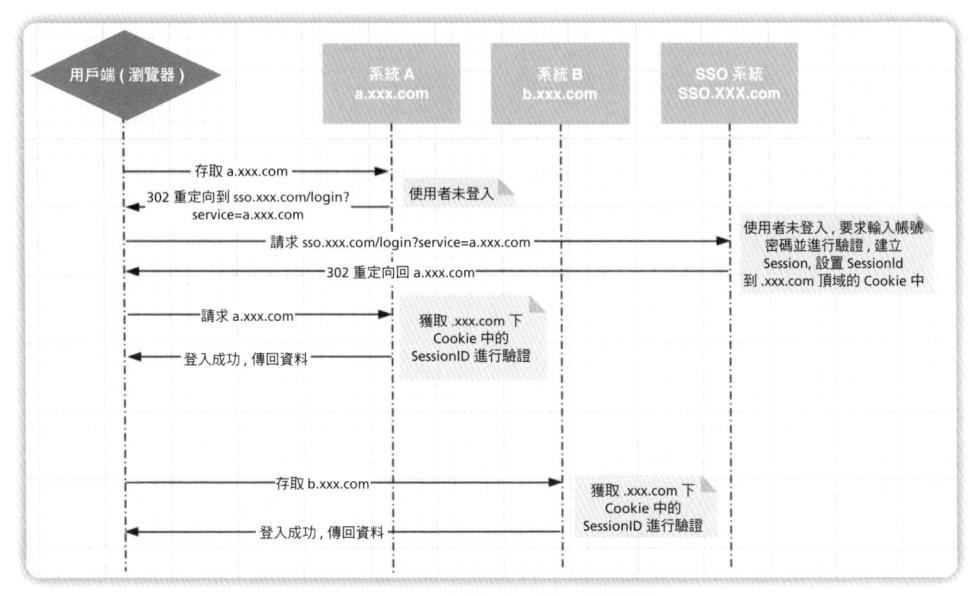

▲ 圖 6-6 單點登入的工作流程

圖 6-6 中的流程詳細描述如下：

（1）用戶端存取系統 A：使用者嘗試存取系統 A，由於在頂層網域名 xxx.com 下未檢測到使用者的 Cookie 資訊，系統觸發重定向流程。

（2）重定向至 SSO 系統：使用者被重定向到 SSO 登入頁面（sso.xxx.com），並附上系統 A 的標識或重定向 URL。使用者此時需輸入帳號和密碼進行登入。登入成功後，系統在頂層網域名下設置 Cookie 資訊。

（3）重定向回系統 A：SSO 系統根據使用者輸入的重定向 URL，將使用者帶回系統 A，並攜帶 SessionID 的 Cookie。系統 A 透過 Cookie 中的 SessionID 驗證使用者身份，確認無誤後允許使用者登入並存取資料。

（4）用戶端存取系統 B：使用者存取系統 B 時，由於頂層網域名 xxx.com 下的 Cookie 中已存在 SessionID，系統 B 透過該 SessionID 查詢並驗證使用者的 Session 資訊。身份驗證通過後，使用者即可成功登入並獲取資料。

（5）退出登入：使用者在任一系統中登出時，系統需清除頂層網域名下的 Cookie 資訊，以確保使用者在所有相關系統中都已登出。

需要注意的是，SSO、系統 A 和 B 雖然是不同的應用，但它們的 Session 儲存在自己的伺服器上，並不預設共用。當使用者登入 SSO 系統後，若要存取系統 A，系統 A 需要能夠辨識和驗證 SSO 系統設置的 Cookie 中的 SessionID。為解決這一問題，可以採用如 Redis 這樣的工具來實現 Session 的共用。此外，也可以使用 Token 機制來維護使用者的認證資訊，Token 由用戶端儲存，從而使服務端達到無狀態化。

理解了同域名下的 SSO 原理後，我們可以進一步探討不同域名下的 SSO 實現方式。

2. 不同域名下的單點登入

在相同域名下實現單點登入時，可以利用頂層網域名 Cookie 的特性來共用階段資訊。然而，不同域名的網站由於瀏覽器的相同來源策略限制，通常無法直接共用 Cookie。要解決這一問題，我們可以採用中央認證服務（Central Authentication Service，CAS）來集中管理身份認證。跨域 SSO 的工作流程如圖 6-7 所示。

圖 6-7 中的流程詳細描述如下：

（1）使用者存取應用 A：使用者在瀏覽器中輸入應用 A 的網址並嘗試存取。

（2）檢查使用者登入狀態：應用 A 檢查使用者是否已登入。如果使用者未登入，則跳躍到統一 CAS 認證中心進行登入。

（3）重定向到 CAS 的登入頁面：應用 A 將使用者重定向到 CAS 的登入頁面，並傳遞應用 A 的標識資訊（例如應用 A 的識別字或回呼 URL）。

（4）使用者登入：使用者在 CAS 系統的登入頁面輸入其憑證（例如使用者名稱和密碼），然後進行身份驗證。

（5）生成權杖：驗證成功後，CAS 系統將生成一個 TGT（Ticket Granted Ticket，俗稱大權杖或票根），該權杖可以用於簽發 ST（Service Ticket，小權杖）。為了隔離不同應用的驗證過程，系統會根據 TGT 再生成 ST，並將該小權杖綁定到重定向 URL 中，隨後傳回給應用 A。同時，CAS 系統會設置一個名為 CASTGC 的 Cookie，該 Cookie 與 TGT 的關係類似於 Web 階段（Session）與階段 ID（SessionID）之間的關係。透過 CASTGC，系統可以定位到相應的 TGT。

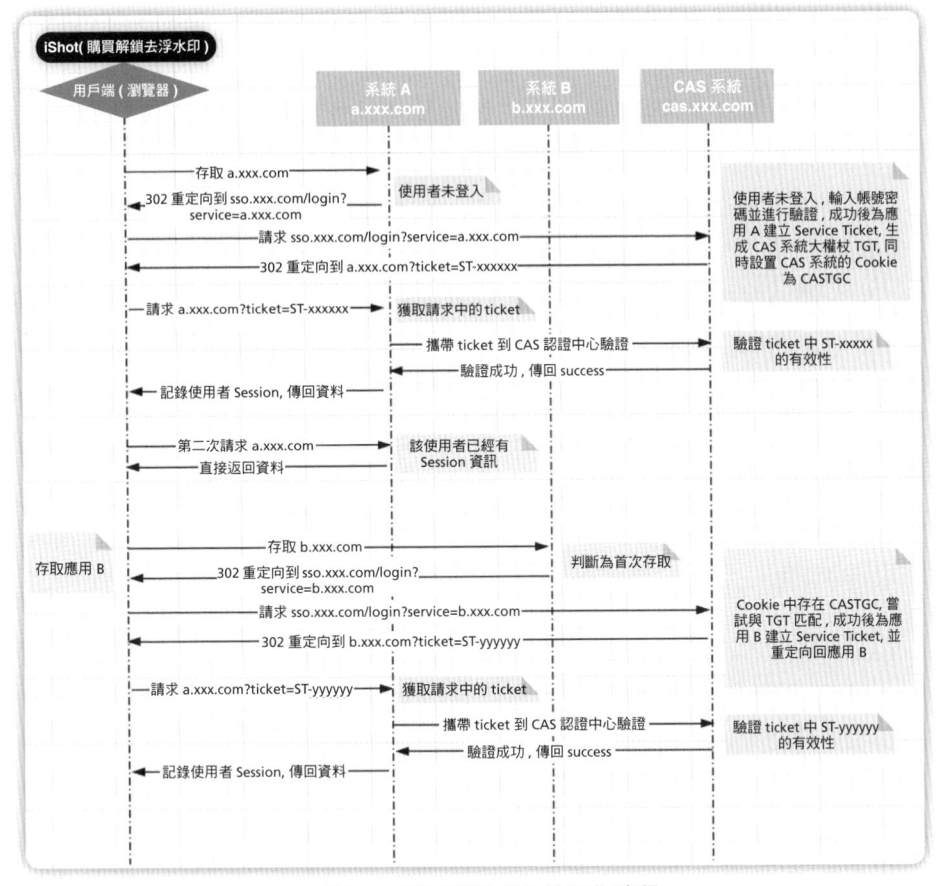

▲ 圖 6-7 跨域單點登入的工作流程

（6）攜帶權杖傳回應用 A：CAS 系統使用者重定向到應用 A 的回呼 URL。

（7）應用 A 驗證權杖：應用 A 獲取 URL 後面攜帶的 Ticket，向 CAS 伺服器發起 HTTP 請求，以驗證 Ticket 的有效性。如果權杖有效，則建立使用者階段，並將使用者標記為已登入狀態。

（8）再次存取應用 A 的資源：使用者現在被視為已登入應用 A，並且可以存取其受保護的資源。

（9）使用者存取應用 B：使用者在同一瀏覽器階段中嘗試存取應用 B。

（10）檢查使用者登入狀態：應用 B 會在守衛或攔截器中（以 Nest 為例）檢查使用者是否已登入，發現使用者首次存取，於是發起重定向，去 CAS 系統登入。

（11）重定向到 CAS 系統：應用 B 將使用者重定向到 CAS 系統，並傳遞應用 B 的回呼 URL。

（12）CAS 系統驗證階段：CAS 系統檢查使用者的階段狀態，發現之前已經存取過一次了，因此 Cookie 中會攜帶一個 CASTGC。

（13）生成新權杖：CAS 系統生成一個新權杖，並將其與使用者的身份相關資訊連結起來。

（14）傳回權杖：CAS 系統透過 TGC 查詢到對應的 TGT，於是用 TGT 簽發一個 ST，回呼 URL 攜帶上 ST，並將使用者重定向到應用 B。

（15）應用 B 驗證權杖：應用 B 獲取到 URL 後面的 Ticket 後，驗證 Ticket 權杖的有效性。如果權杖有效，則建立使用者階段，並將使用者標記為已登入狀態。

（16）存取應用 B 的資源：使用者現在被視為已登入應用 B，並且可以存取其受保護的資源。

到目前為止，我們已經學習了身份驗證中常見的概念，如 Cookie、Session、Token、JWT 以及 SSO 等，相信讀者對這些知識已經有了更加系統和深入的理解。下一節將透過案例演示如何在 Nest 中使用身份驗證。

6.2 基於 Passport 和 JWT 實現身份驗證

本節將介紹如何在 Nest 中結合 Passport（通行證）和 JWT 進行身份驗證。首先，詳細介紹 Passport 的基本概念，包括其作用、使用方式及其與 JWT 的關係。然後，建立一個 Nest 專案，安裝並整合 Passport 與 JWT，用 Passport 提供的本地策略來實現帳號和密碼的登入驗證，並結合 JWT 策略實現介面的驗證功能。

6.2.1 基本概念

Passport 是最流行的 Node.js 身份驗證中介軟體，廣泛應用於許多生產應用中。它支援多種身份驗證策略，包括本地認證（使用者名稱和密碼）、OAuth 策略、OAuth2 策略、OpenID Connect 策略、JWT 策略等，可以滿足不同場景下的身份認證需求。

Passport 基於策略模式擴充不同的驗證方式，它不直接提供 JWT 的具體實現，而是擴充了一個名為 passport-jwt 的策略套件，用於在 Node.js 應用中實現基於 JWT 的無狀態身份驗證機制。JWT 被用作驗證權杖（Token），在用戶端與服務端之間傳遞使用者身份資訊，而 Passport 負責驗證和解析 JWT 中的 Payload，將使用者資訊增加到請求物件中，以便在後續的路由處理常式中透過請求物件來獲取使用者資訊，進行存取控制和許可權判斷。

需要注意的是，在 Nest 中依舊可以單獨用 JWT 進行身份認證，使用 Passport 是因為它提供了進一步的抽象，以簡化身份認證流程。此外，Passport 與 Nest 框架的分層理念相契合，這使得在開發大型應用時，身份認證的維護和擴充變得更加容易。

6.2.2 專案準備

為了更進一步地說明 Passport 與 JWT 的用法，我們將建立一個 Nest 專案來實踐應用它。首先，執行「nest new nest-passport-jwt -p pnpm」命令，建立完成後的結果如圖 6-8 所示。

```
$ nest new nest-passport-jwt -p pnpm
⚡  We will scaffold your app in a few seconds..

CREATE nest-passport-jwt/.eslintrc.js (663 bytes)
CREATE nest-passport-jwt/.prettierrc (51 bytes)
CREATE nest-passport-jwt/README.md (3347 bytes)
CREATE nest-passport-jwt/nest-cli.json (171 bytes)
CREATE nest-passport-jwt/package.json (1958 bytes)
CREATE nest-passport-jwt/tsconfig.build.json (97 bytes)
CREATE nest-passport-jwt/tsconfig.json (546 bytes)
CREATE nest-passport-jwt/src/app.controller.spec.ts (617 bytes)
CREATE nest-passport-jwt/src/app.controller.ts (274 bytes)
CREATE nest-passport-jwt/src/app.module.ts (249 bytes)
CREATE nest-passport-jwt/src/app.service.ts (142 bytes)
CREATE nest-passport-jwt/src/main.ts (208 bytes)
CREATE nest-passport-jwt/test/app.e2e-spec.ts (630 bytes)
CREATE nest-passport-jwt/test/jest-e2e.json (183 bytes)

✔ Installation in progress... 🍥

🚀  Successfully created project nest-passport-jwt
```

▲ 圖 6-8 建立專案

要實現使用者登入驗證，需要使用 Passport 提供的本地策略套件（passport-local），它實現了使用者名稱和密碼的驗證機制。同時，還需安裝 Passport 的其他相依套件。執行以下命令：

```
pnpm add --save @nestjs/passport passport passport-local
pnpm add --save-dev @types/passport-local
```

然後建立兩個功能模組：auth 和 user。執行以下命令：

```
nest g mo auth
nest g s auth --no-spec
nest g mo user
nest g s user --no-spec
```

6.2.3 用本地策略實現使用者登入

建立成功後，auth 模組用於實現驗證使用者身份的相關邏輯，而 user 模組用於使用者的增刪改查操作。為了方便說明，本節將使用模擬資料來代替從資料庫中查詢使用者。下面是 user.service.ts 的程式：

```
import { Injectable } from '@nestjs/common';

export type User = {
  id: number;
  userName: string;
  password: string;
};

@Injectable()
export class UserService {
  private users: User[] = [
    { id: 1, userName: 'user1', password: 'user111' },
    { id: 2, userName: 'user2', password: 'user222' },
    { id: 3, userName: 'user3', password: 'user333' },
  ];

  async findOneByUserName(userName: string): Promise<User | undefined> {
    return this.users.find((item) => item.userName === userName);
  }
}
```

模擬資料庫中有三筆資料，分別為 user1、user2 和 user3 三個使用者，並且定義了 findOneByUserName 方法，用於根據使用者名稱查詢記錄。

在查詢到使用者記錄後，在 auth.service.ts 中定義 validateUser 方法，用於驗證請求中使用者傳遞的密碼是否匹配。如果匹配，則傳回除密碼之外的其他欄位，否則傳回 null。範例程式如下：

```
import { Injectable } from '@nestjs/common';
import { UserService } from '../user/user.service';
```

```typescript
@Injectable()
export class AuthService {
  constructor(private usersService: UserService) {}

  async validateUser(userName: string, passord: string): Promise<any> {
    const user = await this.usersService.findOneByUserName(userName);
    if (user && user.password === password) {
      const { password, ...result } = user;
      return result;
    }
    return null;
  }
}
```

有了驗證方法，現在我們可以實現本地身份驗證策略了。在 auth 資料夾下建立一個名為 local.strategy.ts 的檔案，定義 LocalStrategy 類別並繼承 PassportStrategy。範例程式如下：

```typescript
import { Strategy } from 'passport-local';
import { PassportStrategy } from '@nestjs/passport';
import { Injectable, UnauthorizedException } from '@nestjs/common';
import { AuthService } from './auth.service';

@Injectable()
export class LocalStrategy extends PassportStrategy(Strategy) {
  constructor(private authService: AuthService) {
    super({
      usernameField: 'userName',
    });
  }

  async validate(userName: string, password: string): Promise<any> {
    const user = await this.authService.validateUser(userName, password);
    if (!user) {
      throw new UnauthorizedException();
    }
    return user;
  }
}
```

無論是定義本地策略還是即將實現的 JWT 策略，都需要繼承 PassportStrategy 類別，並實現其 validate 方法。預設情況下，本地策略需要從請求中獲取名為 username 和 password 的兩個屬性。然而，在上述範例程式中，筆者採用了駝峰命名法，將欄位名稱改為 userName。因此，需要在呼叫 super 方法時傳遞一個選項物件，例如 { usernameField: 'userName' }，以重寫預設的欄位名稱。這種做法在電子郵件密碼登入場景下特別有用。

此外，透過呼叫先前定義的 validateUser 方法來驗證使用者身份。如果使用者不存在，則拋出 UnauthorizedException 例外。

然後，在 auth.module.ts 檔案中引入本地策略，並匯入 PassportModule 模組。相關程式如下：

```
import { Module } from '@nestjs/common';
import { AuthService } from './auth.service';
import { LocalStrategy } from './local.strategy';
import { UserModule } from 'src/user/user.module';
import { PassportModule } from '@nestjs/passport';

@Module({
  imports: [UserModule, PassportModule],
  providers: [AuthService, LocalStrategy]
})
export class AuthModule {}
```

最後，在 app.controller.ts 中實現登入方法（login），使用 Nest 中的守衛（Guard）進行切面判斷，並指定路由程式的認證策略為 local，程式如下：

```
import { Controller, Post, UseGuards } from '@nestjs/common';
import { Req } from '@nestjs/common/decorators';
import { AuthGuard } from '@nestjs/passport';
import { AppService } from './app.service';

@Controller()
export class AppController {
  constructor(private readonly appService: AppService) {}

  @UseGuards(AuthGuard('local'))
  @Post('auth/login')
  async login(@Req() req) {
    return req.user;
  }
}
```

接下來，對實現的身份驗證功能進行測試。首先，執行「pnpm start:dev」命令以啟動開發服務。如果服務啟動成功，將繼續進行測試。在之前的章節中，我們使用了 .http 檔案來發送請求進行測試。現在，使用 macOS 系統附帶的 curl 工具來進行測試（對於 Windows 系統使用者，需要先下載並安裝 curl 工具），執行以下命令：

```
curl -X POST http://localhost:3000/auth/login -d '{"userName": "user1", "password": "user111"}' -H "Content-Type: application/json"
```

執行結果如圖 6-9 所示。

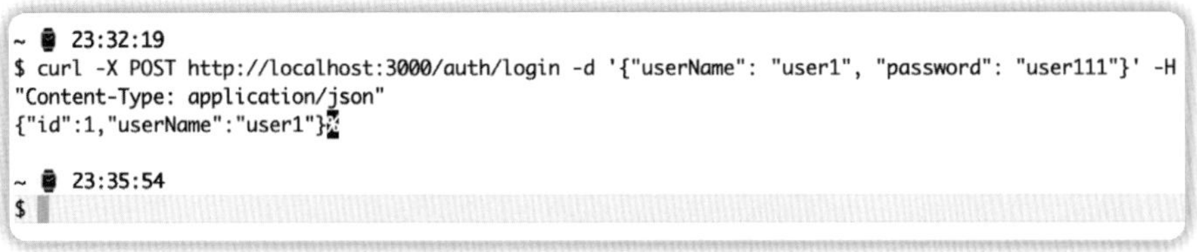

```
~  23:32:19
$ curl -X POST http://localhost:3000/auth/login -d '{"userName": "user1", "password": "user111"}' -H
"Content-Type: application/json"
{"id":1,"userName":"user1"}
~  23:35:54
$
```

▲ 圖 6-9 使用者登入認證成功

此時輸入錯誤的密碼 userxxx，再次執行命令，執行結果如圖 6-10 所示，傳回 401 授權認證失敗提示。

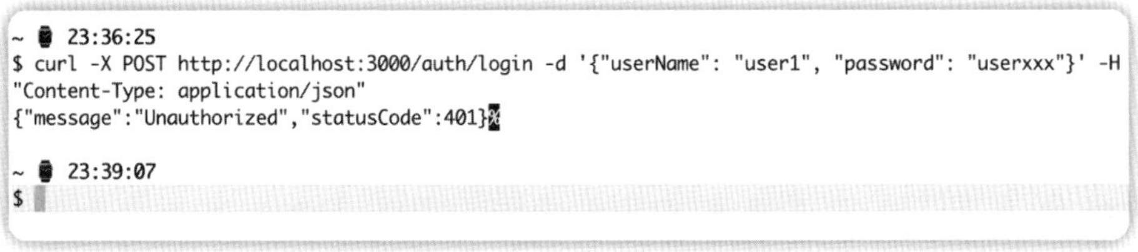

```
~  23:36:25
$ curl -X POST http://localhost:3000/auth/login -d '{"userName": "user1", "password": "userxxx"}' -H
"Content-Type: application/json"
{"message":"Unauthorized","statusCode":401}
~  23:39:07
$
```

▲ 圖 6-10 使用者登入認證失敗

我們已經實現了使用者的身份驗證功能，但在實際應用中，通常只在登入時使用使用者名稱和密碼進行驗證。對於其他介面的請求，我們使用權杖（Token）來進行身份驗證，這時就需要引入 JWT 驗證機制。

6.2.4 用 JWT 策略實現介面驗證

要想在 Passport 的基礎上實現 JWT 驗證，我們還需要安裝其對應的相依，執行以下命令：

```
pnpm add --save @nestjs/jwt passport-jwt
pnpm add --save-dev @types/passport-jwt
```

其中，passport-jwt 是實現 JWT 策略的 Passport 套件，而 Nest 整合了 JWT 相關操作，方便實現相依注入。

現在，我們在 AuthModule 中引入 JwtModule，程式如下：

```
import { Module } from '@nestjs/common';
import { AuthService } from './auth.service';
import { LocalStrategy } from './local.strategy';
import { UserModule } from 'src/user/user.module';
import { PassportModule } from '@nestjs/passport';
import { JwtModule } from '@nestjs/jwt';
import { JwtStrategy } from './jwt.strategy';
```

```
@Module({
  imports: [
    UserModule,
    PassportModule,
    // 新增 JWT 模組
    JwtModule.register({
      // 這裡應該讀取配置中的 secret
      secret: 'jwt-secret',
      signOptions: { expiresIn: '7d' },
    }),
  ],
  providers: [AuthService, LocalStrategy, JwtStrategy],
  exports: [AuthService]
})
export class AuthModule {}
```

有了 JWT 模組，還需要有對應的 JWT 策略，在 auth 資料夾下建立一個名為 jwt-strategy.ts 的檔案，同樣定義 JwtStrategy 類別並繼承 PassportStrategy。範例程式如下：

```
import { ExtractJwt, Strategy } from 'passport-jwt';
import { PassportStrategy } from '@nestjs/passport';
import { Injectable } from '@nestjs/common';

@Injectable()
export class JwtStrategy extends PassportStrategy(Strategy) {
  constructor() {
    super({
      // 表示從 header 中的 Authorization 的 Bearer 標頭中獲取 token 值
      jwtFromRequest: ExtractJwt.fromAuthHeaderAsBearerToken(),
      // 不忽視 token 過期的情況，過期會傳回 401
      ignoreExpiration: false,
      // 這裡應該讀取配置中的 secret
      secretOrKey: 'jwt-secret',
    });
  }

  async validate(payload: any) {
    return { id: payload.sub, username: payload.userName };
  }
}
```

在上述程式中，jwtFromRequest 表示從 Request 物件中獲取 Token 的方式，包含 header、body、URLparameter 等，secretOrKey 是 JWT 金鑰，ignoreExpiration 設置為 false 會在 Token 過期時傳回 401 Unauthorized 提示。最後實現 validate 方法並傳回一個 user 物件。

策略實現完成後，接下來在 AppController 中定義一個名為 getUserInfo 的路由方法。這個方法明顯是受保護的資源，需要進行身份驗證才能存取。同時，我們需要修改 login 方法，以便在使用者登入成功後傳回一個 access_token 給前端進行儲存。範例程式如下：

```
import { Controller, Post, UseGuards } from '@nestjs/common';
import { Get, Req } from '@nestjs/common/decorators';
import { AuthGuard } from '@nestjs/passport';
import { AuthService } from './auth/auth.service';

@Controller()
export class AppController {
  constructor(private readonly authService: AuthService) {}

  @UseGuards(AuthGuard('local'))
  @Post('auth/login')
  async login(@Req() req) {
    // 呼叫 auth 中 login 方法傳回 access_token
    return this.authService.login(req.user);
  }

  // 新增獲取使用者資訊的方法
  @UseGuards(AuthGuard('jwt'))
  @Get('getUserInfo')
  getUserInfo(@Req() req) {
    return req.user;
  }
}
```

可見，我們使用了 AuthGuard('jwt') 來指定該路由方法應用 JWT 認證策略，並且呼叫了 AuthService 中的方法來傳回 access_token。AuthService 中 login 方法的實現程式如下：

```
import { Injectable } from '@nestjs/common';
import { UserService, User } from '../user/user.service';
import { JwtService } from '@nestjs/jwt';

@Injectable()
export class AuthService {
  constructor(
    private usersService: UserService,
    private jwtService: JwtService,
  ) {}

  async validateUser(userName: string, password: string): Promise<any> {
    const user = await this.usersService.findOneByUserName(userName);
```

```
    if (user && user.password === password) {
      const { password, ...result } = user;
      return result;
    }
    return null;
  }
  // 新增 login 方法，生成 token
  async login(user: User) {
    const payload = { username: user.userName, sub: user.id };
    return {
      access_token: this.jwtService.sign(payload),
    };
  }
}
```

我們最終呼叫 JwtService 的 sign 方法生成了一個 token，其中定義 username 和 sub 是為了與 JWT 標準保持一致。

至此，我們已經完成了所有的認證開發過程。接下來測試效果，執行以下命令嘗試獲取使用者資訊：

```
curl -X GET http://localhost:3000/getUserInfo
```

顯然，在未登入情況下，如果沒有攜帶 access_token，伺服器將傳回 401 錯誤，如圖 6-11 所示。

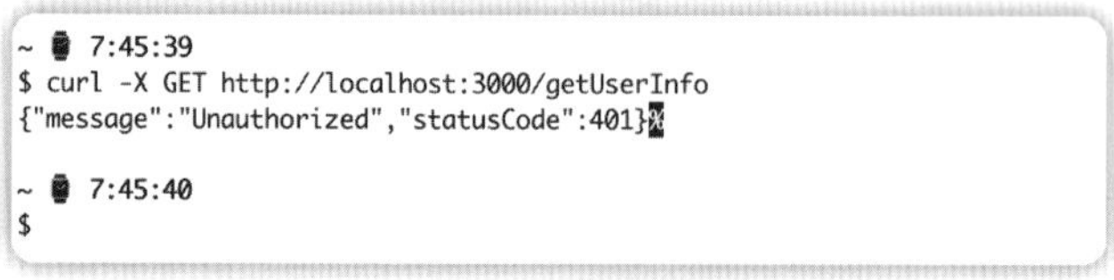

```
~  7:45:39
$ curl -X GET http://localhost:3000/getUserInfo
{"message":"Unauthorized","statusCode":401}
~  7:45:40
$
```

▲ 圖 6-11 獲取使用者資訊失敗

接下來執行使用者登入操作，獲取 access_token，執行以下命令：

```
curl -X POST http://localhost:3000/auth/login -d '{"userName": "user1", "password": "user111"}' -H "Content-Type: application/json"
```

傳回了一串格式為 xxx.xxx.xxx 的 JWT 加密字串，如圖 6-12 所示。

```
~  7:55:26
$ curl -X POST http://localhost:3000/auth/login -d '{"userName": "user1", "password": "user111"}' -H
"Content-Type: application/json"
{"access_token":"eyJhbGciOiJIUzI1NiIsInR5cCI6IkpXVCJ9.eyJ1c2VybmFtZSI6InVzZXIxIiwic3ViIjoxLCJpYXQiOjE
3MDgzODY5MzIsImV4cCI6MTcwODk5MTczMn0.JPJIAYxKD3fhcz9SmQhc5Gz8vgq4VS9bTVtZQK6f7TY"}
```

▲ 圖 6-12 獲取 access_token 成功

我們在請求標頭中增加這串 access_token，執行以下命令：

```
curl -X GET http://localhost:3000/getUserInfo -H "Authorization: Bearer
eyJhbGciOiJIUzI1NiIsInR5cCI6IkpXVCJ9.eyJ1c2VybmFtZSI6InVzZXIxIiwic3ViIjoxLCJpYXQiOjE3MDg
zODY5MzIsImV4cCI6MTcwODk5MTczMn0.JPJIAYxKD3fhcz9SmQhc5Gz8vgq4VS9bTVtZQK6f7TY"
```

可以看到許可權驗證成功，傳回了正確的資料，如圖 6-13 所示。

```
~   7:55:32
$ curl -X GET http://localhost:3000/getUserInfo -H "Authorization: Bearer eyJhbGciOiJIUzI1NiIsInR5cCI
6IkpXVCJ9.eyJ1c2VybmFtZSI6InVzZXIxIiwic3ViIjoxLCJpYXQiOjE3MDgzODY5MzIsImV4cCI6MTcwODk5MTczMn0.JPJIAYx
KD3fhcz9SmQhc5Gz8vgq4VS9bTVtZQK6f7TY"
{"id":1,"username":"user1"}

~   7:56:22
$
```

▲ 圖 6-13 獲取使用者資訊成功

至此，我們基本完成了 JWT 身份驗證的核心過程。但前面的部分程式仍不夠優雅，下一小節跟隨筆者來最佳化它們。

6.2.5 程式最佳化

在AppController中，在路由方法上定義認證策略時，我們使用AuthGuard('local')或AuthGuard('jwt')來指定使用哪個策略，事實上這無形之中引入了魔術字串，並不好維護，可以把它抽離成單獨的類別來管理。在auth資料夾下新建jwt-auth.guard.ts檔案，程式如下：

```
import { Injectable } from '@nestjs/common';
import { AuthGuard } from '@nestjs/passport';

@Injectable()
export class JwtAuthGuard extends AuthGuard('jwt') {}
```

同樣，修改本地策略，在同級目錄下新建 local-auth.guard.ts 檔案，程式如下：

```
import { Injectable } from '@nestjs/common';
import { AuthGuard } from '@nestjs/passport';

@Injectable()
export class LocalAuthGuard extends AuthGuard('local') {}
```

接下來，在 AppController 中引入它們，程式如下：

```
import { LocalAuthGuard } from './auth/local-auth.guard';
import { JwtAuthGuard } from './auth/jwt-auth.guard';

@Controller()
```

```
export class AppController {
  ...
  @UseGuards(LocalAuthGuard)
  ...

  @UseGuards(JwtAuthGuard)
  ...
}
```

除此之外，細心的讀者可能會留意到，在配置 JWT 金鑰時，我們把 jwt-secret 字串強制寫入到程式中。然而，在生產環境中並不建議這樣做，而是應該使用金鑰庫、環境變數或配置服務來管理此金鑰（類似管理資料庫帳號和密碼）。為此，我們可以在專案根目錄下建立一個 .env 檔案來維護它：

```
JWT_SECRET=my-jwt-secret
JWT_EXPIRE_TIME=7d
```

為了獲取配置中的資訊，還需要安裝 @nestjs/config 套件。在 AppModule 中引入 ConfigModule，並將配置模組注入全域，程式如下：

```
import { Module } from '@nestjs/common';
import { AppController } from './app.controller';
import { AppService } from './app.service';
import { AuthModule } from './auth/auth.module';
import { UserModule } from './user/user.module';
import { JwtAuthGuard } from './auth/jwt-auth.guard';
import { ConfigModule } from '@nestjs/config';

@Module({
  imports: [
    AuthModule,
    UserModule,
    // 將配置模組注入全域
    ConfigModule.forRoot({
      isGlobal: true,
    }),
  ],
  controllers: [AppController],
  providers: [
    AppService,
    JwtAuthGuard
  ],
})
export class AppModule {}
```

此時，為了更進一步地獲取配置資訊，我們需要把 JWT 模組改為使用工廠函數方式進行註冊，並注入 ConfigService 獲取配置資訊。範例程式如下：

```
import { Module } from '@nestjs/common';
import { AuthService } from './auth.service';
import { LocalStrategy } from './local.strategy';
import { UserModule } from 'src/user/user.module';
import { PassportModule } from '@nestjs/passport';
import { JwtModule } from '@nestjs/jwt';
import { JwtStrategy } from './jwt.strategy';
import { ConfigModule, ConfigService } from '@nestjs/config';

@Module({
  imports: [
    UserModule,
    PassportModule,
    // 新增 JWT 模組
    JwtModule.registerAsync({
      useFactory: async (configService: ConfigService) => (
        {
          // 讀取配置中的 secret
          secret: configService.get<string>('JWT_SECRET'),
          signOptions: {
            expiresIn: configService.get<string>('JWT_EXPIRE_TIME')
          },
        }
      ),
      // 將 ConfigService 注入工廠函數中
      inject: [ConfigService],
    }),
  ],
  providers: [AuthService, LocalStrategy, JwtStrategy],
  exports: [AuthService]
})
export class AuthModule {}
```

另外，我們還需要在 JWT 策略中修改金鑰獲取方式，將建構元程式修改如下：

```
import { ExtractJwt, Strategy } from 'passport-jwt';
import { PassportStrategy } from '@nestjs/passport';
import { Injectable } from '@nestjs/common';
// 新增相依
import { ConfigService } from '@nestjs/config';

@Injectable()
export class JwtStrategy extends PassportStrategy(Strategy) {
  // 注入配置服務
  constructor(configService: ConfigService) {
    super({
      // 表示從 header 中的 Authorization 的 Bearer 標頭獲取 token 值
      jwtFromRequest: ExtractJwt.fromAuthHeaderAsBearerToken(),
      // 不忽視 token 過期的情況，過期會傳回 401
```

```
      ignoreExpiration: false,
      // 讀取配置中的 secret
      secretOrKey: configService.get<string>('JWT_SECRET'),
    });
  }

  async validate(payload: any) {
    return { id: payload.sub, username: payload.username };
  }
}
```

至此，我們已經完成了所有的最佳化工作。最後來測試一下，登入並成功獲取使用者資訊，效果如圖 6-14 所示。

```
~           11:47:50
$ curl -X POST http://localhost:3000/auth/login -d '{"userName": "user1", "password": "user111"}' -H
"Content-Type: application/json"
{"access_token":"eyJhbGciOiJIUzI1NiIsInR5cCI6IkpXVCJ9.eyJ1c2VybmFtZSI6InVzZXIxIiwic3ViIjoxLCJpYXQiOjE
3MDg0MDA4NzYsImV4cCI6MTcwOTAwNTY3Nn0.q_vJYgG89jTxJfi_sYv-hX4HELGBypP8chglXED_tCQ"}

~           11:47:56
$ curl -X GET http://localhost:3000/getUserInfo -H "Authorization: Bearer eyJhbGciOiJIUzI1NiIsInR5cCI
6IkpXVCJ9.eyJ1c2VybmFtZSI6InVzZXIxIiwic3ViIjoxLCJpYXQiOjE3MDg0MDA4NzYsImV4cCI6MTcwOTAwNTY3Nn0.q_vJYgG
89jTxJfi_sYv-hX4HELGBypP8chglXED_tCQ"
{"id":1,"username":"user1"}

~           11:48:38
$
```

▲ 圖 6-14 執行成功

綜上所述，我們詳細介紹了如何結合 Nest 中的 Passport 模組和 JWT 實現身份驗證。透過 Passport 提供的策略和中介軟體，我們能夠輕鬆實現多種身份驗證策略，而 JWT 提供了一種安全可靠的身份認證方式，使得在無狀態的分散式環境下進行身份驗證更加便捷和高效。

6.3 基於 RBAC 實現許可權控制

在實際應用中，基於角色的存取控制（Role Based Access Control，RBAC）是當前應用最廣泛的角色控制模型。RBAC 透過將使用者分配到不同的角色，並為每個角色分配不同的許可權來管理使用者對系統的存取權。

本節將介紹在 Nest 中如何使用 RBAC 模型，在介面等級實現許可權控制，以確保每個使用者只能存取其具備許可權的介面，從而實現一個簡單可擴充的許可權系統。

6.3.1 基本概念

在 RBAC 模型中，通常包含以下幾個核心概念。

- 使用者（User）：系統中的操作實體，是系統的使用者，被賦予一個或多個角色來獲得對應的許可權。
- 角色（Role）：角色是一組許可權的集合，用於定義使用者在系統中的身份或地位。使用者可以被分配一個或多個角色，不同的角色有不同的許可權，例如店長擁有刪除資料的許可權，而普通員工沒有。
- 許可權（Permission）：許可權是指使用者在系統中執行某個操作的能力，如刪除、編輯、查看等。

它們之間的關係概括如圖 6-15 所示。

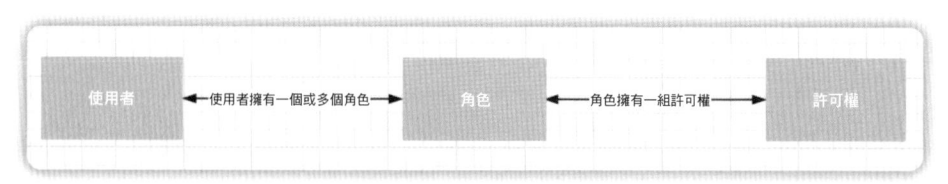

▲ 圖 6-15 RBAC 的基本關係

可見，使用者與角色之間是一對多或多對多的關係，角色與許可權之間是多對多的關係。

有讀者可能會有疑惑，為什麼不直接把許可權分配給使用者呢？

實際上，是可以直接把許可權分配給使用者的，只是這樣做會少一層關係，這在一定程度上削弱了系統的擴充性、維護性、安全性等，因此適用於使用者量少、角色類型少的平臺。

透過角色來管理許可權的分配，這種設計有以下 3 個優點。

（1）擴充性：透過角色來分配許可權，在大型複雜的系統中，可以擴充角色層次，如部門經理、技術總監、組長以及組員，不同角色之間存在包含關係。

（2）安全性：有了角色管理，系統超級管理員只需要為角色分配一組許可權集，這表示無須操作龐大的使用者為其分配許可權，減少了操作失誤的風險，同時也能夠提升操作效率。

（3）維護性：當使用者的角色許可權發生變化時，只需更新角色與許可權之間的關係，而無須為每個使用者單獨修改許可權設置，從而降低維護成本。

現在我們對 RBAC 有了基本的認識，接下來在 Nest 中演示如何使用它實現許可權控制。

6.3.2 資料表設計

以常用的 RBAC 管理系統（Admin）為例，角色與許可權相連結，使用者透過擁有合適的角色獲得相應的許可權。我們透過圖 6-16 來加深理解。

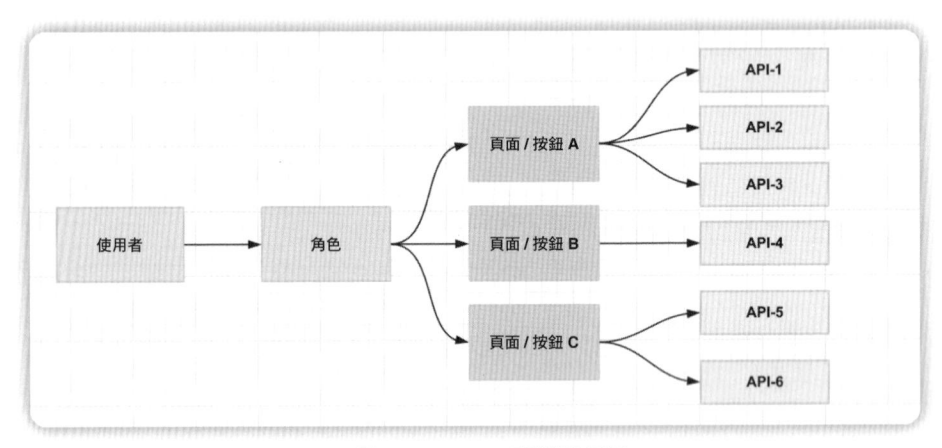

▲ 圖 6-16 RBAC 實際關係圖

可見，角色會得到一組頁面或按鈕的許可權集，每個頁面或按鈕背後可能包含一個或多個介面，前端透過控制頁面 UI 來限制使用者是否有許可權呼叫相關 API，後端同樣需要根據使用者角色判斷是否有指定介面的許可權，這樣才能成功呼叫 API。安全是雙方的事情。

理清關係之後，接下來設計資料庫表結構，如圖 6-17 所示。

分別建立 user 使用者表、role 角色表、permission 許可權表、permission_api 許可權表。另外，user_role 表用於連結 user 表與 role 表，使用 role_permission 表來連結 role 表與 permission 表。為了契合實際開發，本節將不使用外鍵，而是採用業務邏輯來管理表與表之間的連結。

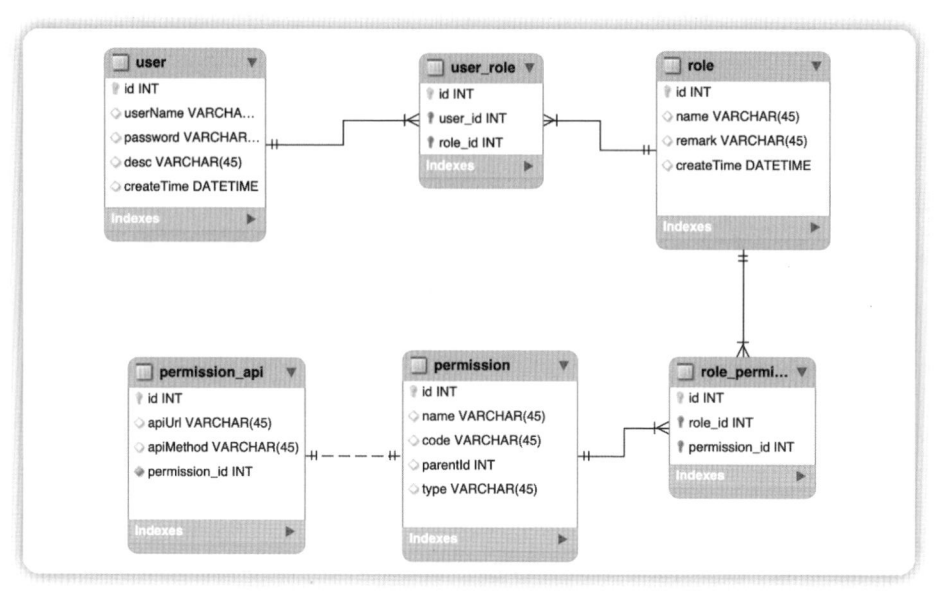

▲ 圖 6-17 RBAC 關係表

6.3.3 專案準備

前面已經實現了登入和介面身份驗證。為了方便起見，本小節基於 nest-passport-jwt 專案實現 RBAC 許可權控制，我們複製專案並將其重新命名為 nest-rbac。

有了 user 模組，我們還需要建立 role 和 permission 模組。執行以下命令：

```
nest g resource permission --no-spec
nest g resource role --no-spec
```

接下來安裝相依，需要用到 typeorm 和 mysql2 @nestjs/typeorm path-to-regexp 套件。執行「pnpm add path-to-regexp typeorm mysql2 @nestjs/typeorm -S」命令安裝所需相依。

安裝完成後，需要初始化資料庫連接。在 AppModule 中匯入 TypeOrmModule，核心程式如下：

```
imports: [
  // 初始化 MySQL 連接
  TypeOrmModule.forRootAsync({
    inject: [ConfigService],
    useFactory: (config: ConfigService) => ({
      type: 'mysql',
```

```
      host: config.get<string>('MYSQL_HOST'),
      port: config.get<number>('MYSQL_PORT'),
      username: config.get<string>('MYSQL_USER'),
      password: config.get<string>('MYSQL_PASSWORD'),
      database: config.get<string>('MYSQL_DATABASE'),
      entities: [__dirname + '/**/*.entity{.ts,.js}'],
      autoLoadEntities: true,
      // 在生產環境中禁止開啟，應該使用資料移轉
      synchronize: true
    })
  }),
  AuthModule,
  UserModule,
  // 將配置模組注入全域
  ConfigModule.forRoot({
    isGlobal: true,
  }),
  PermissionModule,
  RoleModule,
],
```

在上述程式中，透過 ConfigService 獲取資料庫配置，這些配置已經抽離到 .env 環境變數中統一管理。程式如下：

```
# mysql 配置
MYSQL_HOST=localhost
MYSQL_PORT=3306
MYSQL_USER=root
MYSQL_PASSWORD=xxxxx
MYSQL_DATABASE=nest_rbac
```

6.3.4 建立實體

配置完畢後，接下來按照圖 6-16 的結構來設計表結構。每個模組下都有一個 entities 資料夾（user 模組需要手動建立），需要完善對應的 entity 實體。啟動服務時，系統會自動建立資料庫表。

在 user/entities 中分別建立 user.entity.ts 和 user-role.entity.ts 檔案。user-role 實體用於連結使用者與角色，是範例程式如下：

```
// user.entity.ts
import { Entity, PrimaryGeneratedColumn, Column } from "typeorm";

@Entity()
export class User{

    @PrimaryGeneratedColumn()
```

```
    id: number;

    @Column()
    userName: string;

    @Column()
    password: string;

    @Column()
    desc: string;

    @Column()
    createTime: Date;
}

// user-role.entity.ts
import { Entity, PrimaryGeneratedColumn, Column } from "typeorm";

@Entity()
export class UserRole{

    @PrimaryGeneratedColumn()
    id: number;

    @Column()
    userId: number;

    @Column()
    roleId: number;
}
```

同樣，在 role/entities 下的 role.entity.ts 和 role-permission 檔案中分別定義角色、連結角色與許可權實體。範例程式如下：

```
// role.entity.ts
import { Entity, PrimaryGeneratedColumn, Column } from "typeorm";

@Entity()
export class Role{

    @PrimaryGeneratedColumn()
    id: number;

    @Column()
    name: string;
```

```
    @Column()
    remark: string;

    @Column()
    createTime: Date;

}
// role-permission.entity.ts
import { Entity, PrimaryGeneratedColumn, Column } from «typeorm»;

@Entity()
export class RolePermission{

    @PrimaryGeneratedColumn()
    id: number;

    @Column()
    role_id: number;

    @Column()
    permission_id: number;

}
```

最後,在 permission/entities 下完善許可權相關的物理定義。這包括 permission.entity.ts 和 permission-api.entity.ts 兩個檔案。範例程式如下:

```
// permission.entity.ts
import { Entity, PrimaryGeneratedColumn, Column } from "typeorm";

@Entity()
export class Permission{

    @PrimaryGeneratedColumn()
    id: number;

    @Column()
    name: string;

    @Column()
    code: string;

    @Column()
    parentId: number;

    @Column()
    type: string;
```

```
}
// permission-api.entity.ts
import { Entity, PrimaryGeneratedColumn, Column } from «typeorm»;

@Entity()
export class PermissionApi{

    @PrimaryGeneratedColumn()
    id: number;

    @Column()
    apiUrl: string;

    @Column()
    apiMethod: string;

    @Column()
    permission_id: number;

}
```

6.3.5 啟動服務

我們已經設計並建立好了資料庫表。執行「pnpm start:dev」命令啟動服務後，可以在 VS Code 的視覺化資料庫工具中看到成功建立的資料表，如圖 6-18 所示。

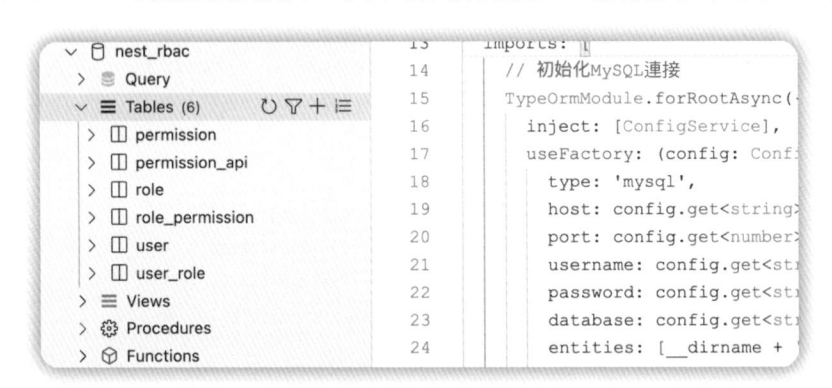

▲ 圖 6-18 成功啟動服務

6.3.6 實現角色守衛控制

為了實現不同使用者角色管理不同介面許可權的功能，我們首先考慮透過守衛（Guard）來實現這一需求。守衛的工作流程如下：

（1）判斷當前請求的介面是否需要進行 Token 驗證（舉例來說，登入和註冊介面通常不需要 Token 驗證）。

（2）判斷該介面是否需要進行許可權驗證（一些向外部開放的介面可能不需要進行許可權驗證）。

（3）檢查當前使用者是否具有存取當前介面的許可權。

在明確了這一邏輯之後，我們可以在 auth 資料夾下建立一個新的守衛檔案 role-auth.guard.ts。範例程式如下：

```typescript
import { CanActivate, Inject, ExecutionContext, Injectable, ForbiddenException }
        from '@nestjs/common'
import { Reflector } from '@nestjs/core'
import { pathToRegexp } from 'path-to-regexp'
import { ALLOW_NO_PERMISSION } from '../decorators/permission.decorator'
import { PermissionService } from 'src/permission/permission.service'
import { ALLOW_NO_TOKEN } from 'src/decorators/noToken.decorator'

@Injectable()
export class RoleAuthGuard implements CanActivate {
  constructor(
    private readonly reflector: Reflector,
    @Inject(PermissionService)
    private readonly permissionService: PermissionService,
  ) {}

  async canActivate(ctx: ExecutionContext): Promise<boolean> {
    // 若函數請求標頭配置 @AllowNoToken() 裝飾器，則無須驗證 token 許可權
    const allowNoToken = this.reflector.getAllAndOverride<boolean>(ALLOW_NO_TOKEN, [ctx.getHandler(), ctx.getClass()])
    if (allowNoToken) return true

    // 若函數請求標頭配置 @AllowNoPermission() 裝飾器，則無須驗證許可權
    const allowNoPerm = this.reflector.getAllAndOverride<boolean>(ALLOW_NO_PERMISSION, [ctx.getHandler(), ctx.getClass()])
    if (allowNoPerm) return true

    const req = ctx.switchToHttp().getRequest()
    const user = req.user
    // 若沒有攜帶 token，則直接傳回 false
    if (!user) return false

    // 獲取該使用者所擁有的介面許可權
    const userApis = await this.permissionService.findUserApis(user.id)
    console.log(< 當前使用者擁有的 URL 許可權集:',userApis);

    const index = userApis.findIndex((route) => {
      // 請求方法類型相同
```

```
      if (req.method.toUpperCase() === route.method.toUpperCase()) {
        // 比較當前請求 url 是否存在於使用者介面許可權集中
        const reqUrl = req.url.split('?')[0]
        console.log('當前請求 URL：', reqUrl);

        return !!pathToRegexp(route.path).exec(reqUrl)
      }
      return false
    })
    if (index === -1) throw new ForbiddenException('您無許可權存取該介面')
    return true
  }
}
```

同時，在 PermissionService 中，我們需要實現一個介面，用於查詢使用者所擁有的許可權集，程式如下：

```
import { Injectable } from '@nestjs/common';
import { CreatePermissionDto } from './dto/create-permission.dto';
import { UpdatePermissionDto } from './dto/update-permission.dto';
import { DataSource } from 'typeorm'
import { RouteDto } from './dto/route.dto';

@Injectable()
export class PermissionService {
  constructor(
    private dataSource: DataSource
  ) {}
  // 查詢使用者介面許可權集
  async findUserApis(userId: string): Promise<RouteDto[]> {
    const permsResult = await this.dataSource
      .createQueryBuilder()
      .select(['pa.apiUrl', 'pa.apiMethod'])
      .from('user_role', 'ur')
      .leftJoin('role_permission', 'rp', 'ur.roleId = rp.role_id')
      .leftJoin('permission_api', 'pa', 'rp.permission_id = pa.permission_id')
      .where('ur.userId = :userId', { userId })
      .groupBy('pa.apiUrl')
      .addGroupBy('pa.apiMethod')
      .getRawMany()
    const perms = permsResult.map(item => ({path: item.pa_apiUrl, method: item.pa_apiMethod}));
    return perms
  }
  create(createPermissionDto: CreatePermissionDto) {
    return 'This action adds a new permission';
  }

  findAll() {
```

```
    return `This action returns all permission`;
  }

  findOne(id: number) {
    return `This action returns a #${id} permission`;
  }

  update(id: number, updatePermissionDto: UpdatePermissionDto) {
    return `This action updates a #${id} permission`;
  }

  remove(id: number) {
    return `This action removes a #${id} permission`;
  }
}
```

由於角色判斷是通用邏輯，因此我們應該把它註冊為全域守衛，而無須在每個路由方法中綁定，核心程式如下：

```
providers: [
  AppService,
  {
    provide: APP_GUARD,
    useClass: JwtAuthGuard
  },
  {
    provide: APP_GUARD,
    useClass: RoleAuthGuard,
  },
],
```

別忘記，在前面的 RoleAuthGuard 中呼叫了 PermissionService 服務。接下來，在 PermissionModule 中將其匯出：

```
@Module({
  controllers: [PermissionController],
  providers: [PermissionService],
  exports: [PermissionService]
})
```

此外，細心的讀者可能會發現，在上面的程式中，我們需要實現 @AllowNoToken() 和 @AllowNoPermission() 裝飾器。這些裝飾器用於對特定介面進行放行，允許它們繞過 Token 驗證和許可權驗證。

為了組織這些裝飾器，我們可以在 src 目錄下新增一個 decorators 資料夾，並在其中建立 token.decorator.ts 和 permission.decorator.ts 檔案。實現程式如下：

```typescript
// token.decorator.ts
import { SetMetadata } from '@nestjs/common'

/**
 * 介面允許 Token 存取
 */
export const ALLOW_NO_TOKEN = 'allowNoToken'

export const AllowNoToken = () => SetMetadata(ALLOW_NO_TOKEN, true)
// permission.decorator.ts
import { SetMetadata } from '@nestjs/common'

/**
 * 介面允許無許可權存取
 */
export const ALLOW_NO_PERMISSION = 'allowNoPerm'

export const AllowNoPermission = () => SetMetadata(ALLOW_NO_PERMISSION, true)
```

然後，在 AppController 的路由程式中使用它們，程式如下：

```typescript
@Post('auth/login')
// 登入介面無須 Token 驗證
@AllowNoToken()
@UseGuards(LocalAuthGuard)
async login(@Req() req) {
  // 呼叫 auth 中的 login 方法傳回 access_token
  return this.authService.login(req.user);
}
// 開放 API 無須角色驗證
@Get('xxx')
@AllowNoPermission()
OpenXxxApi() {
  return []
}
```

至此，我們透過守衛實現了不同場景下的介面許可權控制。

6.3.7 生成測試資料

開始測試之前，先在資料庫各表中插入一些資料，以便進行測試。執行以下 SQL 敘述：

```sql
-- Active: 1705225958726@@127.0.0.1@3306@nest_rbac
-- 新增許可權資料
INSERT INTO `permission` VALUES (1, '新增員工', 'user_add', 1, 2);
INSERT INTO `permission` VALUES (2, '刪除員工', 'user_delete', 1, 2);
INSERT INTO `permission` VALUES (3, '編輯員工', 'user_edit', 1, 2);
```

```sql
INSERT INTO `permission` VALUES (4, '員工列表', 'user_list', 1, 1);
INSERT INTO `permission` VALUES (5, '角色管理', 'role_list', 2, 1);
INSERT INTO `permission` VALUES (6, '編輯角色', 'role_edit', 2, 2);
INSERT INTO `permission` VALUES (7, '刪除角色', 'role_delete', 2, 2);

-- 新增許可權 API
INSERT INTO `permission_api` VALUES (1, '/user/add', 'POST', 1);
INSERT INTO `permission_api` VALUES (2, '/user/delete', 'GET', 2);
INSERT INTO `permission_api` VALUES (3, '/user/edit', 'POST', 3);
INSERT INTO `permission_api` VALUES (4, '/user/list', 'GET', 4);
INSERT INTO `permission_api` VALUES (5, '/role/list', 'GET', 5);
INSERT INTO `permission_api` VALUES (6, '/role/edit', 'GET', 6);
INSERT INTO `permission_api` VALUES (7, '/role/delete', 'GET', 7);

-- 新增角色
INSERT INTO `role` VALUES (1, '管理員', '系統管理員,擁有全部許可權', '2024-02-23');
INSERT INTO `role` VALUES (2, '普通使用者', '普通使用者,擁有部分許可權', '2024-02-23');

-- 新增使用者
INSERT INTO `user` VALUES (1, 'admin', 'admin123', '我是管理員', '2024-02-23');
INSERT INTO `user` VALUES (2, 'user1', 'user111', '我是 user1', '2024-02-23');
INSERT INTO `user` VALUES (3, 'user2', 'user222', '我是 user2', '2024-02-23');

-- 新增使用者角色連結資料：1 個管理員和 2 個使用者

INSERT INTO `user_role` VALUES (1, 1, 1);
INSERT INTO `user_role` VALUES (2, 2, 2);
INSERT INTO `user_role` VALUES (3, 3, 2);

-- 新增角色許可權連結資料
--- 管理員擁有全部許可權,使用者擁有讀寫許可權,沒有刪除許可權
-- 管理員【1,2,3,4,5,6,7】
-- 使用者 1【1,3,4】
-- 使用者 2【1,5,6】
INSERT INTO `role_permission` VALUES (1, 1, 1);
INSERT INTO `role_permission` VALUES (2, 1, 2);
INSERT INTO `role_permission` VALUES (3, 1, 3);
INSERT INTO `role_permission` VALUES (4, 1, 4);
INSERT INTO `role_permission` VALUES (5, 1, 5);
INSERT INTO `role_permission` VALUES (6, 1, 6);
INSERT INTO `role_permission` VALUES (7, 1, 7);
INSERT INTO `role_permission` VALUES (8, 2, 1);
INSERT INTO `role_permission` VALUES (9, 2, 3);
INSERT INTO `role_permission` VALUES (10, 2, 4);
INSERT INTO `role_permission` VALUES (11, 3, 1);
INSERT INTO `role_permission` VALUES (12, 3, 5);
```

```
INSERT INTO `role_permission` VALUES (13, 3, 6);
```

有了真實資料後，我們就可以替換 6.2 節中 UserService 遺留的模擬資料邏輯，改用 Repository 來查詢使用者資料，程式如下：

```
import { Injectable } from '@nestjs/common';
import { InjectRepository } from '@nestjs/typeorm';
import { User as UserEntity } from './entities/user.entity';
import { Repository } from 'typeorm';

export type User = {
  id: number;
  userName: string;
  password: string;
};

@Injectable()
export class UserService {
  constructor(
    @InjectRepository(UserEntity)
    private readonly userRepo: Repository<UserEntity>,
  ) {}
  async findOneByUserName(userName: string): Promise<User | undefined> {
    return this.userRepo.findOne({where: { userName }});
  }
}
```

同時，在 AppController 中新增路由方法，用於模擬管理員和普通使用者的操作請求，程式如下：

```
// 新增獲取使用者列表的方法
@Get('user/list')
getUserList() {
  return [];
}

// 新增刪除使用者的方法
@Get('user/delete')
deleteUser() {
  return " 刪除成功 ";
}
```

6.3.8 測試效果

接下來，我們透過 curl 工具登入管理員帳號獲取 access_token，先後請求 user/list 介面和需要管理員許可權的 user/delete 介面，如圖 6-19 所示。

```
~  20:22:55
$ curl -X POST http://localhost:3000/auth/login -d '{"userName": "admin", "password": "admin123"}' -H
"Content-Type: application/json"
{"access_token":"eyJhbGciOiJIUzI1NiIsInR5cCI6IkpXVCJ9.eyJ1c2VybmFtZSI6ImFkbWluIiwic3ViIjoxLCJpYXQiOjE
3MDg2OTA5ODMsImV4cCI6MTcwOTI5NTc4M30.SUYhJYhnFg3zrxLdqrNvQnGbygNy2eljVITOuNvdDIs"}

~  20:23:03
$ curl -X GET http://localhost:3000/user/list -H "Authorization: Bearer eyJhbGciOiJIUzI1NiIsInR5cCI6I
kpXVCJ9.eyJ1c2VybmFtZSI6ImFkbWluIiwic3ViIjoxLCJpYXQiOjE3MDg2OTA5ODMsImV4cCI6MTcwOTI5NTc4M30.SUYhJYhnF
g3zrxLdqrNvQnGbygNy2eljVITOuNvdDIs"
[]

~  20:23:30
$ curl -X GET http://localhost:3000/user/delete -H "Authorization: Bearer eyJhbGciOiJIUzI1NiIsInR5cCI
6IkpXVCJ9.eyJ1c2VybmFtZSI6ImFkbWluIiwic3ViIjoxLCJpYXQiOjE3MDg2OTA5ODMsImV4cCI6MTcwOTI5NTc4M30.SUYhJYh
nFg3zrxLdqrNvQnGbygNy2eljVITOuNvdDIs"
刪除成功
```

▲ 圖 6-19 管理員請求結果

從圖中可以明顯看出，管理員帳號存取時能夠正常傳回預期的結果。同樣地，如果使用普通使用者帳號登入並發送相同的請求，其效果將如圖 6-20 所示。

```
~  20:35:02
$ curl -X POST http://localhost:3000/auth/login -d '{"userName": "user1", "password": "user111"}' -H
"Content-Type: application/json"
{"access_token":"eyJhbGciOiJIUzI1NiIsInR5cCI6IkpXVCJ9.eyJ1c2VybmFtZSI6InVzZXIxIiwic3ViIjoyLCJpYXQiOjE
3MDg2OTE3MjIsImV4cCI6MTcwOTI5NjUyMn0.OW4_x1UqHC8jUsc2YnwUXD9DgxM_FEEKLgIJ53lxZ3I"}

~  20:35:22
$ curl -X GET http://localhost:3000/user/list -H "Authorization: Bearer eyJhbGciOiJIUzI1NiIsInR5cCI6I
kpXVCJ9.eyJ1c2VybmFtZSI6InVzZXIxIiwic3ViIjoyLCJpYXQiOjE3MDg2OTE3MjIsImV4cCI6MTcwOTI5NjUyMn0.OW4_x1UqH
C8jUsc2YnwUXD9DgxM_FEEKLgIJ53lxZ3I"
[]

~  20:36:14
$ curl -X GET http://localhost:3000/user/delete -H "Authorization: Bearer eyJhbGciOiJIUzI1NiIsInR5cCI
6IkpXVCJ9.eyJ1c2VybmFtZSI6InVzZXIxIiwic3ViIjoyLCJpYXQiOjE3MDg2OTE3MjIsImV4cCI6MTcwOTI5NjUyMn0.OW4_x1U
qHC8jUsc2YnwUXD9DgxM_FEEKLgIJ53lxZ3I"
{"message":"您無限權存取該介面","error":"Forbidden","statusCode":403}
```

▲ 圖 6-20 普通使用者請求的結果

從圖中可以明顯看出，當普通使用者請求 /user/list 時，能夠傳回預期的結果。然而，當嘗試請求許可權管理員存取的 /user/delete 介面時，系統傳回了 403 錯誤，提示缺少許可權。這表明我們的守衛成功攔截了未授權的請求。

透過這一過程，我們成功實現了基於角色的存取控制（RBAC）系統。我們透過設置守衛和自訂裝飾器來區分不同場景下的介面存取控制，並根據不同角色的許可權集來管理和分發系統許可權。掌握了本節內容後，相信你將完全有能力設計出一套包括超級管理員、總監、財務、行政、銷售、倉管等角色的複雜許可權系統。

第 7 章
系統部署與擴充

系統部署是許多讀者關注的重要環節，同時也是非運行維護（後端）人員相對較少涉足的領域。特別是在涉及前後端統一部署時，關於這一主題的網路資料通常不夠充分。本章將深入探討系統部署所涉及的容器技術—Docker。我們將學習如何在 Nest 中撰寫自訂的 Dockerfile 來建立鏡像，並最終透過 Docker Compose 來實操部署一個完整的 Nest 應用程式。

7.1 快速上手 Docker

在後端應用的部署中，我們通常需要部署多種服務，例如 MySQL 資料庫、Redis 快取服務、Nginx 作為反向代理閘道等，同時還包括我們自己開發的業務服務。如果採用微服務架構，可能還需要部署數十甚至數百個微服務。要將這些應用部署到多台伺服器上時，需要對它們執行一系列相同的環境配置和依賴安裝操作。這不僅工作量巨大，而且容易出錯。

Docker 容器技術可以幫助我們高效率地解決這些問題。本節將介紹 Docker 的安裝和基本使用方法，涵蓋透過命令列介面（CLI）和視覺化工具兩種方式操作 Docker，為後續 Nest 專案的應用提供更好的技術支援。

7.1.1 初識 Docker

在 Docker 普及之前，將後端服務如 Redis、Nginx、MySQL 以及訊息佇列服務部署到多台伺服器時，通常使用虛擬機器（Virtual Machine，VM）實現服務部署和隔離。然而，虛擬機器存在一些明顯的缺點，包括資源消耗大、啟動速度慢以及隔離和擴充的複雜性較高。Docker 等容器技術的出現有效地解決了這些問題，因此迅速獲得了業界的廣泛認可和採用。

Docker 的設計理念圖如圖 7-1 所示。

▲ 圖 7-1 Docker 設計理念

接下來，介紹 Docker 中的三個核心概念。

- 鏡像（Image）：鏡像就像靜止等待運送的集裝箱，它包含了執行某個程式所需的一切相依項、函數庫檔案和配置資訊等，被看作容器的範本。
- 容器（Container）：容器就像運送中的集裝箱，是鏡像執行時期的實例。每個容器都擁有自己的檔案系統、CPU、記憶體和網路資源，是一個獨立的小型虛擬機器，它比傳統虛擬機器更輕量、啟動更快。通俗地說，每個容器都是一個獨立隔離的空間，彼此互不影響。
- 倉庫（Registry）：倉庫類似於集中儲存和分發集裝箱的超級碼頭，匯聚了世界各地的集裝箱，可以把集裝箱分發到世界各地。Docker 中最流行的是 Docker Hub，我們可以從倉庫中拉取（Pull）所需的鏡像或將鏡像推送（Push）到倉庫中以供他人使用。

了解以上概念後，接下來探討 Docker 的工作原理。

Docker 的執行過程如下：首先，從遠端倉庫中拉取（Pull）所需的鏡像檔案到本地環境。然後，利用這個鏡像檔案，Docker 可以建立並啟動一個輕量級的虛擬容器。每個容器都可以獨立執行一個或多個服務，類似於我們在自己的電腦上可以同時執行 MySQL、Redis 等不同的服務。這表示可以在任何支援 Docker 的機器上執行指定的鏡像，從而確保不同環境中都能使用相同配置和依賴的服務。

對 Docker 有了基本的認識之後，下一步我們將安裝並學習如何使用 Docker。

7.1.2 安裝 Docker

Docker 是開放原始碼的商業產品，提供社區版和企業版，個人開發者選擇社區版即可。以 Mac 系統為例（M1 晶片 -Apple Chip），讀者可以選擇適合自身系統的 Docker Desktop 版本進行下載和安裝，如圖 7-2 所示。

▲ 圖 7-2　下載 Docker Desktop

安裝完成後，在命令列視窗輸入「docker -v」或「docker –h」命令，查看是否安裝成功，如圖 7-3 所示。

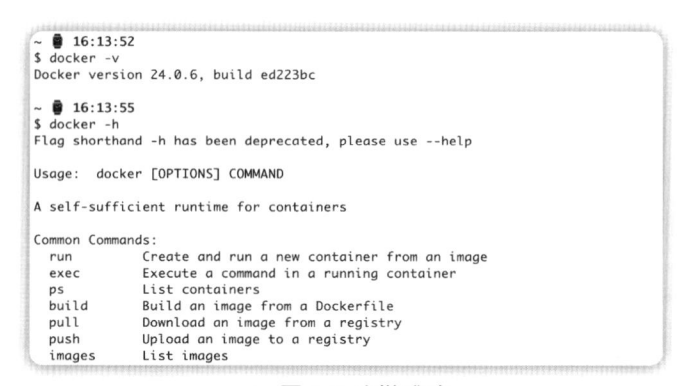

▲ 圖 7-3　安裝成功

接下來，開啟 Docker Desktop 視覺化操作介面，如圖 7-4 所示。

▲ 圖 7-4　Docker 視覺化操作介面

其中，Images 是 Docker 在本地的所有鏡像，Containers 是執行鏡像之後的容器，Volumes 是持久化容器內的應用程式執行時期產生的資料，儲存在宿主機檔案系統中。

7.1.3 Docker 的使用

為了演示 Docker 的使用方法，我們從倉庫中拉取 Nginx 進行測試，在搜尋框中輸入 nginx，如圖 7-5 所示。

點擊 Pull 按鈕將所需的鏡像拉取到本地，成功後，你可以在 Images 中看到這個鏡像，如圖 7-6 所示。

▲ 圖 7-6 本地鏡像列表

接下來，執行這個鏡像。點擊 Run 按鈕之後，填寫容器的執行配置資訊，如圖 7-7 所示。

▲ 圖 7-7 填寫容器的執行配置資訊

其中，docker-nginx-test 是自訂的容器名稱。如果沒有設置，Docker 會自動生成隨機的名稱。容器中的 Nginx 通訊埠預設是 80，這裡需要把宿主機上的通訊埠映射為容器的通訊埠才能進行存取。Volumes 在前文介紹過了，表示資料卷冊。容器執行後的資料會儲存在這個資料卷冊中，並持久化到宿主機的某個目錄下（筆者掛載到 /tmp/docker-nginx-test 目錄下，讀者可以自訂）。在下一次容器執行時期，依舊可以使用這個資料。最後，我們需要為容器設置環境變數，這些環境變數包含容器執行過程中所需的參數。這個過程類似於在執行 Node.js 應用程式時為其設置參數。

點擊 Run 按鈕，可以看到 Nginx 服務成功執行，如圖 7-8 所示。

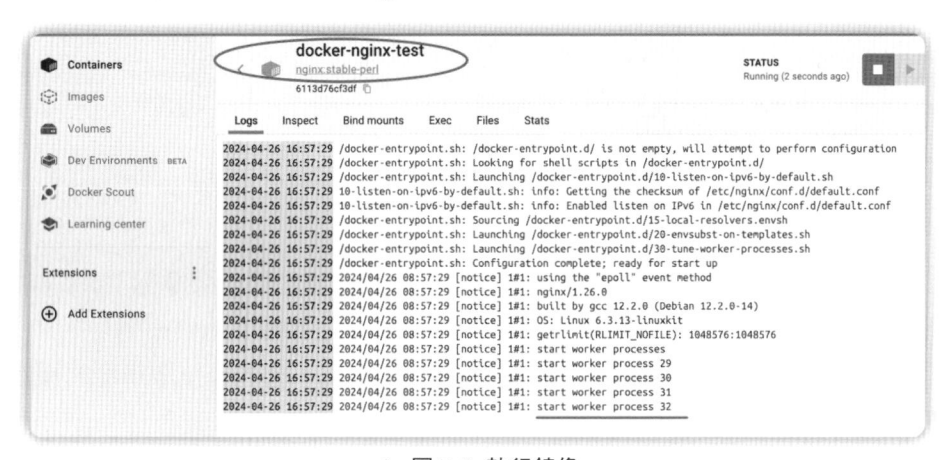

▲ 圖 7-8 執行鏡像

Nginx 通常用於託管靜態資源，我們在 /tmp/docker-nginx-test 目錄下建立一個 index.html 檔案，執行 echo 'hello docker' > ./index.tml 命令，隨後嘗試造訪 http://localhost，結果如圖 7-9 所示。

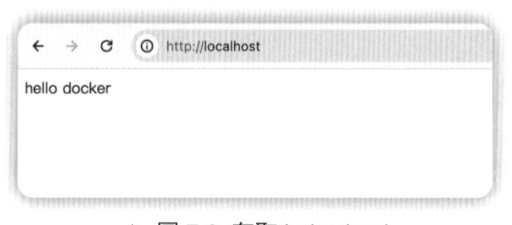

▲ 圖 7-9 存取 index.html

此時，執行在 Docker 容器上的 Nginx 服務可以正常託管我們的靜態資源。切換至 Files 標籤頁，在容器目錄（/usr/share/nginx/html）下可以看到剛剛增加在宿主機目錄（/tmp/docker-nginx-test）下的 index.html 檔案，這表明檔案已經成功被掛載到容器中，如圖 7-10 所示。

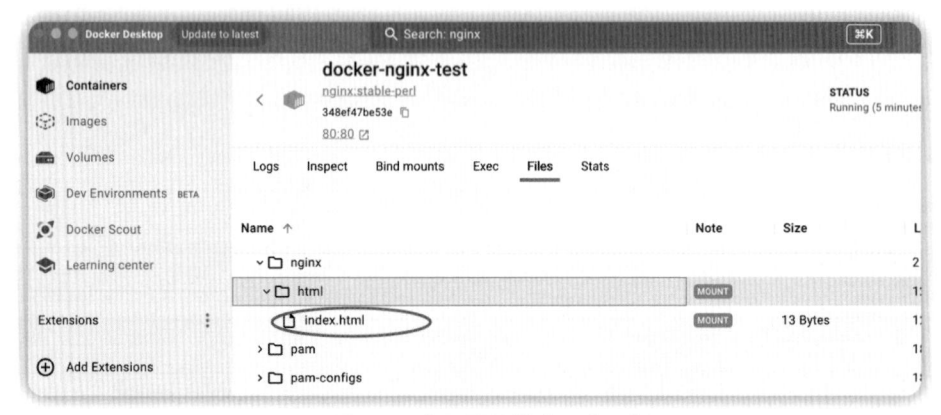

▲ 圖 7-10 容器目錄檔案同步更新

雖然使用視覺化工具操作 Docker 非常方便，但我們仍需要掌握 Docker 的基本命令列操作，以便在沒有視覺化工具的伺服器環境中進行 Docker 管理。

下面透過命令列方式對前面的流程進行類比說明。首先是拉取鏡像的操作，這等於執行以下命令：

```
docker pull nginx:latest
```

執行鏡像前需要配置資訊，執行以下命令：

```
docker run --name docker-nginx-test -p 80:80 -v /tmp/docker-nginx-test:/usr/share/nginx/html -e key1=value1
```

其中，--name 用於設置容器名稱，-p 用於指定映射通訊埠，-v 用於設置資料卷冊掛載目錄，-e 用於設置容器環境變數。

執行成功後會建立一個容器，我們可以執行「docker ps -a」命令獲取本地建立的所有容器列表，如圖 7-11 所示。

▲ 圖 7-11 查看容器列表

然後執行「docker images」或「docker image ls」命令，獲取本地所有的鏡像列表，如圖 7-12 所示。

```
$ docker images
REPOSITORY   TAG                    IMAGE ID       CREATED        SIZE
nginx        stable-perl            ff53ed93b7c9   2 days ago     243MB
nginx        stable-alpine3.17-slim ab6510b890a4   6 months ago   16.6MB
```

▲ 圖 7-12 查看鏡像列表

接著執行「docker logs docker-nginx-test」命令，查看容器的執行日誌，如圖 7-13 所示。

```
2024/04/26 09:10:56 [notice] 1#1: worker process 25 exited with code 0
2024/04/26 09:10:56 [notice] 1#1: worker process 26 exited with code 0
2024/04/26 09:10:56 [notice] 1#1: signal 29 (SIGIO) received
/docker-entrypoint.sh: /docker-entrypoint.d/ is not empty, will attempt to perform configuration
/docker-entrypoint.sh: Looking for shell scripts in /docker-entrypoint.d/
/docker-entrypoint.sh: Launching /docker-entrypoint.d/10-listen-on-ipv6-by-default.sh
10-listen-on-ipv6-by-default.sh: info: IPv6 listen already enabled
/docker-entrypoint.sh: Launching /docker-entrypoint.d/20-envsubst-on-templates.sh
/docker-entrypoint.sh: Launching /docker-entrypoint.d/30-tune-worker-processes.sh
/docker-entrypoint.sh: Configuration complete; ready for start up
2024/04/26 10:10:12 [notice] 1#1: using the "epoll" event method
2024/04/26 10:10:12 [notice] 1#1: nginx/1.24.0
2024/04/26 10:10:12 [notice] 1#1: built by gcc 12.2.1 20220924 (Alpine 12.2.1_git20220924-r4)
2024/04/26 10:10:12 [notice] 1#1: OS: Linux 6.3.13-linuxkit
2024/04/26 10:10:12 [notice] 1#1: getrlimit(RLIMIT_NOFILE): 1048576:1048576
2024/04/26 10:10:12 [notice] 1#1: start worker processes
2024/04/26 10:10:12 [notice] 1#1: start worker process 22
2024/04/26 10:10:12 [notice] 1#1: start worker process 23
2024/04/26 10:10:12 [notice] 1#1: start worker process 24
2024/04/26 10:10:12 [notice] 1#1: start worker process 25
```

▲ 圖 7-13 查看容器日誌

接著執行「docker volume」命令，管理容器的資料卷冊，如圖 7-14 所示。

```
$ docker volume

Usage:  docker volume COMMAND

Manage volumes

Commands:
  create      Create a volume
  inspect     Display detailed information on one or more volumes
  ls          List volumes
  prune       Remove unused local volumes
  rm          Remove one or more volumes

Run 'docker volume COMMAND --help' for more information on a command.
```

▲ 圖 7-14 管理資料卷冊

最後，對容器進行啟動、停止和刪除操作，分別對應的命令為「docker start docker-nginx-test」「docker stop docker-nginx-test」「docker rm docker-nginx-test」，如圖 7-15 所示。

```
~  18:04:29
$ docker start docker-nginx-test
docker-nginx-test

~  18:06:54
$ docker stop docker-nginx-test
docker-nginx-test

~  18:07:36
$ docker rm docker-nginx-test
docker-nginx-test
```

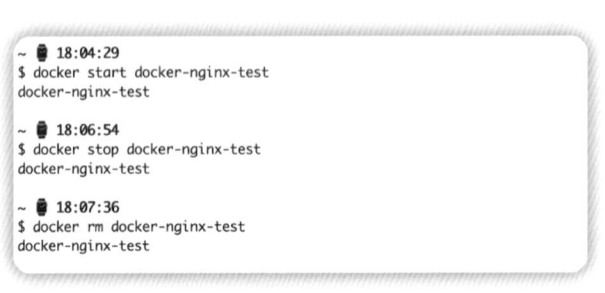

▲ 圖 7-15 容器的啟動、停止與刪除

以上就是 Docker 中常用的命令。在後續使用 Docker 時還會涉及其他命令，在後續的章節中會介紹。

至此，我們對 Docker 有了一個全新的認識。本節透過「Docker Desktop」拉取遠端倉庫的鏡像演示了 Docker 的基本使用方法。然而在實際業務中，我們需要根據自身的需求自訂鏡像。接下來的 7.2 節將介紹如何建構一個自訂的 Docker 鏡像。

7.2 快速上手 Dockerfile

遠端倉庫中擁有豐富的 Docker 鏡像，如 Nginx、MySQL 和 Kafka 等。我們可以直接拉取並使用它們。除此之外，我們還需要學會自訂業務鏡像，以便在開發、測試和生產各階段使用同一份鏡像，確保環境的高度一致，從而減少因環境差異而引起的執行問題。本節將介紹 Dockerfile 的基本概念和語法，同時建立一個實際的 Nest 專案，並為其建構一個自訂的 Docker 鏡像。

7.2.1 Docker 的基本概念

在 Docker 中，我們所說的鏡像是一個二進位檔案。當鏡像執行起來時，它變成一個容器實例，容器實例本身也是一個檔案，稱為容器檔案。鏡像可以從遠端倉庫拉取，使用者也可以自訂鏡像，這時就需要用到 Dockerfile。Dockerfile 是一個文字檔，用於描述如何建構一個 Docker 鏡像，裡面包含若干指令，這些指令按照一定的順序排序，Docker 根據這些指令一步一步生成鏡像檔案。簡單來說，鏡像是基於 Dockerfile 中定義的指令建構生成的產物。

7.2.2 Dockerfile 的基本語法

以某個專案的 Dockerfile 為例，程式如下：

```
FROM node:18
CMD [ "mkdir", "/upload" ]
WORKDIR /servers
COPY . .
ENV TZ=Asia/guangzhou
RUN npm i --registry=https://registry.npmmirror.com -g pnpm && pnpm i && pnpm run build
EXPOSE 8080
```

上面一共有 7 行指令，這些指令的含義解析如下。

- FROM node:18：該 image 檔案指定了基礎鏡像，基於 node:18 鏡像進行建構。
- CMD ["mkdir","/upload"]：該 CMD 指令用於指定容器啟動時需要執行的命令，當容器啟動時，在容器根目錄下建立一個名為 upload 的目錄。一個 Dockerfile 中只能執行一個 CMD 指令。
- WORKDIR /servers：WORKDIR 指令用於設置容器的工作目錄，這表示後續的命令都會在 /servers 目錄下執行。
- COPY ..：COPY 指令用於將建構上下文中的檔案或目錄複寫到鏡像中。
- ENV TZ=Asia/Guangzhou：ENV 指的是設置環境變數，其中 TZ（Time Zone，時區）指定了執行該鏡像容器時預設使用廣州時區。
- RUN npm：RUN 指令用於在鏡像中執行命令。這裡首先使用 npm 全域安裝了 pnpm 套件管理器，然後使用 pnpm 安裝相依並進行打包建構，這些相依會被一同打包到鏡像中。一個 Dockerfile 中可以執行多筆 RUN 指令。
- EXPOSE 8080：EXPOSE 指令宣告了容器在執行時期監聽的網路通訊埠，這裡宣告容器將監聽 8080 通訊埠。

7.2.3 Dockerfile 實踐

接下來，我們建立一個名為 nest-docker 的 Nest 專案進行演示，執行「nest n nest-docker -p pnpm」命令，結果如圖 7-16 所示。

```
$ nest n nest-docker -p pnpm
⚡  We will scaffold your app in a few seconds..
CREATE nest-docker/.eslintrc.js (663 bytes)
CREATE nest-docker/.prettierrc (51 bytes)
CREATE nest-docker/README.md (3347 bytes)
CREATE nest-docker/nest-cli.json (171 bytes)
CREATE nest-docker/package.json (1952 bytes)
CREATE nest-docker/tsconfig.build.json (97 bytes)
CREATE nest-docker/tsconfig.json (546 bytes)
CREATE nest-docker/src/app.controller.spec.ts (617 bytes)
CREATE nest-docker/src/app.controller.ts (274 bytes)
CREATE nest-docker/src/app.module.ts (249 bytes)
CREATE nest-docker/src/app.service.ts (142 bytes)
CREATE nest-docker/src/main.ts (208 bytes)
CREATE nest-docker/test/app.e2e-spec.ts (630 bytes)
CREATE nest-docker/test/jest-e2e.json (183 bytes)

✔ Installation in progress... ☕

🚀 Successfully created project nest-docker
```

▲ 圖 7-16 建立專案

在根目錄下建立 .dockerignore 檔案，排除不需要打包的檔案或目錄，範例如下：

```
.git
node_modules
dist
package-lock.json
npm-debug.log
```

在同一目錄下建立 Dockerfile 檔案，輸入 Docker 鏡像建構指令，程式如下：

```
FROM node:18
WORKDIR /servers
COPY . .
ENV TZ=Asia/Guangzhou
RUN npm set registry=https://registry.npmmirror.com
RUN npm i -g pnpm && pnpm i && pnpm run build
EXPOSE 3000
CMD node dist/main.js
```

在上面的程式中，與前面範例不同的是，將暴露通訊埠設置為 3000，並且在容器啟動時自動執行 CMD 命令「node dist/main.js」以啟動 Nest 服務。

Dockerfile 撰寫完畢之後，執行「docker build -t nest-docker:0.0.1 .」命令建構鏡像，結果如圖 7-17 所示。

```
$ docker build -t nest-docker:0.0.1 .
[+] Building 42.8s (10/10) FINISHED                                    docker:desktop-linux
 => [internal] load .dockerignore                                                       0.0s
 => => transferring context: 2B                                                         0.0s
 => [internal] load build definition from Dockerfile                                    0.0s
 => => transferring dockerfile: 238B                                                    0.0s
 => [internal] load metadata for docker.io/library/node:18                              2.1s
 => [1/5] FROM docker.io/library/node:18@sha256:98218110d09c63b72376137860d1f30a4f61ce029d7de4caf2e8c00f3  0.0s
 => [internal] load build context                                                       0.6s
 => => transferring context: 2.79MB                                                     0.5s
 => CACHED [2/5] WORKDIR /servers                                                       0.0s
 => [3/5] COPY . .                                                                      2.0s
 => [4/5] RUN npm set registry=https://registry.npmmirror.com                           0.5s
 => [5/5] RUN npm i -g pnpm && pnpm i && pnpm run build                                36.5s
 => exporting to image                                                                  1.0s
 => => exporting layers                                                                 1.0s
 => => writing image sha256:06669be0aac7c0caac5dc8d62e88c8e22182b8c1ff95c2ffe58fad882843387b  0.0s
 => => naming to docker.io/library/nest-docker:0.0.1                                    0.0s
```

▲ 圖 7-17 建構鏡像

其中，-t 參數用來指定鏡像檔案的名稱，後面可以用冒號指定標籤，如果不指定，預設表示 latest。命令中的最後一個點（.）指示 Docker 建構上下文的位置，即 Dockerfile 所在的目錄。在上例中，點（.）表示 Dockerfile 位於執行「docker build」命令的目前的目錄。

執行成功之後，執行「docker images」命令，可以看到生成了新的 docker-test 鏡像檔案，如圖 7-18 所示。

```
$ docker images
REPOSITORY      TAG                      IMAGE ID        CREATED         SIZE
nest-docker     0.0.1                    06669be0aac7    2 hours ago     1.24GB
nginx           stable-perl              ff53ed93b7c9    3 days ago      243MB
nginx           stable-alpine3.17-slim   ab6510b890a4    6 months ago    16.6MB
```

▲ 圖 7-18 成功建構鏡像

圖 7-18 中包含鏡像名稱、標籤、ID 和體積等資訊。接下來執行鏡像檔案以啟動一個新的容器，執行以下命令：

```
docker container run -p 8000:3000 -it nest-docker:0.0.1
```

此時可以看到 Nest 服務被啟動了。這表示容器在啟動時成功執行了 CMD 命令，如圖 7-19 所示。

```
$ docker container run -p 8000:3000 -it nest-docker:0.0.1
[Nest] 8  - 04/27/2024, 3:32:53 AM     LOG [NestFactory] Starting Nest application...
[Nest] 8  - 04/27/2024, 3:32:53 AM     LOG [InstanceLoader] AppModule dependencies initialized +5ms
[Nest] 8  - 04/27/2024, 3:32:53 AM     LOG [RoutesResolver] AppController {/}: +5ms
[Nest] 8  - 04/27/2024, 3:32:53 AM     LOG [RouterExplorer] Mapped {/, GET} route +1ms
[Nest] 8  - 04/27/2024, 3:32:53 AM     LOG [NestApplication] Nest application successfully started +2ms
```

▲ 圖 7-19 執行容器並自動啟動服務

在瀏覽器中輸入 http://localhost:8000/ 存取 Nest 服務，可以看到正常傳回了「Hello World！」文字，說明 Nest 服務已成功部署到 Docker 上，如圖 7-20 所示。

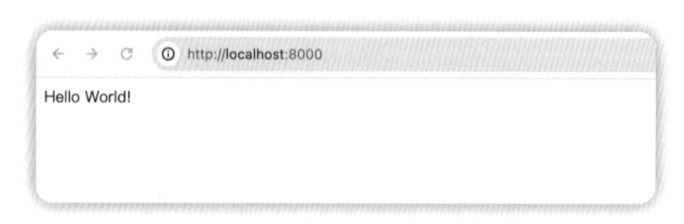

▲ 圖 7-20 存取 Nest 介面服務

透過執行「docker container ls」命令可以看到正在執行中的容器，如圖 7-21 所示。

▲ 圖 7-21 查看執行中的容器

當然，如果你想查看所有容器列表，包含終止執行的容器，可以執行「docker container ls –all」命令。

執行上述命令後，Docker 會輸出容器的 ID、名稱等資訊。我們可以透過這個 ID 來終止容器執行，命令格式為「docker container kill [containerID]」。終止執行的容器仍然會佔用磁碟空間。此時可以使用「docker container rm [containerID]」命令將其刪除。或在容器執行時期加上 --rm 參數，這樣容器在終止執行後將自動刪除其檔案系統和網路設置。對應的命令為「docker container run --rm -p 8000:3000 -it nest-docker」。

除此之外，如果不想在容器啟動時自動啟動 Nest 服務，第一種方式是在 Dockerfile 中刪除 CMD 命令，第二種方式是在啟動容器時增加參數覆蓋 CMD 命令，例如：

```
docker container run --rm -p 8000:3000 -it nest-docker /bin/bash
```

執行上述命令後，會傳回一個命令提示符號，這表示系統當前已經處於容器內的 Bash Shell 環境，並且提示符號是容器內部的 Shell 提示符號，如圖 7-22 所示。

```
$ docker container run -p 8000:3000 -it nest-docker:0.0.1 /bin/bash
root@0ae9ce4bf945:/servers#
root@0ae9ce4bf945:/servers#
root@0ae9ce4bf945:/servers#
```

▲ 圖 7-22 容器 Shell 介面

手動執行「node dist/main」命令即可啟動 Nest 服務，執行結果如圖 7-23 所示。

```
root@0ae9ce4bf945:/servers# node dist/main
[Nest] 7  - 04/27/2024, 4:09:22 AM     LOG [NestFactory] Starting Nest application...
[Nest] 7  - 04/27/2024, 4:09:22 AM     LOG [InstanceLoader] AppModule dependencies initialized +5ms
[Nest] 7  - 04/27/2024, 4:09:22 AM     LOG [RoutesResolver] AppController {/}: +5ms
[Nest] 7  - 04/27/2024, 4:09:22 AM     LOG [RouterExplorer] Mapped {/, GET} route +1ms
[Nest] 7  - 04/27/2024, 4:09:22 AM     LOG [NestApplication] Nest application successfully started +1ms
```

▲ 圖 7-23 手動啟動 Nest 服務

注意：在實際的自動化部署（CI/CD）流程中，如果希望容器一旦啟動就立即提供服務而不需要手動干預，可以設置自動啟動以確保容器啟動後立即進入工作狀態。

最後，透過圖 7-18 可以看到鏡像體積高達 1.24GB，導致鏡像體積過大，這影響了建構性能。為了提升性能，建議採用體積更小的 alpine 版本鏡像。有關 alpine 版本鏡像的詳細資訊，可以參考圖 7-24 所示的網站進行查詢。

https://hub.docker.com/_/node

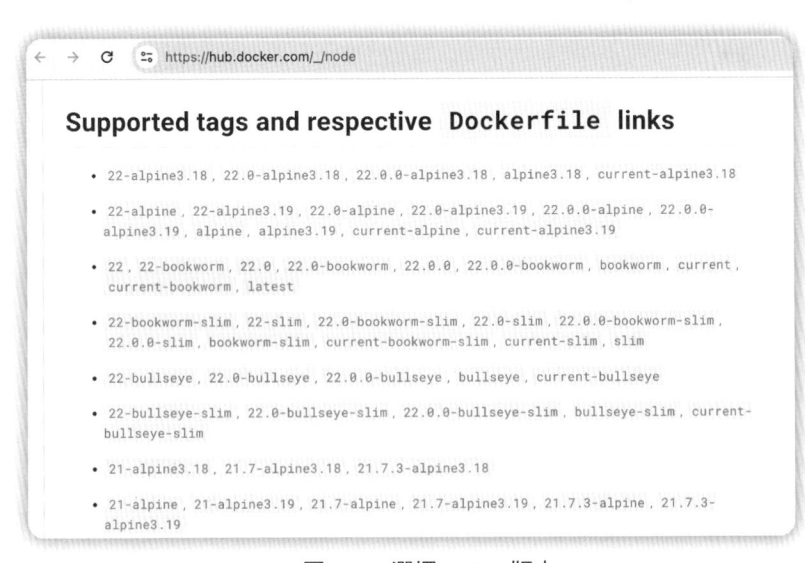

Supported tags and respective Dockerfile links

- 22-alpine3.18, 22.0-alpine3.18, 22.0.0-alpine3.18, alpine3.18, current-alpine3.18
- 22-alpine, 22-alpine3.19, 22.0-alpine, 22.0-alpine3.19, 22.0.0-alpine, 22.0.0-alpine3.19, alpine, alpine3.19, current-alpine, current-alpine3.19
- 22, 22-bookworm, 22.0, 22.0-bookworm, 22.0.0, 22.0.0-bookworm, bookworm, current, current-bookworm, latest
- 22-bookworm-slim, 22-slim, 22.0-bookworm-slim, 22.0-slim, 22.0.0-bookworm-slim, 22.0.0-slim, bookworm-slim, current-bookworm-slim, current-slim, slim
- 22-bullseye, 22.0-bullseye, 22.0.0-bullseye, bullseye, current-bullseye
- 22-bullseye-slim, 22.0-bullseye-slim, 22.0.0-bullseye-slim, bullseye-slim, current-bullseye-slim
- 21-alpine3.18, 21.7-alpine3.18, 21.7.3-alpine3.18
- 21-alpine, 21-alpine3.19, 21.7-alpine, 21.7-alpine3.19, 21.7.3-alpine, 21.7.3-alpine3.19

▲ 圖 7-24 選擇 alpine 版本

在本案例中，我們使用 18-alpine3.18 作為基礎鏡像。在 Dockerfile 中，我們將基礎鏡像修改為 FROM node:18-alpine3.18，並重新建構標籤為 1.0.0 的鏡像。鏡像建構前後的體積大小如圖 7-25 所示。

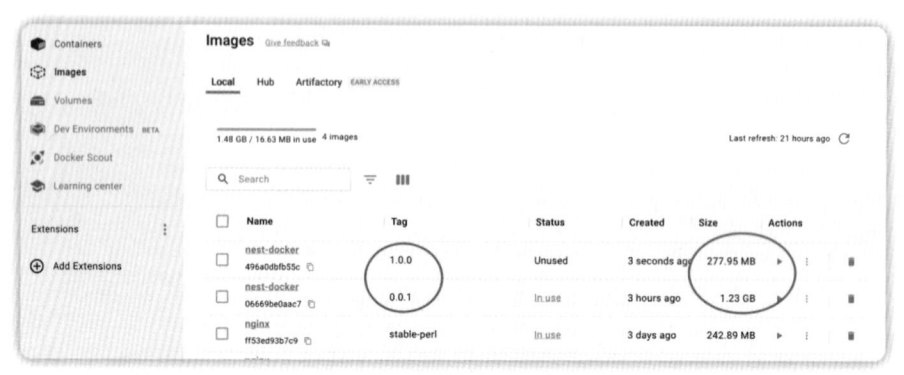

▲ 圖 7-25 最佳化前後體積對比

可以看出，使用不同版本的基礎鏡像建構出的產物大小差異顯著。

透過本節的學習，我們不僅掌握了 Dockerfile 的基本概念和語法，還完成了 Nest 專案的 Dockerfile 檔案撰寫，成功建構了鏡像，並部署執行了 Nest 服務。此外，選擇使用 alpine 版本的基礎鏡像可以顯著提升建構速度並減少鏡像體積，從而最佳化整體的部署效率。

第 3 部分

擴 充 篇

　　第 3 部分我們將深入學習 Nest 框架生態系統中的各種測試方法，探討如何透過單元測試、整合測試和點對點測試來提升專案品質。同時，我們也將了解如何建構高效的日誌系統和完整的錯誤處理機制，以提升系統的穩定性。透過全面掌握這些技術，我們可以打造出更健壯、更可靠的應用程式。

第 8 章
單元測試與點對點測試

本章將深入探討軟體測試中的三個關鍵測試方法：單元測試、整合測試和點對點測試。首先介紹單元測試，這是一種針對程式中最小可測試單元的測試方法，通常由開發人員撰寫，用於驗證程式的行為是否符合預期。接著擴大測試範圍，進行模組的整合測試，這涉及將各個獨立測試完成的模組聚合到一起，確保它們作為一個整體協作工作。最後，探討點對點測試，這是一種模擬真實使用者操作流程的測試方法，旨在驗證整個軟體系統在實際環境中的行為。透過學習這幾種測試方法，讀者將能夠更進一步地保證軟體品質，並確保系統在不同層面上的穩定性和可靠性。

8.1 重新認識單元測試

單元測試是一個經常被討論的話題，但許多人可能對它持有一種敬而遠之的態度。儘管單元測試的重要性被廣泛認可，但在實際應用中，它可能沒有得到充分的利用。本節將從多個角度深入探討單元測試。首先介紹單元測試的基本概念，包括它的定義和重要性。然後探討為什麼我們應該撰寫單元測試，以及它能夠為我們帶來哪些好處。接下來討論在實踐中是應該先撰寫測試還是先撰寫程式，因為這兩種方法在效果上存在顯著差異。最後介紹測試驅動開發（Test-Driven Development，TDD），這是一種在撰寫功能程式之前先撰寫測試程式的程式設計實踐方法，有助提高程式品質和開發效率。

8.1.1 什麼是單元測試

所謂單元，通常指的是程式中的最小可測試部分，例如一個函數、方法或類別中的方法。單元測試是一種軟體測試方法，其目的是獨立地測試應用程式中的每個程式單元，以驗證它們是否按照開發階段的預期設計正常執行。這個過程包括為每個單元撰寫測試用例，執行這些測試，並驗證結果是否符合預期。單元測試在軟體開發生命週期中扮演著至關重要的角色，它有助在開發早期發現和修復錯誤，從而提高程式品質並減少後期的維護成本。

8.1.2 為什麼大部分公司沒有進行單元測試

根據筆者的從業經驗及任職過的公司來看，絕大多數公司對程式設計師撰寫單元測試的要求並不高。筆者認為主要有以下幾點原因。

1. 沒有真正感受到單元測試帶來的好處

很多人可能聽說單元測試很有作用，事實上，甚至一些公司的管理層也未體驗過單元測試帶來的價值。潛意識中可能仍然認為這會影響需求評估的耗時，並增加了單元測試的成本。時常會出現這種聲音：「這個版本把需求完成已經十分困難了，更不用說加上單元測試。」

2. 潛意識認為發現 BUG 是測試人員的工作

在網際網路軟體公司中，測試團隊通常承擔著產品品質的最終把關責任，他們面臨的壓力並不亞於開發團隊。開發人員可能沒有意識到，不撰寫單元測試會增加測試階段的工作量和壓力，因為他們可能認為測試團隊會負責發現並修復 BUG。這種做法導致開發人員在完成基本需求後匆忙進入測試階段，而測試過程中的 BUG 修復可能引發更多的問題，即所謂的「改崩了」現象。這種情況通常源於開發人員在理解需求和進行程式設計時花費了大量時間，而沒有足夠的時間進行單元測試。

為了改善這一狀況，公司可以鼓勵開發團隊在撰寫程式的同時進行單元測試，以提前發現和修復 BUG，減少後期測試階段的壓力。此外，透過提高開發人員對單元測試重要性的認識，可以幫助他們更高效率地完成需求，並提高產品品質。

3. 投入產出比（Return on Investment，ROI）考量

在一些公司中，單元測試可能被認為不如專案交付和滿足客戶需求重要，因此在面臨高產量需求時，可能會被認為不那麼重要。一旦忽視了單元測試，重新開機並實施這項工作將變得更加具有挑戰性。

8.1.3 為什麼要撰寫單元測試

需要明確指出，撰寫單元測試是一項需要投入大量精力和時間的工作。許多人撰寫單元測試是因為公司的要求，目的是達到一定的測試覆蓋率，從而使相關資料看起來更加令人滿意。此外，有些人認為撰寫單元測試的主要目的是減少 BUG 的產生，但在面對複雜功能時，他們可能會選擇性地撰寫一些測試，以求得心理上的安慰，畢竟有品質保證（Quality Assurance，QA）團隊作為最終的測試保障。然而，許多優秀的程式設計師深刻意識到單元測試的巨大作用，並從內心重視其撰寫。單元測試究竟能為我們帶來哪些好處呢？以下是筆者認為的幾個關鍵點。

1. 驗證程式的正確性

通常情況下，我們開發完成後需要自己測試一遍才能交付給測試團隊。最常見的方式是執行程式，簡單測試主要分支場景。如果測試成功，我們可能會認為自己的程式沒有問題。然而，這種方式很難覆蓋所有特殊場景或邊界情況。單元測試可以輕鬆建構各種測試場景，特別是在後續的維護中，直接執行測試用例，以確保我們的程式可以交付測試。

2. 確保重構的可行性

在實際工作中，我們往往不敢隨意重構舊系統，因為不確定它的影響範圍，擔心改動邏輯後影響其他模組的正常執行。單元測試可以幫助我們了解程式改動後的影響，只需執行測試用例，從而讓我們自信地進行程式修改。

3. 加深對業務的理解

單元測試為我們提供了設計測試用例時考慮各種邊界條件和業務場景的機會。換句話說，如果我們不清楚業務的具體實現，就無法知道程式應該如何執行，應該輸出什麼，也就無法撰寫出完整的測試用例。

4. 完善研發流程

在研發閉環流程中，我們應盡可能透過單元測試讓 BUG 在方案設計階段、開發階段和測試階段暴露出來。因為 BUG 發現得越晚，修復成本越高。舉例來說，當 BUG 出現在預上線環境中時，可能涉及環境、資料相容、配置等因素，修復成本隨之增大。

5. 單元測試是最好的開發文件

單元測試覆蓋了介面的所有使用方法和邊界條件，是最佳的範例程式。它清晰地告訴我們輸入什麼資料會得到什麼輸出。範例程式如下：

```
describe('numeral.add', () => {
  test('加法', () => {
    expect(numeral.add(1, 1)).toBe(2);
    expect(numeral.add(0.1, 0.2)).toBe(0.3);
    expect(numeral.add(0.7, 0.1)).toBe(0.8);
    expect(numeral.add(0.2, 0.4)).toBe(0.6);
    expect(numeral.add(35.41, 19.9)).toBe(55.31);
  });
});
```

在上述程式中，numeral.add 方法解決了 JavaScript 中浮點數運算的精度問題，如 0.1 加 0.2 在 JavaScript 中可能會得到 0.30000000000000004，而 numeral.add 確保了結果為 0.3，從而解決了精度遺失的問題。

8.1.4 先撰寫單元測試還是先撰寫程式

先撰寫程式後撰寫單元測試是一種常見的做法，目的是確保所撰寫的程式能夠正常執行。而測試驅動開發（Test-Driven Development，TDD）則是一種不同的方法，它從業務需求出發，先設計測試用例再撰寫程式，這種方法鼓勵開發者從不同的角度思考問題。

採用先撰寫程式後撰寫單元測試的方式，雖然能夠驗證程式的基本功能，但可能無法全面覆蓋所有邊界條件和場景。這種方法可能導致開發者在撰寫測試用例時受到原有程式邏輯的限制，形成思維定勢，從而難以發現潛在的問題。我們通常將這種方法視為一種「防守」策略，即在程式撰寫後透過測試來防止錯誤。

相比之下，TDD 是一種「進攻」策略，它要求開發者在撰寫任何程式之前先從業務場景出發設計測試用例。這種方法與 QA 工程師的工作方式相似，他們透過各種測試手段來驗證程式的正確性，目標是發現並修復潛在的錯誤。TDD 鼓勵開發者在撰寫程式之前就深入思考業務需求和可能的測試場景，從而提高程式的品質和可維護性。

8.1.5 測試驅動開發

測試驅動開發是有步驟的，通常分為紅、綠、重構三部分，圖解流程如圖 8-1 所示。

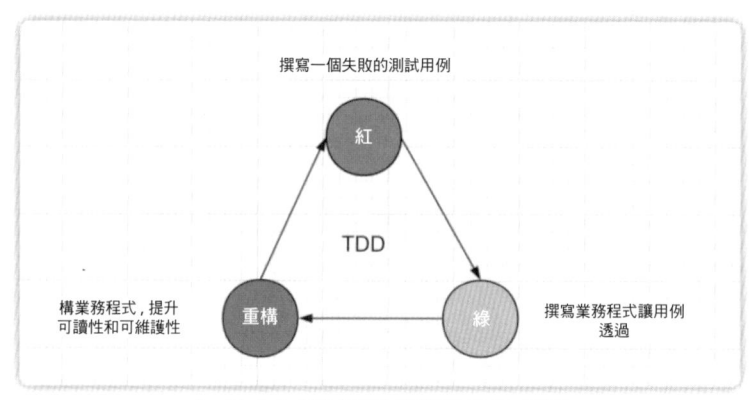

▲ 圖 8-1 測試驅動開發的 TDD 原理

圖 8-1 中的流程詳細說明如下：

（1）撰寫失敗的測試用例：根據業務邏輯撰寫一個新的測試用例，由於未撰寫實際程式，這個測試用例通常無法透過。

（2）撰寫實際程式：撰寫足夠的程式讓測試用例透過。

（3）重構實際程式：重構剛寫的程式，以提高程式的品質和可維護性。如果在重構的過程中測試失敗，我們應該修復問題，使測試再次透過並恢復到綠色狀態。只有當測試全部透過，我們才能認為重構流程成功完成。

有些人可能不太習慣或接受先測試後開發的開發模式，認為在專案初期投入時間撰寫測試並不值得。這種觀點可能源於對 TDD 優勢的不了解。實際上，TDD 在特定場景下可以顯著提高開發效率和程式品質。

TDD 特別適用於以下幾種場景：

（1）撰寫純函數：例如前面提到的 add 方法，它是一個純函數，具有明確的輸入/輸出關係，使得撰寫測試用例變得相對簡單。這類函數通常用於工具方法中。

（2）撰寫新的模組：以 Nest 框架為例，TDD 可以幫助我們實現 API 介面、資料庫讀寫、中介軟體及守衛等功能的開發，確保輸入/輸出的一致性。

（3）修復 BUG：面對 BUG 時，TDD 允許我們透過撰寫測試用例來複現問題，並在測試成功後確保 BUG 被修復。

從這些場景中，我們可以得出結論：TDD 特別適合那些邏輯複雜但輸入輸出關係明確的開發任務。

透過重新檢查單元測試，我們更深入地理解了撰寫單元測試的目的、方法和流程。在下一節中，我們將透過 Nest 框架的實際案例，演示如何遵循 TDD 流程撰寫有效的測試用例，以提升 API 介面的可靠性和穩定性。

8.2 在 Nest 中使用 Jest 撰寫單元測試

在深入理解單元測試的重要性之後，本節將首先介紹主流的單元測試函數庫—Jest。然後，將應用測試驅動開發（TDD）流程，以撰寫一個常見的功能—登入帳號密碼驗證的測試用例。在撰寫測試用例後，我們將根據測試用例的要求來實現具體的業務邏輯程式。最後，為了進一步提升程式的品質和可維護性，我們將對實現的程式進行重構。

透過本節的學習，讀者將了解如何利用 Jest 和 TDD 流程來提高軟體開發的品質和效率。

8.2.1 初識 Jest

1. 基本概念

Jest 是由 Facebook 開發的開放原始碼 JavaScript 測試框架，主要用於自動化測試以確保程式按預期執行。它不僅廣泛應用於前端和後端的測試場景，還支援單元測試、整合測試和端對端測試等多種測試類型。

Jest 提供了包括豐富的斷言函數庫、模擬功能和詳盡的測試報告在內的多種工具，幫助開發者進行有效的測試。此外，它還內建了程式覆蓋率分析、快照測試和並行測試等高級功能，以全面檢測程式品質並提高開發效率。

2. 常用的 API 介紹

Jest 提供了豐富的 API 供開發者使用，下面簡單介紹一下常用的 API 及其作用，以便讓讀者有一個初步的認識。

1）describe

describe 函數用於對測試用例進行分組，以幫助我們更進一步地組織用例程式。我們可以用它來管理同一類型的用例。範例程式如下：

```
const myBeverage = {
  delicious: true,
  sour: false,
};
```

```
describe('my beverage', () => {
  test('is delicious', () => {
    expect(myBeverage.delicious).toBeTruthy();
  });

  test('is not sour', () => {
    expect(myBeverage.sour).toBeFalsy();
  });
});
```

2）test 和 it

test 和 it 函數的作用一致，用於撰寫單一測試用例，多個 test 或 it 函數可以使用 describe 進行包裹。範例程式如下：

```
test('adds 1 + 2 to equal 3', () => {
  expect(1 + 2).toBe(3);
});
```

3）expect

expect 函數用於斷言，對實際的結果進行判斷，即期望結果應該是什麼樣的。範例程式如下：

```
expect(sum(1, 2)).toBe(3);
expect(sum(0.1, 0.2)).toBe(0.3);
```

4）mock

jest.mock 函數用於模擬相依項，例如模擬呼叫外部 API 或模擬模組引用。假設有一個 userManager 使用者管理模組，我們需要測試其中的 getUserName 方法，但它相依於 userService 中的 getUser 方法，用於從資料庫中獲取使用者資訊。在這種情況下，我們就可以用 mock 函數來模擬 userService.getUser 方法，以便在測試中獨立驗證 userManager.getUserName 函數的行為。範例程式如下：

```
// userService.js
export const userService = {
  getUser: (userId) => {
    // 這裡進行資料庫查詢操作
    // 傳回使用者資訊
  }
};

// userManager.js
import { userService } from './userService';
// 獲取使用者名稱
export const getUserName = (userId) => {
  const user = userService.getUser(userId);
```

```js
    return user.name;
};

// userManager.test.js
import { userService } from './userService';
import { getUserName } from './userManager';

// 用 Jest 的 mock 函數來模擬 userService.getUser 方法
jest.mock('./userService', () => ({
  userService: {
    getUser: jest.fn().mockReturnValue({ name: '小銘同學' })
  }
}));

test('getUserName returns the correct user name', () => {
  const userName = getUserName(123);
  expect(userName).toBe('小銘同學');
  expect(userService.getUser).toHaveBeenCalledWith(123);
});
```

5）spyOn

jest.spyOn 函數用於建立一個被監視的函數，追蹤函數的呼叫和傳回結果。依舊使用上面的案例，我們可以監視是否呼叫了 userService.getUser 方法以及是否傳回預期結果。範例程式如下：

```js
// userService.js
export const userService = {
  getUser: (userId) => {
    // 實際的資料庫查詢操作
    // 傳回使用者資訊
    return { name: '小銘同學' }
  }
};

// userManager.js
import { userService } from './userService';

export const getUserName = (userId) => {
  const user = userService.getUser(userId);
  return user.name;
};

// userManager.test.js
import { userService } from './userService';
import { getUserName } from './userManager';

test('getUserName calls userService.getUser and returns the correct user name', () => {
  // 使用 jest.spyOn 來建立一個對 userService.getUser 方法的監視器
```

```
    const spy = jest.spyOn(userService, 'getUser');

    // 呼叫 getUserName 方法
    const userName = getUserName(123);

    // 驗證 userService.getUser 方法是否被呼叫
    expect(spy).toHaveBeenCalledWith(123);

    // 驗證 getUserName 方法是否傳回了正確的使用者名稱
    expect(userName).toBe('小銘同學');
});
```

6）beforeEach 和 afterEach

beforeEach 和 afterEach 函數用於在每個測試用例執行前後執行特定的操作，例如初始化資料庫和清理資料庫，確保每個測試用例都處於相同的初識狀態。範例程式如下：

```
beforeEach(() => {
  // 在每個測試運行之前初始化資料庫
  initializeDatabase();
});

afterEach(() => {
  // 在每個測試運行之後清理資料庫
  clearDatabase();
});
```

7）beforeAll 和 afterAll

beforeAll 和 afterAll 函數用於在所有測試用例執行前後執行特定的操作，例如設置全域測試資料和清理全域資料。範例程式如下：

```
beforeAll(() => {
  // 在所有測試運行之前執行設置全域測試資料的操作
  initializeData();
});

afterAll(() => {
  // 在所有測試運行之後執行清理全域測試資料的操作
  clearData();
});
```

8.2.2 專案準備

現在，我們對 Jest 已經有了大概的了解，接下來在 Nest 中建立使用者測試。在此之前，先明確需要實現的需求，使用 TypeOrm 實現透過帳號和密碼建立使用者，其中密碼必須經過雜湊處理，並且使用者資訊需儲存到資料庫中。

執行「nest n nest-jest -p pnpm」命令建立專案，指定套件管理器為 pnpm，建立成功後如圖 8-2 所示。

```
$ nest n nest-jest -p pnpm
⚡  We will scaffold your app in a few seconds..

CREATE nest-jest/.eslintrc.js (663 bytes)
CREATE nest-jest/.prettierrc (51 bytes)
CREATE nest-jest/README.md (3347 bytes)
CREATE nest-jest/nest-cli.json (171 bytes)
CREATE nest-jest/package.json (1950 bytes)
CREATE nest-jest/tsconfig.build.json (97 bytes)
CREATE nest-jest/tsconfig.json (546 bytes)
CREATE nest-jest/src/app.controller.spec.ts (617 bytes)
CREATE nest-jest/src/app.controller.ts (274 bytes)
CREATE nest-jest/src/app.module.ts (249 bytes)
CREATE nest-jest/src/app.service.ts (142 bytes)
CREATE nest-jest/src/main.ts (208 bytes)
CREATE nest-jest/test/app.e2e-spec.ts (630 bytes)
CREATE nest-jest/test/jest-e2e.json (183 bytes)

✔ Installation in progress... ☕

🚀  Successfully created project nest-jest
```

▲ 圖 8-2　建立專案

接著執行「nest g resource user」命令，建立 user 模組。細心的讀者會發現，Nest CLI 在初始化專案和生成指定模組時，已經為使用者自動建立了各種以 .spec 命名的單元測試檔案，並提供了部分測試用例，讀者可以在此基礎上快速擴充程式邏輯，如圖 8-3 所示。

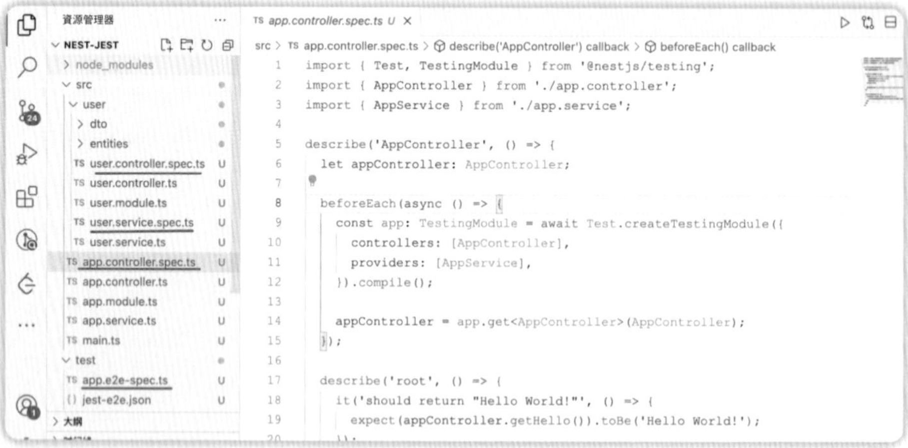

▲ 圖 8-3　自動生成測試檔案

8.2.3 撰寫測試用例

在 user.service.spec.ts 檔案中新增測試用例，程式如下：

```
// 測試 createUser 函數
it('createUser 方法中必須用 hash 對使用者密碼進行加密', async () => {
  // 建立 DTO
  const createUserDto: CreateUserDto = {
    username: 'mouse',
    password: 'mouse123'
  }
  // 儲存使用者
  const saveUser: User = await service.createUser(createUserDto)
  const passFlag: boolean = await bcrypt.compare(
    createUserDto.password,
    saveUser.password
  )
  // 儲存動作應該成功
  expect(saveUser).toBeDefined()
  // 儲存前後帳號應該相同
  expect(saveUser.username).toBe(createUserDto.username)
  // 比較輸入密碼與資料庫中的密碼的雜湊值，一致則為 true
  expect(passFlag).toBeTruthy()
})
```

在上述程式中，我們建立了 CreateUserDto 類別來模擬用戶端請求的資料結構。然後，呼叫 service.createUser 方法以儲存使用者資訊，並獲取傳回的已儲存資料。接下來，對使用者輸入的密碼執行雜湊處理，並將其與資料庫中儲存的雜湊密碼進行比較。最後，對儲存結果、使用者名稱以及雜湊密碼的比較結果進行斷言。

由於尚未實現業務邏輯，包括 createUser 方法的定義、bcrypt 套件的安裝、CreateUserDto 類別的定義以及 User 實體類別的定義，預期該測試用例將失敗。執行「pnpm test:watch」命令以啟動測試的監視模式，測試結果如圖 8-4 所示。

```
FAIL  src/user/user.service.spec.ts
  ● Test suite failed to run

  src/user/user.service.spec.ts:28:36 - error TS2339: Property 'createUser' does not exist on type 'UserService'
  
  28     const saveUser = await service.createUser(createUserDto)
  src/user/user.service.spec.ts:29:37 - error TS2552: Cannot find name 'bcrypt'. Did you mean 'crypto'?
  
  29     const passFlag: boolean = await bcrypt.compare(
  
      node_modules/.pnpm/registry.npmmirror.com+typescript@5.1.3/node_modules/typescript/lib/lib.dom.d.ts:27706:1
  
  27706 declare var crypto: Crypto;
  
  'crypto' is declared here.
  src/user/user.service.spec.ts:30:21 - error TS2339: Property 'password' does not exist on type 'CreateUserDto
  
  30     createUserDto.password,
  src/user/user.service.spec.ts:36:50 - error TS2339: Property 'username' does not exist on type 'CreateUserDto
```

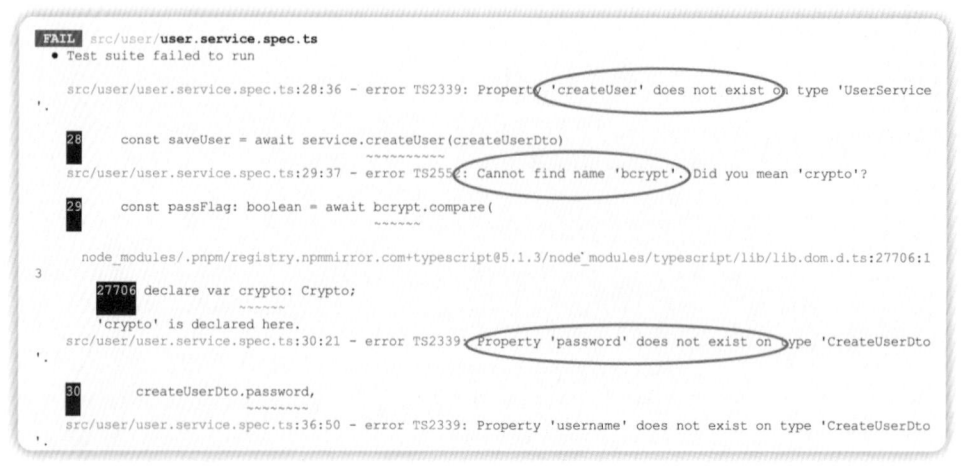

▲ 圖 8-4 執行測試用例

　　根據測試結果的輸出，我們可以看到測試用例未能透過，原因是出現了一系列的相依項未找到或未安裝的錯誤。這通常表示專案中缺少必要的函數庫或模組，需要透過安裝相應的相依來解決。

8.2.4 實現業務程式

　　有了初步執行結果後，接下來撰寫實際業務程式讓測試用例透過。首先安裝 TypeORM、@nestjs/typeorm 和 bcrypt 相依，執行下面的命令：

```
pnpm add typeorm @nestjs/typeorm mysql2 bcryptjs @types/bcryptjs -S
```

　　安裝完成後，使用 typeorm 初始化資料庫，在 AppModule 中引入 TypeOrmModule，程式如下：

```
import { Module } from '@nestjs/common';
import { AppController } from './app.controller';
import { AppService } from './app.service';
import { UserModule } from './user/user.module';
import { TypeOrmModule } from '@nestjs/typeorm'

@Module({
  imports: [
    UserModule,
    // 初始化 MySQL 連接
    TypeOrmModule.forRoot({
      type: 'mysql',
      host: 'localhost',
      port: 3306,
```

```
      username: 'root',
      password: 'jminjmin',
      database: 'nest_jest',
      entities: [__dirname + '/**/*.entity{.ts,.js}'],
      autoLoadEntities: true,
      // 在生產環境中禁止開啟，應該使用資料移轉
      synchronize: true
    })
  ],
  controllers: [AppController],
  providers: [AppService],
})
export class AppModule {}
```

在 VS Code 中的視覺化資料庫外掛程式中建立名為 nest_jest 的資料庫，如圖 8-5 所示。

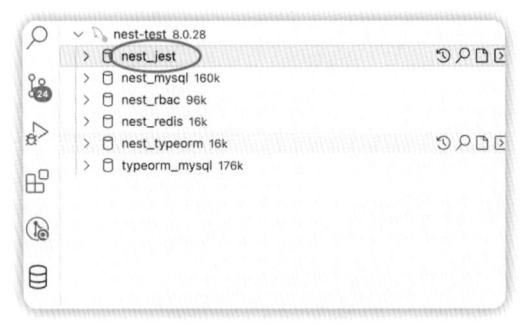

▲ 圖 8-5 建立資料庫

然後，在 user.entity.ts 實體檔案中設計資料表結構，程式如下：

```
import { Column, Entity, PrimaryGeneratedColumn } from "typeorm";

@Entity()
export class User {
  @PrimaryGeneratedColumn()
  id: number;

  @Column()
  username: string;

  @Column()
  password: string;
}
```

同時，別忘記完善建立使用者的 DTO 類別，程式如下：

```
export class CreateUserDto {
  username: string;
```

```
  password: string;
}
```

最後，在 user.service.ts 中實現 createUser 函數邏輯。由於前面我們撰寫了測試用例，這時讀者應該清楚在 createUser 方法中將對密碼進行雜湊加密，然後將資料儲存到資料庫並傳回儲存結果，程式如下：

```
import { Injectable } from '@nestjs/common';
import { CreateUserDto } from './dto/create-user.dto';
import { InjectRepository } from '@nestjs/typeorm';
import { User } from './entities/user.entity';
import { Repository } from 'typeorm';
import bcrypt from 'bcryptjs'

@Injectable()
export class UserService {
  @InjectRepository(User)
  private userRepository: Repository<User>

  async createUser(createUserDto: CreateUserDto) {
    createUserDto.password = await bcrypt.hash(createUserDto.password, 10)
    const saveUser = await this.userRepository.save(createUserDto)
    return saveUser;
  }
}
```

別忘記在 UserController 中定義 createUser 路由方法：

```
@Post()
create(@Body() createUserDto: CreateUserDto) {
  return this.userService.createUser(createUserDto);
}
```

執行 pnpm start:dev 命令啟動服務，程式正常執行，如圖 8-6 所示。

```
[23:34:03] Starting compilation in watch mode...

[23:34:04] Found 0 errors. Watching for file changes.

[Nest] 97642  - 2024/02/26 23:34:05     LOG [NestFactory] Starting Nest application...
[Nest] 97642  - 2024/02/26 23:34:05     LOG [InstanceLoader] TypeOrmModule dependencies initialized +49ms
[Nest] 97642  - 2024/02/26 23:34:05     LOG [InstanceLoader] AppModule dependencies initialized +0ms
[Nest] 97642  - 2024/02/26 23:34:05     LOG [InstanceLoader] TypeOrmCoreModule dependencies initialized +135m
s
[Nest] 97642  - 2024/02/26 23:34:05     LOG [InstanceLoader] TypeOrmModule dependencies initialized +0ms
[Nest] 97642  - 2024/02/26 23:34:05     LOG [InstanceLoader] UserModule dependencies initialized +0ms
[Nest] 97642  - 2024/02/26 23:34:05     LOG [RoutesResolver] AppController {/}: +6ms
[Nest] 97642  - 2024/02/26 23:34:05     LOG [RouterExplorer] Mapped {/, GET} route +1ms
```

▲ 圖 8-6 啟動服務成功

服務正常執行只能說明當前執行的業務程式沒有語法錯誤，但並不表示可以透過單元測試，執行「pnpm test」命令來看看效果，如圖 8-7 所示。

```
FAIL  src/user/user.service.spec.ts
  ● Test suite failed to run

    src/user/user.service.spec.ts:30:37 - error TS2552: Cannot find name 'bcrypt'. Did you mean 'crypto'?

    30     const passFlag: boolean = await bcrypt.compare(
                                           ~~~~~~

      node_modules/.pnpm/registry.npmmirror.com+typescript@5.1.3/node_modules/typescript/lib/lib.dom.d.ts:27706:13
        27706 declare var crypto: Crypto;
                          ~~~~~~
        'crypto' is declared here.

PASS  src/app.controller.spec.ts
FAIL  src/user/user.controller.spec.ts
  ● UserController › should be defined

    Nest can't resolve dependencies of the UserService. Please make sure that the "userRepository" property is available in the current

    Potential solutions:
    - Is RootTestModule a valid NestJS module?
    - If UserRepository is a provider, is it part of the current RootTestModule?
    - If UserRepository is exported from a separate @Module, is that module imported within RootTestModule?
      @Module({
        imports: [ /* the Module containing UserRepository */ ]
      })

       7 |
       8 |   beforeEach(async () => {
    >  9 |     const module: TestingModule = await Test.createTestingModule({
```

▲ 圖 8-7 單元測試結果

在圖 8-7 中，第一個錯誤是由於缺少 bcrypt 相依，可以透過引入 bcrypt 相依即可解決。需要注意的是，這裡需要匯入 bcrypt 的整個模組，使用「import * as bcrypt from 'bcryptjs'」匯入即可。

第二個錯誤表示在 UserController 和 UserService 兩個測試模組中都未找到 userRepository 屬性，原因在於我們使用 TypeOrm 的儲存庫來操作資料庫，而 Jest 無法自動模擬這個相依。為了解決這個問題，我們需要透過模擬來注入外部相依。在兩個測試模組中都加入以下程式：

```
// 模擬外部儲存庫
class UserRepository {
  save(user: User) {
    return user;
  }
}

beforeEach(async () => {
  const module: TestingModule = await Test.createTestingModule({
    providers: [
      UserService,
      // 注入相依
      {
        provide: getRepositoryToken(User),
        useClass: UserRepository
      }
    ],
  }).compile();

  service = module.get<UserService>(UserService);
});
```

此時，主控台中還有一個錯誤，提示接收到的 passFlag 值為 false，與期望值 true 衝突，如圖 8-8 所示。

```
PASS  src/app.controller.spec.ts
PASS  src/user/user.controller.spec.ts
FAIL  src/user/user.service.spec.ts
  ● UserService › createUser方法中必須使用 hash 對使用者密碼進行加密

    expect(received).toBeTruthy()

    Received: false

      53 |       expect(saveUser.username).toBe(createUserDto.username)
      54 |       // 比較輸入密碼與資料庫中的密碼的雜湊值，一致則為 true
    > 55 |       expect(passFlag).toBeTruthy()
         |                        ^
      56 |     })
      57 |   });
      58 |

      at Object.<anonymous> (user/user.service.spec.ts:55:22)

Test Suites: 1 failed, 2 passed, 3 total
Tests:       1 failed, 3 passed, 4 total
Snapshots:   0 total
Time:        2.269 s, estimated 3 s
Ran all test suites related to changed files.
```

▲ 圖 8-8 測試用例錯誤

顯然，這表示使用者輸入的密碼在經過雜湊加密後與儲存在資料庫中的密碼不一致，導致無法透過單元測試。透過審查程式，發現 createUserDto 物件在 createUser 方法呼叫後被修改為資料庫傳回的資料，導致前後加密不匹配。下面來最佳化這段程式：

```
async createUser(createUserDto: CreateUserDto) {
  const password = await bcrypt.hash(createUserDto.password, 10)
  const user: User = new User()
  user.username = createUserDto.username;
  user.password = password

  const saveUser = await this.userRepository.save(user)
  return saveUser;
}
```

透過新建一個 user 實例，我們解決了之前的問題。現在，主控台顯示所有測試用例都已成功透過，如圖 8-9 所示。

```
PASS  src/app.controller.spec.ts
PASS  src/user/user.controller.spec.ts
PASS  src/user/user.service.spec.ts

Test Suites: 3 passed, 3 total
Tests:       4 passed, 4 total
Snapshots:   0 total
Time:        2.86 s, estimated 3 s
Ran all test suites related to changed files.

Watch Usage
 › Press a to run all tests.
 › Press f to run only failed tests.
 › Press p to filter by a filename regex pattern.
 › Press t to filter by a test name regex pattern.
 › Press q to quit watch mode.
 › Press Enter to trigger a test run.
```

▲ 圖 8-9 單元測試成功

可以看到測試用例全部透過，筆者感到非常欣慰，因為這表示我們的程式現在更加健壯，更接近最終的目標。

8.2.5 重構程式

透過不斷迭代修改業務程式和測試用例，我們最終通過了單元測試。這個過程雖然反覆，但確保了程式的品質和功能的正確性。現在，我們可以開始重構（最佳化）業務程式。幸運的是，單元測試可以隨時告訴我們程式是否存在問題，並幫助我們快速定位問題所在。

此時，可能會有讀者問：我的功能已經實現了，接下來應該如何重構或最佳化程式呢？實際上，重構和最佳化並沒有絕對的標準。對於簡單的功能，一旦測試用例透過，我們可能就完成了 TDD 流程。然而，對於複雜的功能，即使測試用例透過，我們仍然可以審查並最佳化程式。

這裡提供一種想法：當測試用例透過後，我們可以審查程式是否符合當前企業要求的開發標準，包括 JavaScript/TypeScript 開發標準、性能標準、介面標準、日誌標準、資料庫標準以及快取標準等。這正是許多企業中 CodeReview 的標準。

舉例來說，為了驗證使用者透過 DTO 發送的資料，我們可以安裝 class-validator 和 class-transformer（版本 0.3.1）相依，並為 username 和 password 增加驗證器。範例程式如下：

```
import { IsNotEmpty, IsString } from "class-validator";

export class CreateUserDto {
  @IsString({message: "username 必須是字串類型 "})
```

```
@IsNotEmpty()
username: string;

@IsString({message: "password 必須是字串類型 "})
@IsNotEmpty()
password: string;
}
```

　　至此，我們已經完成了功能的單元測試部分。透過遵循測試驅動開發（TDD）流程，我們首先分析了需求並進行了功能設計。接下來，我們撰寫了第一個註定失敗的測試用例，這是 TDD 流程的起點。測試失敗提供了回饋，指導我們開始完善業務程式，引入所需的相依，並實現缺失的方法。

　　我們執行了「pnpm test:watch」命令來啟動熱測試服務，它為我們提供了即時的回饋，讓我們知道測試用例何時透過。一旦開發完成並透過所有測試用例，我們根據企業的開發標準進行了程式的重構和最佳化，以提高程式的可讀性、性能和可維護性。

　　在整個 TDD 流程中，我們不斷迭代，直到功能模組開發完成。這個過程不僅幫助我們確保程式品質，也讓我們在開發過程中保持清晰的想法和高效的問題解決能力。TDD 是一種有效的開發實踐，它透過先撰寫測試用例來指導開發，從而提高軟體的可靠性和開發效率。

8.3 整合測試

　　整合測試是一種關鍵的軟體測試方法，它超越了單元測試的範圍，專注於驗證多個單元或元件在合併後的功能是否能夠協作工作。與我們熟悉的 Vue 或 React 元件化開發類似，整合測試可以看作是將獨立開發的元件組合到一起的過程。在元件化開發中，我們分別開發每個元件，然後將它們組裝起來，形成完整的使用者介面。同理，整合測試將不同的模組或服務融合在一起，測試它們作為一個整體時是否能夠正常互動和執行。

　　本節將繼續上一節的程式，透過撰寫整合測試來進一步完善我們的應用程式。我們將採用真實資料庫進行測試，以確保使用者資訊能夠正確地儲存和檢索。這不僅驗證了模組間的互動，也確保了我們的應用程式在實際執行環境中的可靠性。

整合測試的目的是確保應用程式的各個部分在合併後能夠作為一個協調一致的整體執行，從而提高軟體品質和使用者滿意度。透過這種測試，我們可以發現並修復那些在單元測試階段可能被忽略的問題。

8.3.1 撰寫測試用例

為了方便讀者閱讀程式，我們對上一節中的 nest-jest 進行備份，並將其命名為 nest-jest-integration。我們將 UserModule 與 TypeOrmModule 模組結合起來，以測試新使用者是否儲存在資料庫中。

同時，將 user.service.spec.ts 檔案重新命名為 user.integration-spec.ts，以明確這是整合測試檔案。修改後的程式如下：

```
import { Test, TestingModule } from '@nestjs/testing';
import { UserService } from './user.service';
import { CreateUserDto } from './dto/create-user.dto';
import { User } from './entities/user.entity';
import { TypeOrmModule } from '@nestjs/typeorm';
import * as bcrypt from 'bcryptjs'
import { UserModule } from './user.module';

describe('測試 user 服務與資料庫互動', () => {
  let service: UserService;

  beforeEach(async () => {
    const module: TestingModule = await Test.createTestingModule({
      imports: [
        TypeOrmModule.forRoot({
          type: 'mysql',
          host: 'localhost',
          port: 3306,
          username: 'root',
          password: 'jminjmin',
          database: 'nest_jest',
          entities: [__dirname + '/**/*.entity{.ts,.js}'],
          autoLoadEntities: true,
          // 在生產環境中禁止開啟，應該使用資料移轉
          synchronize: true
        }),
        UserModule,
      ],
    }).compile();

    service = module.get<UserService>(UserService);
  });
```

```typescript
it('should be defined', () => {
  expect(service).toBeDefined();
});

// 測試 createUser 函數
it('createUser 方法中必須使用 hash 對使用者密碼進行加密', async () => {
  // 建立 DTO
  const createUserDto: CreateUserDto = {
    username: 'mouse',
    password: 'mouse123'
  }
  // 儲存使用者
  const saveUser: User = await service.createUser(createUserDto)
  const passFlag: boolean = await bcrypt.compare(
    createUserDto.password,
    saveUser.password
  )
  // 儲存動作應該成功
  expect(saveUser).toBeDefined()
  // 儲存前後帳號應該相同
  expect(saveUser.username).toBe(createUserDto.username)
  // 比較輸入密碼與資料庫中的密碼的雜湊值比較，一致則為 true
  expect(passFlag).toBeTruthy()
})
});
```

上述程式建立了一個測試模組，匯入 TypeOrmModule 與 UserModule 並連接了測試資料庫 nest_jest，將兩者組合到一起進行測試。

8.3.2 測試效果

執行「pnpm test」測試命令。當測試用例透過後，我們可以看到指定資料已成功插入到資料庫中，如圖 8-10 所示。

▲ 圖 8-10 成功插入資料

可見，使用者密碼成功被雜湊加密，並且使用者資訊已經成功儲存到資料庫中。這種模組組合稱為整合測試，使用這種方式可以不斷豐富整合測試用例，例如引入 Redis 快取等。

然而，Nest CLI 在生成專案和模組的時候，並不會生成類似 xxx.integration-spec.ts 這樣的模組測試檔案。筆者認為這可能有兩個原因：

- 其一是 Nest 框架的核心邏輯主要由控制器、服務和其他提供者進行處理，而模組作為管理相依的入口，通常並不包含太多業務邏輯。
- 其二是 Nest 提供了端對端測試檔案（xxx.e2e-spec.ts），用於模擬真實使用者的系統行為，提供更全面的測試回饋。

關於端對端測試的更多細節，我們將在下一節介紹。

8.4 點對點測試

與測試單一類或模組的單元測試和整合測試不同，端對端（e2e）測試涵蓋類別和模組在更高聚合程度上的互動，即我們需要知道真實使用者在操作應用系統時會發生什麼事情，端對端測試提供了這種手段，讓我們可以模擬真實使用者的行為，以確保專案在真實場景下正常執行。

本節依舊使用帳號和密碼建立使用者功能，模擬使用者發送 HTTP 請求，對使用者輸入的資訊進行合法性驗證測試，並傳回相應的資訊提示。

8.4.1 撰寫測試用例

沿用 8.2 節的 nest-jest 專案，將其備份並重新命名為 nest-jest-e2e。在 test 目錄下，新建 user.e2e-spec.ts 檔案，用於對 user 模組進行 e2e 測試。我們將使用 supertest 提供的 request 方法來模擬對 createUser 方法的請求，並期望傳回 201 狀態碼。以下是測試用例的程式：

```
import { NestApplication } from '@nestjs/core'
import { Test, TestingModule } from '@nestjs/testing'
import { TypeOrmModule } from '@nestjs/typeorm'
import { AppController } from '../src/app.controller'
import { UserModule } from '../src/user/user.module'
import * as request from 'supertest'
import { AppService } from '../src/app.service'

describe('user module e2e test, () => {
  let app: NestApplication

  beforeAll(async () => {
```

```typescript
    const moduleFixture: TestingModule = await Test.createTestingModule({
      imports: [
        TypeOrmModule.forRoot({
          type: 'mysql',
          host: 'localhost',
          port: 3306,
          username: 'root',
          password: 'jminjmin',
          database: 'nest_jest',
          entities: [__dirname + '/**/*.entity{.ts,.js}'],
          autoLoadEntities: true,
          // 在生產環境中禁止開啟，應該使用資料移轉
          synchronize: true
        }),
        UserModule,
      ],
      controllers: [AppController],
      providers: [AppService]
    }).compile()

    app = moduleFixture.createNestApplication()
    await app.init()
  })

  afterAll(async () => {
    await app.close()
  })

  describe('/user/create (POST)', () => {
    it('should return 201', async () => {
      const requestBody = {
        username: 'mouse100',
        password: '12345678',
      }
      await request(app.getHttpServer())
        .post('/user/create')
        .send(requestBody)
        .expect(201)
    })
  })
})
```

執行「pnpm test:e2e」命令啟動端對端測試，結果如圖 8-11 所示。

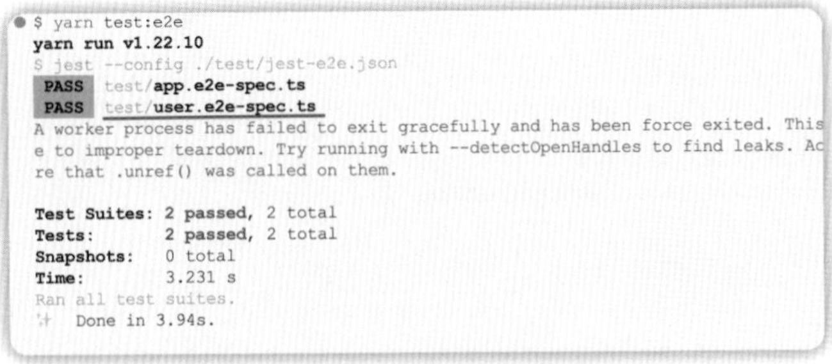

▲ 圖 8-11 執行 e2e 測試

測試用例成功執行並透過測試，在資料庫的視覺化介面中可以看到使用者資訊已成功插入，如圖 8-12 所示。

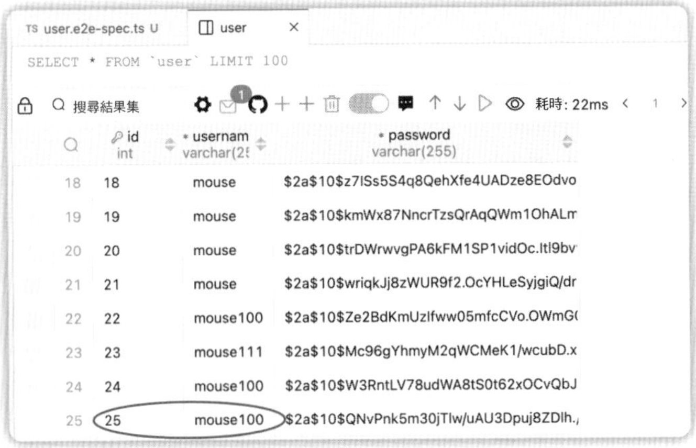

▲ 圖 8-12 e2e 測試插入資料

到目前為止，我們的使用者建立功能還不完善，需要增加更多的場景，比如當帳號已存在時或在根據帳號查詢使用者時帳號輸入錯誤，都應傳回 400 錯誤。接下來我們來補充測試用例：

```
// 建立使用者
describe('/user/create (POST)', () => {
  // 正常建立使用者
  it('should return 201', async () => {
    const requestBody = {
      username: 'mouse200',
      password: '12345678',
    }
    await request(app.getHttpServer())
      .post('/user/create')
```

```
      .send(requestBody)
      .expect(201)
  })

  // 傳入已存在的使用者名稱，傳回 400
  it('should return 400 given exist username', async () => {
    const requestBody = {
      username: 'mouse100',
      password: '12345678',
    }
    await request(app.getHttpServer())
      .post('/user/create')
      .send(requestBody)
      .expect(400)
  })
})

// 查詢使用者
describe('/user/findOne (GET)', () => {
  // 正常查詢使用者
  it('should return 200', async () => {
    await request(app.getHttpServer())
      .get('/user/findOne')
      .query({username: 'mouse111'})
      .expect(200)
  })
  // 查詢不存在的使用者，傳回 400
  it('should return 400 given not exist username', async () => {
    await request(app.getHttpServer())
      .get('/user/findOne')
      .query({username: 'mouse000'})
      .expect(400)
  })
})
```

在原有測試的基礎上，我們新增了三個測試用例，分別用於驗證以下場景：使用者名稱重複時的行為、正常條件下查詢使用者的功能，以及查詢一個不存在的使用者。這些測試用例的設計旨在全面覆蓋使用者管理功能的不同方面。

執行這些測試後，按預期結果應該會看到 1 個測試成功（在測試報告中通常以綠色顯示），而另外 3 個測試未透過（通常以紅色顯示）。出現這種結果是因為我們尚未實現這三個新增測試用例所對應的業務邏輯。

觀察到預期的測試結果後，接下來我們將實現這些測試用例所對應的業務邏輯，並重新執行測試，以驗證實現的正確性。

8.4.2 實現業務程式

我們依然按照 TDD 的「紅 - 綠 - 重構」流程來完善程式。接下來，我們將撰寫業務程式以透過測試（即讓測試變「綠」）。在 createUser 方法中，我們增加了使用者名稱重複的判斷邏輯，程式如下：

```
async createUser(createUserDto: CreateUserDto) {
  // 判斷使用者是否已經存在
  let existUser: User = await this.userRepository.findOneBy({username: createUserDto.username})
  if (existUser && existUser.username) throw new BadRequestException('使用者名稱已存在')

  const password = await bcrypt.hash(createUserDto.password, 10)
  const user: User = new User()
  user.username = createUserDto.username;
  user.password = password

  const saveUser = await this.userRepository.save(user)
  return saveUser;
}
```

接著實現查詢使用者功能，在 UserController 中增加 findOne 路由方法，接收 username 作為查詢參數，並呼叫 UserService 中對應的 findOne 方法進行查詢，程式如下：

```
// user.controller.ts
@Get('/findOne')
findOne(@Query('username') username: string) {
  return this.userService.findOne(username);
}

// user.service.ts
async findOne(username: string) {
  const user: User = await this.userRepository.findOneBy({username: username});
  if (!user || user.username !== username) throw new BadRequestException('當前使用者不存在')
  return user
}
```

再次執行測試，可以看到新增的所有測試用例都通過了，如圖 8-13 所示。

```
$ yarn test:e2e
yarn run v1.22.10
$ jest --config ./test/jest-e2e.json
PASS  test/app.e2e-spec.ts
PASS  test/user.e2e-spec.ts
A worker process has failed to exit gracefully and has been force exited. This is
due to improper teardown. Try running with --detectOpenHandles to find leaks. Act
ensure that .unref() was called on them.

Test Suites: 2 passed, 2 total
Tests:       5 passed, 5 total
Snapshots:   0 total
Time:        3.205 s
Ran all test suites.
✨  Done in 3.79s.
```

▲ 圖 8-13 成功透過 e2e 測試

　　至此，我們已經實現了使用者模組的端對端測試。在實際專案中，我們可以將測試命令增加到 CI/CD 管線中。當檢測到分支更新時，自動執行端對端測試並生成測試覆蓋率，以實現持續整合。

第9章
日誌與錯誤處理

本章將深入探討後端開發中的日誌記錄和錯誤處理機制，這些是建構穩定系統的關鍵組成部分。本章首先強調日誌記錄在監控系統執行和捕捉潛在問題中的重要性，並介紹記錄日誌的最佳實踐，將會幫助使用者確保系統的健康和可維護性。然後，深入探討 Nest 框架提供的內建日誌器。透過學習其基本用法，讀者將了解如何有效地在應用程式中實現日誌記錄功能。為了進一步提升日誌管理的靈活性和功能，最後將介紹 Winston，這是一個功能強大的第三方日誌管理器。我們將探討 Winston 的優勢，包括其高度可配置性、支援多種日誌等級和傳輸機制，以及如何將其整合到 Nest 應用程式中。

透過本章的學習，讀者將獲得必要的知識和工具，以建立一個健壯的日誌記錄系統，這對於維護現代後端服務的穩定性和可靠性至關重要。

9.1 如何在 Nest 中記錄日誌

在日常開發過程中，我們經常使用瀏覽器環境提供的 console 物件來輸出資訊，這有助偵錯工具和記錄日誌。同樣，在伺服器端開發中，Nest 框架也提供了一套強大的基於文字的日誌系統。它不僅可以記錄程式執行中的錯誤、警告和偵錯資訊，還支援訂製化功能，如自訂日期格式、日誌加密、壓縮，以及將日誌上報至外部儲存系統等，滿足多樣化的業務需求。

本節內容不僅教讀者如何使用 Nest 的日誌記錄功能，還將深入探討日誌記錄的重要性，並介紹一系列最佳實踐方法。這些方法包括但不限於日誌等級的合理使用、日誌的可讀性與結構化以及日誌的監控和警告策略。掌握這些方法後，讀者可以有效地將它們應用到實際專案中，提高專案的可維護性和穩定性。

透過本節的學習，讀者將更加深入地理解日誌記錄的價值，並掌握如何透過 Nest 框架的日誌系統來最佳化應用程式的日誌記錄策略。

9.1.1 為什麼要記錄日誌

作為開發人員，在開發環境中偵錯和定位問題相對容易。然而，在生產環境中，通常無法方便地附加偵錯器來追蹤 BUG。此時，日誌記錄的作用就顯得至關重要。

日誌記錄就像一部相機，能夠捕捉程式執行時期的各個方面的事件。如果能夠正確地使用和維護，它將成為我們進行故障排除和程式診斷的強大工具。透過日誌，我們可以回溯問題發生時的具體情況，從而快速定位並解決問題。

在執行 Node 應用程式時，通常需要多個伺服器和元件，如資料庫或 Redis 服務，它們需要相互協作。如果這些服務因某些問題而未能正常執行，沒有日誌記錄，我們可能永遠無法得知伺服器失敗的真正原因。這就像是公路上發生了交通事故，如果沒有監控裝置記錄下事故的經過，那麼事故的調查和責任認定將變得極為困難。

因此，建立一個全面且有效的日誌記錄系統對於任何應用程式的生產執行都是必不可少的。它不僅幫助我們監控應用程式的狀態，還能在出現問題時提供關鍵的資訊，幫助我們快速回應並恢復服務。

9.1.2 內建日誌器 Logger

在開始探索 Nest 內建的 Logger 日誌器之前，我們首先需要建立一個新的 Nest 專案。可以透過執行「nest n nest-logger -p pnpm」命令來完成，結果如圖 9-1 所示。

```
$ nest n nest-logger -p pnpm
⚡  We will scaffold your app in a few seconds..

CREATE nest-logger/.eslintrc.js (663 bytes)
CREATE nest-logger/.prettierrc (51 bytes)
CREATE nest-logger/README.md (3347 bytes)
CREATE nest-logger/nest-cli.json (171 bytes)
CREATE nest-logger/package.json (1952 bytes)
CREATE nest-logger/tsconfig.build.json (97 bytes)
CREATE nest-logger/tsconfig.json (546 bytes)
CREATE nest-logger/src/app.controller.spec.ts (617 bytes)
CREATE nest-logger/src/app.controller.ts (274 bytes)
CREATE nest-logger/src/app.module.ts (249 bytes)
CREATE nest-logger/src/app.service.ts (142 bytes)
CREATE nest-logger/src/main.ts (208 bytes)
CREATE nest-logger/test/app.e2e-spec.ts (630 bytes)
CREATE nest-logger/test/jest-e2e.json (183 bytes)

✔ Installation in progress... ☕

🚀  Successfully created project nest-logger
```

▲ 圖 9-1 建立專案

Nest 中預設開啟了日誌記錄（Logger），當我們執行「pnpm start:dev」命令啟動 Nest 應用時，Logger 會記錄啟動過程中發生的事情，如圖 9-2 所示。

```
[Nest] 72290  - 2024/03/11 08:56:14     LOG [NestFactory] Starting Nest application...
[Nest] 72290  - 2024/03/11 08:56:14     LOG [InstanceLoader] AppModule dependencies initialized +5ms
[Nest] 72290  - 2024/03/11 08:56:14     LOG [RoutesResolver] AppController {/}: +7ms
[Nest] 72290  - 2024/03/11 08:56:14     LOG [RouterExplorer] Mapped {/, GET} route +1ms
[Nest] 72290  - 2024/03/11 08:56:14     LOG [NestApplication] Nest application successfully started +1ms
```

▲ 圖 9-2 開機記錄記錄

可見，日誌資訊包含時間戳記、日誌等級和上下文等重要內容。根據這些資訊，可以確認程式執行的狀態。預設格式如下：

```
[AppName] [PID] [Timestamp] [LogLevel] [Context] Message [+ms]
```

其中，AppName 為應用程式名稱，固定為 Nest；PID 為系統分配的處理程序編號；Timestamp 為輸出的當前系統時間。

Logger 分為多種等級，包括 'log'、'error'、'warn'、'debug' 和 'verbose'。使用者可以指定任意組合來啟用記錄，例如只在出現錯誤或警告時列印日誌。範例程式如下：

```
async function bootstrap() {
  const app = await NestFactory.create(AppModule, {
    logger: ['error', 'warn'],
  });
  await app.listen(3000);
}
```

Context 上下文表示當前日誌的產生階段，例如應用啟動階段或路由程式執行時。

在許多情況下，我們需要使用 Logger 手動記錄自訂事件或訊息，Nest 會以相同的格式和行為把它記錄下來。舉例來說，在 AppService 的 getHello 方法中記錄錯誤，程式如下：

```
import { Injectable, Logger } from '@nestjs/common';

@Injectable()
export class AppService {

  private logger = new Logger(AppService.name);

  getHello(): string {
    this.logger.error('getHello error')
```

```
    return 'Hello World!';
  }
}
```

在上述範例程式中，我們透過向 Nest 的 Logger 類別傳遞上下文來建立了一個 logger 實例。這樣做可以幫助我們在記錄檔項目中標識它們所屬的模組或功能區。接著，我們呼叫了 logger 實例的 error 方法來記錄錯誤資訊。

為了模擬對伺服器的 GET 請求，我們執行了「curl -X GET http://localhost:3000/」命令。這個請求觸發了應用程式中的 getHello 方法，主控台中記錄了一筆格式化的錯誤日誌訊息，如圖 9-3 所示。

```
[Nest] 73613   - 2024/03/11 09:39:26    ERROR [AppService] getHello error
```

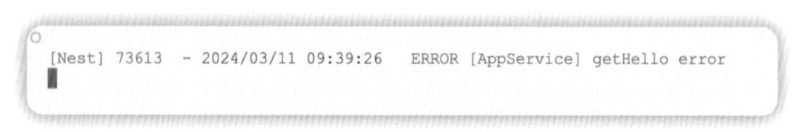

▲ 圖 9-3 記錄錯誤日誌

有些讀者可能會注意到，使用這種方式記錄日誌存在一些問題，例如：

- 每次使用 Logger 都需要先實例化它。
- 無法記錄其他服務資訊，如 ConfigService 環境配置。

那麼，能否進行相依注入呢？答案是可以的。此時，我們需要擴充內建日誌器。

9.1.3 訂製日誌器

我們知道相依注入的範圍是模組等級的。首先，透過執行「nest g module logger --no-spec」命令來建立一個 LoggerModule 模組，操作十分簡單，範例程式如下：

```
import { Module } from '@nestjs/common';
import { Logger } from './logger';

@Module({
  providers: [Logger],
  exports: [Logger]
})
export class LoggerModule {}
```

還需要建立一個 logger 服務提供者，用於擴充日誌方法並獲取配置服務的資訊，如環境變數等應用資訊。此時有兩種方案：

（1）完全訂製一套屬於自己的日誌器，需要實現更底層的 LoggerService 類別；

（2）繼承 ConsoleLogger 類別，這樣我們依然可以使用 Nest 內建的日誌行為。範例程式如下：

```
import { ConsoleLogger, Injectable, Scope } from '@nestjs/common';

@Injectable()
export class Logger extends ConsoleLogger {
  error(message: any, stack?: string, context?: string) {
    // 在這裡呼叫 configService 方法，例如加上環境標識
    ...
    message = message + ' 環境：dev'
    super.error(message, stack, context);
  }
}
```

此時，在 main.ts 中，透過 useLogger 方法將應用程式的預設日誌器替換為訂製的日誌器實例。這樣，當我們在應用程式的任何地方使用 this.logger.error() 列印日誌時，將呼叫自訂 Logger 實例中定義的方法。範例程式如下：

```
async function bootstrap() {
  const app = await NestFactory.create(AppModule, {
    bufferLogs: true,
  });
  app.useLogger(app.get(Logger));
  await app.listen(3000);
}
```

重新執行「curl -X GET http://localhost:3000/」命令請求介面，可以看到記錄的自訂環境配置資訊，如圖 9-4 所示。

```
[Nest] 81470  - 2024/03/11 13:34:30     LOG [InstanceLoader] LoggerModule dependencies initialized
[Nest] 81470  - 2024/03/11 13:34:30     LOG [InstanceLoader] AppModule dependencies initialized
[Nest] 81470  - 2024/03/11 13:34:30     LOG [RoutesResolver] AppController {/}:
[Nest] 81470  - 2024/03/11 13:34:30     LOG [RouterExplorer] Mapped {/, GET} route
[Nest] 81470  - 2024/03/11 13:34:30     LOG [NestApplication] Nest application successfully started
[Nest] 81470  - 2024/03/11 13:34:34   ERROR [AppService] getHello failed 環境：dev
```

▲ 圖 9-4 自訂日誌資訊

9.1.4 記錄日誌的正確姿勢

在後端服務開發中，通常要求開發者及時對某些行為進行日誌記錄，以便應用上線後能及時追蹤執行狀態。為了確保日誌記錄的高效性和準確性，我們需要注意以下幾點：

（1）日誌行為不應該出現異常。當我們使用日誌記錄系統行為時，首先需要保證日誌行為不能出現異常，如空指標異常等。範例程式如下：

```
this.logger.log('服務操作使用者 id=${userService.getUser().id}')
```

其中，userService.getUser() 的執行結果可能為 null，導致獲取 id 操作拋出例外，應避免出現這種情況。

（2）日誌行為不應該產生副作用。日誌應該是無狀態的，日誌行為不應該對業務邏輯產生任何副作用，否則系統行為可能完全出乎意料。範例程式如下：

```
this.logger.log('儲存使用者成功 ${userRepository.save(user)}')
```

其中，執行 userRepository.save 後會在資料庫中新增記錄，應該杜絕這種情況。

（3）日誌資訊不應該包含敏感資訊。日誌用於記錄程式的執行狀態，例如呼叫了哪些函數、發生了什麼錯誤等，需要確保不記錄使用者的帳號和密碼、銀行卡等財務敏感資訊。

（4）日誌記錄盡可能詳細。日誌是開發人員導向的，當描述錯誤時，應該提及具體操作及失敗原因，並提醒接下來應該怎麼做。範例程式如下：

```
this.logger.error('使用者 ${id} 呼叫 getUserInfo 方法失敗，進行重試中 ', error)
getUserInfo()
```

9.1.5 第三方日誌器 Winston

Nest 內建的日誌器可以在開發過程中記錄系統行為，而在生產環境中通常使用專用的日誌統計模組，如 Winston。它能夠滿足特定的日誌記錄要求，包括高級過濾、格式化和集中日誌記錄。下一節將詳細介紹如何在 Nest 中使用 Winston 進行日誌管理。

9.2 Winston 日誌管理實踐

在實際的生產應用中，日誌記錄扮演著關鍵角色。它不僅用於服務問題排除，還廣泛應用於業務資料分析和監控警告平臺的建設。在這些場景中，我們通常不會使用傳統的 console 模組來列印日誌，因為它不支援持久化。相反，我們需要將日誌寫入資料庫或檔案中，以確保它們的長期儲存和可追溯性。

Winston 是一個在 Node.js 領域廣泛使用的日誌框架，它提供了一套強大的功能來幫助我們實現這一目標。Winston 支援將日誌傳輸到多種目的地，包括檔案系統、主控台以及資料庫等。此外，它還支援定義多種日誌等級（如 DEBUG、INFO、WARN、ERROR 等）和自訂日誌格式，以滿足不同場景下的日誌需求。

使用 Winston，我們可以輕鬆地配置和訂製日誌策略，確保日誌記錄的靈活性和有效性。無論是開發人員進行問題診斷還是運行維護團隊進行系統監控，Winston 都能提供必要的支援。

9.2.1 Winston 的基礎使用

為了更進一步地演示在 Nest 中整合 Winston，我們先來建立一個 Nest 專案。執行「nest n nest-winston -p pnpm」命令，效果如圖 9-5 所示。

```
$ nest n nest-winston -p pnpm
⚡  We will scaffold your app in a few seconds..

CREATE nest-winston/.eslintrc.js (663 bytes)
CREATE nest-winston/.prettierrc (51 bytes)
CREATE nest-winston/README.md (3347 bytes)
CREATE nest-winston/nest-cli.json (171 bytes)
CREATE nest-winston/package.json (1953 bytes)
CREATE nest-winston/tsconfig.build.json (97 bytes)
CREATE nest-winston/tsconfig.json (546 bytes)
CREATE nest-winston/src/app.controller.spec.ts (617 bytes)
CREATE nest-winston/src/app.controller.ts (274 bytes)
CREATE nest-winston/src/app.module.ts (249 bytes)
CREATE nest-winston/src/app.service.ts (142 bytes)
CREATE nest-winston/src/main.ts (208 bytes)
CREATE nest-winston/test/app.e2e-spec.ts (630 bytes)
CREATE nest-winston/test/jest-e2e.json (183 bytes)

✔ Installation in progress... ☕

🚀  Successfully created project nest-winston
```

▲ 圖 9-5 建立專案

執行以下命令，生成 logger 模組和服務提供者，專門用於處理日誌模組：

```
nest g module logger --no-spec
nest g provider logger --no-spec
```

其中，在 LoggerModule 中匯出 Logger，以便在其他模組中共用它，程式如下：

```
import { Module } from '@nestjs/common';
import { Logger } from './logger';

@Module({
  providers: [Logger],
  // 匯出 Logger 供其他模組使用
  exports: [Logger]
})
export class LoggerModule {}
```

接下來,重點是實現 logger.ts 的邏輯。在此之前,需要安裝 winston、winston-daily-rotate-file、dayjs、chalk 這些相依。其中,winston 是日誌框架;winston-daily-rotate-file 可以根據日期和大小限制進行記錄檔的輪轉,舊日誌可以根據計數或已用天數進行刪除;dayjs 用於格式化日期;chalk 用於為主控台文字的著色。執行「pnpm add winston winston-daily-rotate-file dayjs chalk -S」命令即可安裝這些相依。

由於我們要訂製一個全新的日誌模組來替換系統內建的日誌器,這裡選擇實現 Nest 底層的 LoggerService 類別來撰寫 Logger 類別,這表示需要在 LoggerService 類別下實現日誌方法。首先,我們來實現在主控台列印日誌,程式如下:

```
import { Injectable, LoggerService } from '@nestjs/common';
import 'winston-daily-rotate-file';
import {
  Logger as WinstonLogger,
  createLogger,
  transports,
} from 'winston';

@Injectable()
export class Logger implements LoggerService {
  private logger: WinstonLogger;
  constructor() {
    this.logger = createLogger({
      level: 'debug',
      transports: [
        // 列印到主控台,生產環境可關閉
        new transports.Console()
      ],
    });
  }
  log(message: string, context: string) {
    this.logger.log('info', message, { context });
  }
  info(message: string, context: string) {
    this.logger.info(message, { context });
  }
  error(message: string, context: string) {
    this.logger.error(message, { context });
  }
  warn(message: string, context: string) {
    this.logger.warn(message, { context });
  }
}
```

上述程式透過 createLogger 方法建立了一個日誌管理器，並指定傳輸方式為主控台，同時實現了常用的 log、info、error、warn 方法。

Winston 支援以下幾種日誌等級統計，它們是：

```
const levels = {
  error: 0,
  warn: 1,
  info: 2,
  http: 3,
  verbose: 4,
  debug: 5,
  silly: 6
};
```

這裡指定為 debug，表示所有小於數字 5 的日誌等級都會被統計並記錄，只有 silly 等級不會被統計和記錄。

然後，在 main.js 中指定使用 Winston 日誌器，並關閉內建日誌器，程式如下：

```
async function bootstrap() {
  const app = await NestFactory.create(AppModule, {
    logger: false,
  });
  // 使用自訂 Logger
  app.useLogger(app.get(Logger));
  await app.listen(3000);
}
```

在 AppService 中注入 Logger 類別，並在 getHello 路由方法中呼叫日誌方法列印結果，程式如下：

```
import { Injectable } from '@nestjs/common';
import { Logger } from './logger/logger';

@Injectable()
export class AppService {
  constructor(private logger: Logger) {}
  getHello(): string {
    this.logger.info('getHello info', AppService.name);
    this.logger.warn('getHello warn', AppService.name);
    this.logger.error('getHello error', AppService.name);
    return 'Hello World!';
  }
}
```

執行「pnpm start:dev」命令啟動服務後，可以在主控台中看到最基礎的日誌記錄，如圖 9-6 所示。

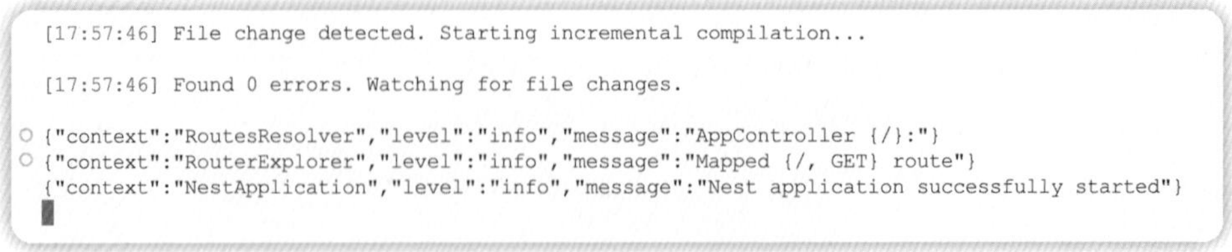

```
[17:57:46] File change detected. Starting incremental compilation...
[17:57:46] Found 0 errors. Watching for file changes.
{"context":"RoutesResolver","level":"info","message":"AppController {/}:"}
{"context":"RouterExplorer","level":"info","message":"Mapped {/, GET} route"}
{"context":"NestApplication","level":"info","message":"Nest application successfully started"}
```

▲ 圖 9-6　在主控台中列印日誌

執行「curl -X GET http://localhost:3000/」命令請求 getHello 介面後，可以在主控台看到成功列印出了日誌，如圖 9-7 所示。

```
{"context":"RoutesResolver","level":"info","message":"AppController {/}:"}
{"context":"RouterExplorer","level":"info","message":"Mapped {/, GET} route"}
{"context":"NestApplication","level":"info","message":"Nest application successfully started"}
{"context":"AppService","level":"info","message":"getHello info"}
{"context":"AppService","level":"warn","message":"getHello warn"}
{"context":"AppService","level":"error","message":"getHello error"}
```

▲ 圖 9-7　自訂日誌記錄

9.2.2 本地持久化日誌

講到這裡，有些讀者可能會有疑惑：Nest 內建的日誌系統缺少時間戳記，並且日誌在服務重新啟動後會遺失，應該如何解決？

為了同時滿足上述需求，我們需要改造 logger.ts 的邏輯，以實現以下幾點要求：

- 日誌記錄加上時間戳記，並格式化日誌輸出。
- 日誌記錄加上顏色進行美化。
- 根據日誌等級進行分類統計，特別是區分錯誤日誌，並按日實現捲動日誌。
- 定期清理記錄檔。

logger.ts 實現邏輯如下：

```typescript
import { Injectable, LoggerService } from '@nestjs/common';
import 'winston-daily-rotate-file';
import {
  Logger as WinstonLogger,
  createLogger,
  format,
  transports,
```

```typescript
} from 'winston';
import * as chalk from 'chalk';
import * as dayjs from 'dayjs';

@Injectable()
export class Logger implements LoggerService {
  private logger: WinstonLogger;
  constructor() {
    this.logger = createLogger({
      level: 'debug',
      transports: [
        // 列印到主控台，生產環境可關閉
        new transports.Console({
          format: format.combine(
            // 顏色
            format.colorize(),
            // 日誌格式
            format.printf(({ context, level, message, timestamp }) => {
              const appStr = chalk.green(`[NEST]`);
              const contextStr = chalk.yellow(`[${context}]`);

              return `${appStr} ${timestamp} ${level} ${contextStr} ${message} `;
            }),
          ),
        }),
        // 儲存到檔案
        new transports.DailyRotateFile({
          // 記錄檔資料夾
          dirname: process.cwd() + '/src/logs',
          // 記錄檔名稱 %DATE% 會自動設置為當前日期
          filename: 'application-%DATE%.info.log',
          // 日期格式
          datePattern: 'YYYY-MM-DD',
          // 壓縮文件，用於定義是否對存檔的記錄檔進行 gzip 壓縮，預設值為 false
          zippedArchive: true,
          // 檔案最大大小，可以是 Bytes、KB、MB、GB
          maxSize: '20M',
          // 最大檔案數，可以是檔案數，也可以是天數，天數加單位 "d"
          maxFiles: '7d',
          // 格式定義，同 winston
          format: format.combine(
            format.timestamp({
              format: 'YYYY-MM-DD HH:mm:ss',
            }),
            format.json(),
          ),
          // 日誌等級，如果不設置，所有日誌將記錄在同一個檔案中
          level: 'info',
```

```
      }),
      // 用上述方法區分 error 日誌和 info 日誌,儲存在不同檔案中,方便排除問題
      new transports.DailyRotateFile({
        dirname: process.cwd() + '/src/logs',
        filename: 'application-%DATE%.error.log',
        datePattern: 'YYYY-MM-DD',
        zippedArchive: true,
        maxSize: '20m',
        maxFiles: '14d',
        format: format.combine(
          format.timestamp({
            format: 'YYYY-MM-DD HH:mm:ss',
          }),
          format.json(),
        ),
        level: 'error',
      }),
    ],
  });
}
log(message: string, context: string) {
  const timestamp = dayjs(Date.now()).format('YYYY-MM-DD HH:mm:ss');
  this.logger.log('info', message, { context, timestamp });
}
info(message: string, context: string) {
  const timestamp = dayjs(Date.now()).format('YYYY-MM-DD HH:mm:ss');
  this.logger.info(message, { context, timestamp });
}
error(message: string, context: string) {
  const timestamp = dayjs(Date.now()).format('YYYY-MM-DD HH:mm:ss');
  this.logger.error(message, { context, timestamp });
}
warn(message: string, context: string) {
  const timestamp = dayjs(Date.now()).format('YYYY-MM-DD HH:mm:ss');
  this.logger.warn(message, { context, timestamp });
}
}
```

在上述程式中,筆者特別強調了日誌系統的關鍵配置部分。為了將日誌輸出到檔案,我們使用了 dailyRotateFile 傳輸方式,這允許日誌按日進行輪轉,從而便於管理和歸檔。此外,我們還使用了 Winston 的 format 方法來定義一組格式化日誌輸出的格式,包括:

```
[AppName] [Timestamp] [LogLevel] [Context] Message
```

輸出的日誌如圖 9-8 所示。

```
[18:31:47] Starting compilation in watch mode...
[18:31:47] Found 0 errors. Watching for file changes.
[NEST] 2024-03-12 18:31:48 info [RoutesResolver] AppController {/}:
[NEST] 2024-03-12 18:31:48 info [RouterExplorer] Mapped {/, GET} route
[NEST] 2024-03-12 18:31:48 info [NestApplication] Nest application successfully started
```

▲ 圖 9-8 格式化日誌

同時，在檔案目錄中可以看到生成的 logs 資料夾，專門用於存放記錄檔，如圖 9-9 所示。

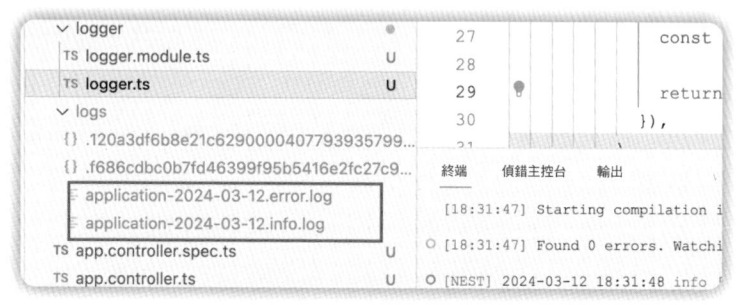

▲ 圖 9-9 生成的 .log 記錄檔

在 .info 記錄檔中可以看到生成了日誌記錄，如圖 9-10 所示。

```
src > logs > ≡ application-2024-03-12.info.log
  1  {"context":"RoutesResolver","level":"info","message":"AppController {/}:","timestamp":"2024-03-12
  2  {"context":"RouterExplorer","level":"info","message":"Mapped {/, GET} route","timestamp":"2024-03
  3  {"context":"NestApplication","level":"info","message":"Nest application successfully started","ti
  4
```

▲ 圖 9-10 生成日誌記錄

再次執行「curl -X GET http://localhost:3000/」命令請求 getHello 介面後，可以在 .error 記錄檔中看到新增了一筆錯誤日誌，如圖 9-11 所示。

```
src > logs > ≡ application-2024-03-12.error.log
  1  {"context":"AppService","level":"error","message":"getHello error","timestamp":
  2
```

▲ 圖 9-11 生成錯誤日誌

至此，我們已經完成了日誌本地持久化的功能。在實際業務中，我們也可以將記錄檔上傳到遠端伺服器，或透過指定 transports.Http 的方式將日誌持久化到資料庫中，以便進行後續的業務資料分析、資料過濾和資料清洗等操作。

9.3 面向切面日誌統計實踐

在實際業務環境中，日誌記錄是一種關鍵的監控和問題診斷工具。雖然在某些特殊業務場景中，開發人員可能會選擇手動記錄日誌，但在處理請求、回應和捕捉服務異常等常見場景時，手動記錄日誌可能效率較低。為了減少容錯的日誌程式並統一日誌格式，通常會採用全域日誌記錄策略，這基於面向切面程式設計（Aspect-Oriented Programming，AOP）的思想來實現。

本節將探討如何在 Nest 應用程式中使用以下全域機制來自動化日誌記錄過程：全域中介軟體，用於自動記錄每個請求的參數、方法和 IP 位址等資訊；全域回應攔截器，用於統一應用程式的回應格式，確保回應的一致性；全域異常篩檢程式，用於捕捉應用程式中的異常，並在合適的位置記錄錯誤日誌。

透過這些機制，我們可以減少日誌記錄的程式量，同時確保關鍵資訊的記錄，提高應用程式的可維護性和可監控性。

9.3.1 中介軟體日誌統計

在 Nest 中，中介軟體是指在路由處理常式之前或之後執行的函數。它們可以操作請求和回應物件，或執行其他執行時期確定的任務。一般來說中介軟體用於收集請求參數、請求本體、請求方法、IP 位址等資訊，這些資訊對於後續的問題排除至關重要。

為了在 nest-winston 專案中實現日誌記錄功能，我們需要新建一個檔案 common/logger.middleware.ts。在這個檔案中，我們將實現一個類別中介軟體，用於記錄每個請求的關鍵資訊。範例程式如下：

```
import { Inject, Injectable, NestMiddleware } from '@nestjs/common';
import { NextFunction, Request, Response } from 'express';
import { Logger } from 'src/logger/logger';

@Injectable()
export class LoggerMiddleware implements NestMiddleware {
  @Inject(Logger)
  private logger: Logger
```

```
  use(req: Request, res: Response, next: NextFunction) {
    const statusCode = res.statusCode;
    const logFormat = `
######################################################################################
####################
RequestOriginal: ${req.originalUrl}
Method: ${req.method}
IP: ${req.ip}
StatusCode: ${statusCode}
Params: ${JSON.stringify(req.params)}
Query: ${JSON.stringify(req.query)}
Body: ${JSON.stringify(req.body)}
######################################################################################
####################
`;

    next();

    if (statusCode >= 500) {
      this.logger.error(logFormat, 'Request LoggerMiddleware');
    } else if (statusCode >= 400) {
      this.logger.warn(logFormat, 'Request LoggerMiddleware');
    } else {
      this.logger.log(logFormat, 'Request LoggerMiddleware');
    }
  }
}
```

上述程式在中介軟體中注入了 Logger 物件，對不同的狀態碼進行判斷，並用 Logger 收集指定格式的日誌記錄，然後在 AppModule 中將中介軟體應用於所有路由上，程式如下：

```
export class AppModule implements NestModule{
  configure(consumer: MiddlewareConsumer) {
    consumer.apply(LoggerMiddleware).forRoutes('*')
  }
}
```

執行「curl -X GET http://localhost:3000/」命令，可以看到在主控台中列印出了上面格式的日誌記錄，如圖 9-12 所示。

```
[NEST] 2024-03-12 23:31:06 info [Request LoggerMiddleware]
######################################################################################
####
RequestOriginal: /
Method: GET
IP: ::ffff:127.0.0.1
StatusCode: 200
Params: {"0":""}
Query: {}
Body: {}
######################################################################################
####
```

▲ 圖 9-12 中介軟體列印日誌

我們成功地使用中介軟體收集了所有的 HTTP 請求資訊,並將其持久化到指定目錄中。由於每個請求都有對應的響應資料,因此通常我們使用攔截器來記錄 HTTP 響應日誌。

9.3.2 攔截器日誌統計

在 common 目錄下建立 response.interceptor.ts 檔案,用於實現使用攔截器記錄 HTTP 回應成功日誌的功能。範例程式如下:

```
import {
  CallHandler,
  ExecutionContext,
  Injectable,
  NestInterceptor,
} from '@nestjs/common';
import { Observable } from 'rxjs';
import { map } from 'rxjs/operators';
import { Logger } from '../logger/logger';

@Injectable()
export class ResponseInterceptor implements NestInterceptor {
  constructor(private readonly logger: Logger) {}

  intercept(context: ExecutionContext, next: CallHandler): Observable<any> {
    const req = context.getArgByIndex(1).req;
    return next.handle().pipe(
      map((data) => {
        const logFormat = `
################################################################################
####################
Request original url: ${req.originalUrl}
Method: ${req.method}
IP: ${req.ip}
Response data: ${JSON.stringify(data)}
################################################################################
####################
`;
        this.logger.info(logFormat, 'Response ResponseInterceptor');
        return data;
      }),
    );
  }
}
```

上述程式收集了請求 URL、請求方式、IP 位址和回應本體資料,並在傳回資料之前列印日誌。然後在 AppModule 中註冊該中介軟體:

```
providers: [
  AppService,
  // 應用攔截器
  {
    provide: APP_INTERCEPTOR,
    useClass: ResponseInterceptor,
  }
],
```

執行「curl -X GET http://localhost:3000/」命令，可以看到在主控台中列印出了回應日誌，如圖 9-13 所示。

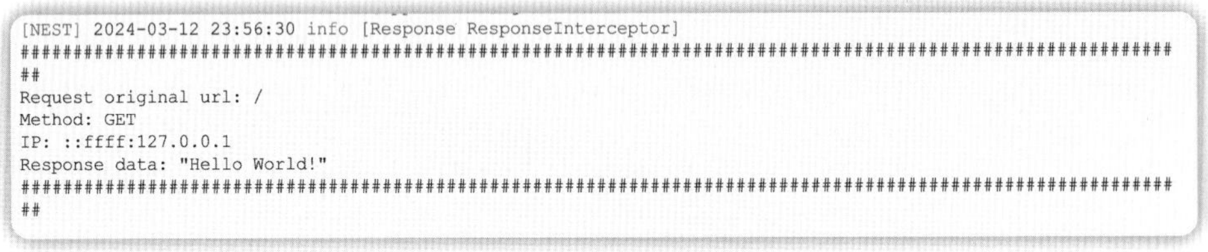

```
[NEST] 2024-03-12 23:56:30 info [Response ResponseInterceptor]
################################################################################
##
Request original url: /
Method: GET
IP: ::ffff:127.0.0.1
Response data: "Hello World!"
################################################################################
##
```

▲ 圖 9-13 攔截器列印日誌

此時可以在 logs 記錄檔中看到最新記錄，如圖 9-14 所示。

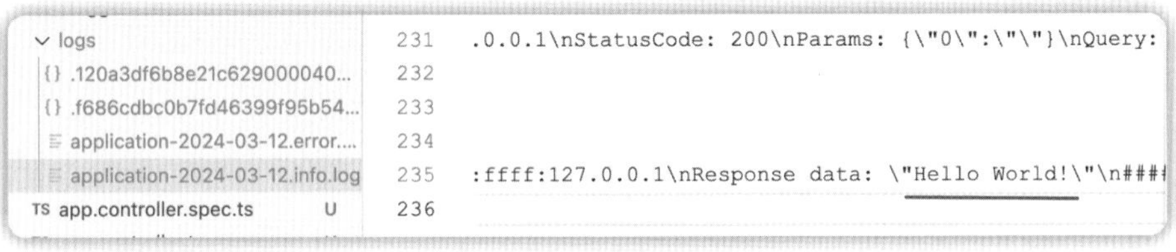

▲ 圖 9-14 成功收集日誌記錄

除正常的 HTTP 請求回應外，我們還需要收集異常日誌資訊。接下來使用篩檢程式來實現這一功能。

9.3.3 篩檢程式日誌統計

在 common 目錄下新建 http-exceptions.filter.ts 檔案，用於實現使用篩檢程式收集 HTTP 異常資訊的功能。範例程式如下：

```
import {
  Catch,
  HttpException,
  ExceptionFilter,
  ArgumentsHost,
  Inject,
```

```
} from '@nestjs/common';
import { Logger } from 'src/logger/logger';

@Catch(HttpException)
export class HttpExceptionsFilter implements ExceptionFilter {
  @Inject(Logger)
  private loggger: Logger;
  catch(exception: any, host: ArgumentsHost) {
    const ctx = host.switchToHttp();
    const response = ctx.getResponse();
    const request = ctx.getRequest();
    const status = exception.getStatus();
    const exceptionResponse = exception.getResponse();
    const logFormat = `
 #######################################################################################
 ####################
Request original url: ${request.originalUrl}
Method: ${request.method}
IP: ${request.ip}
Status code: ${status}
Response: ${
      exception.toString() +
      `(${exceptionResponse?.message || exception.message}) `
    }
 #######################################################################################
 ####################
`;
    this.loggger.error(logFormat, 'HttpException filter');
    response.status(status).json({
      code: status,
      error: exceptionResponse?.message || exception.message,
      msg: `${status >= 500 ? 'Service Error' : 'Client Error'}`,
    });
  }
}
```

上述程式記錄了發生異常的 HTTP 請求的 URL、請求方法、IP 位址、異常狀態碼以及異常資訊，並呼叫 Logger 的 error 方法統計日誌。然後在 AppModule 中註冊該篩檢程式：

```
providers: [
  AppService,
  // 應用攔截器
  {
    provide: APP_INTERCEPTOR,
    useClass: ResponseInterceptor,
  },
  // 應用篩檢程式
```

```
    {
      provide: APP_FILTER,
      useClass: HttpExceptionsFilter,
    },
  ],
```

接下來在 AppService 的 getHello 方法中拋出一個 HTTP 例外：

```
getHello(): string {
  throw new HttpException('getHello 請求異常 ', HttpStatus.EXPECTATION_FAILED)
  return 'Hello World!';
}
```

執行「curl -X GET http://localhost:3000/」命令，在主控台列印出異常日誌，如圖 9-15 所示。

```
[NEST] 2024-03-13 00:30:37 error [HttpException filter ]
################################################################################
##
Request original url: /
Method: GET
IP: ::ffff:127.0.0.1
Status code: 417
Response: HttpException: getHello 請求異常 （getHello 請求異常）
################################################################################
##
```

▲ 圖 9-15 異常篩檢程式列印日誌

在 .error 記錄檔中收集到了異常資訊，可以明確提示開發者異常發生的地方，幫助快速定位問題，如圖 9-16 所示。

```
 8   atus code: 417\nResponse: HttpException: getHello 請求異常（getHello 請求異常）\n##
 9
10   atus code: 417\nResponse: HttpException: getHello 請求異常（getHello 請求異常）\n##
11
12   atus code: 417\nResponse: HttpException: getHello 請求異常（getHello 請求異常）\n##
13
```

▲ 圖 9-16 成功收集異常日誌

至此，我們完成了三種全域日誌的統計和收集。讀者可以利用 Nest 中切面導向的程式設計思想，實現無侵入式的日誌統計效果。請在你的專案中盡情使用這些日誌記錄工具吧！

第 4 部分

Nest 專案實戰篇

　　在本書的最後部分，我們將透過專案實戰把前面學到的知識應用於實際開發。從產品需求分析到技術選型，再到開發實現，你將學習如何從零開始架設一個完整的 Nest 應用。最後，我們將演示如何將專案成功部署到 Docker 環境中，實現從開發到部署的全流程覆蓋。

第 10 章
數字門店管理平臺開發

終於迎來了本書最後的實戰部分。本章將深入探索數字化門店管理系統的開發與應用。如今，隨著數字化工具的廣泛普及，線下門店紛紛採用管理系統以最佳化業務流程，包括營運資料分析、使用者行為視覺化、數字化行銷策略以及人工智慧等。這些工具極大地提升了門店的營運效率和管理效能。

我們將在 Nest 框架中建構一個通用的門店管理平臺，這不僅是對在前面章節中所學知識的綜合運用，也是對專案實戰技能的一次全面展示。透過本專案，讀者將親身體驗到理論知識在實際開發中的應用。

需要特別指出的是，在本章的開發階段，我們將重點講解後端服務的實現細節。雖然前端互動也是系統開發的重要組成部分，但它並非本書的核心內容。因此，我們不會深入展開每個頁面元件的具體實現，尤其是那些功能相似的清單元件、表單元件等。然而，對於系統中的核心功能，筆者仍將不吝篇幅，詳細介紹其實現想法和過程，以確保讀者能夠深刻理解並掌握關鍵技術。

10.1 產品需求分析與設計

許多線下門店都擁有自己的營運管理系統。這些系統中通常包含多種角色，比如在餐飲行業中，有店長、收銀員、服務員、廚師長和廚師等職務角色；而在健身行業中，有管理員、店長、前臺、教練和銷售等職務角色。以餐飲行業為例，管理員擁有最高許可權，可以根據門店人員職位的情況建立角色，並為每個角色分配相應的許可權。不同的角色根據其所擁有的許可權集合來存取和作業系統。

10.1.1 產品需求說明

普通員工可以註冊帳號。在註冊過程中，他們透過郵件獲取驗證碼，並在註冊成功後使用帳號登入系統。普通員工的帳號將被分配最低許可權的功能。管理員帳號是系統內建的，管理員可以為使用者指定角色並分配許可權。

管理員與普通使用者的功能許可權如圖 10-1 所示。

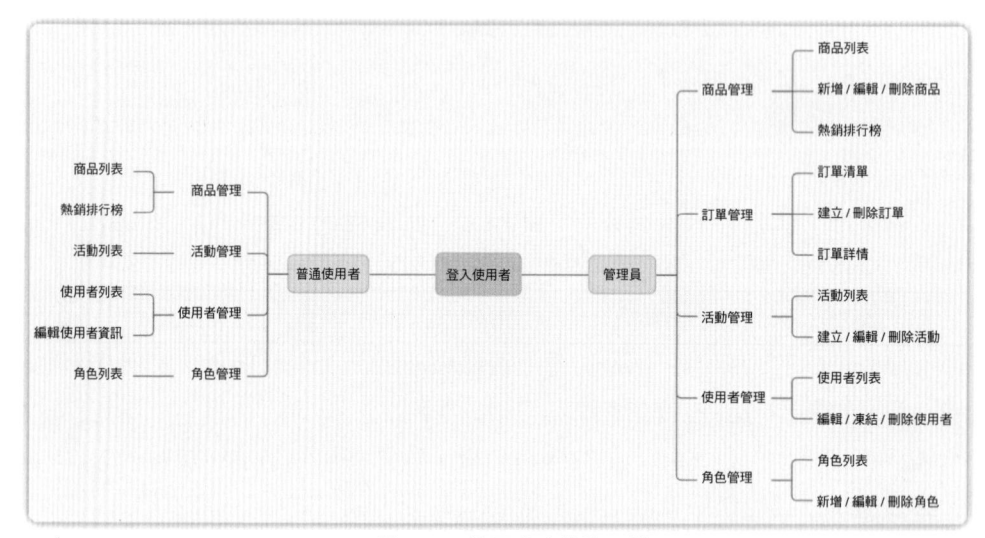

▲ 圖 10-1 使用者功能許可權

系統整體功能可分為五大模組，分別為商品管理、訂單管理、活動管理、使用者管理和角色管理。普通使用者預設只有存取資料列表和編輯個人資訊的許可權，而訂單管理模組包含較為敏感的資訊，普通使用者無權查看該模組。

管理員擁有系統最高許可權，能夠對系統進行新增、修改和編輯等敏感操作。通常情況下，管理員會為指定的普通使用者分配角色，如收銀員，使得擁有這個角色的員工可以在客戶點餐時生成待付款訂單，並在付款完成後將訂單改為已付款。

商品售出後，需要根據銷售量進行統計，並展示出門店熱銷的商品，以便門店進行活動推廣。

值得注意的是，管理員可以凍結指定使用者的帳號，被凍結的使用者不能登入系統，需要管理員解凍後才能繼續存取。

10.1.2 功能原型圖

在分析需求後,接下來對功能模組進行原型設計,以展示系統最基礎的互動效果。

1. 登入註冊

首先在使用者登入頁面輸入帳號和密碼進行登入,如圖 10-2 所示。

▲ 圖 10-2 登入功能

如果忘記密碼,可以透過電子郵件驗證找回密碼,如圖 10-3 所示。

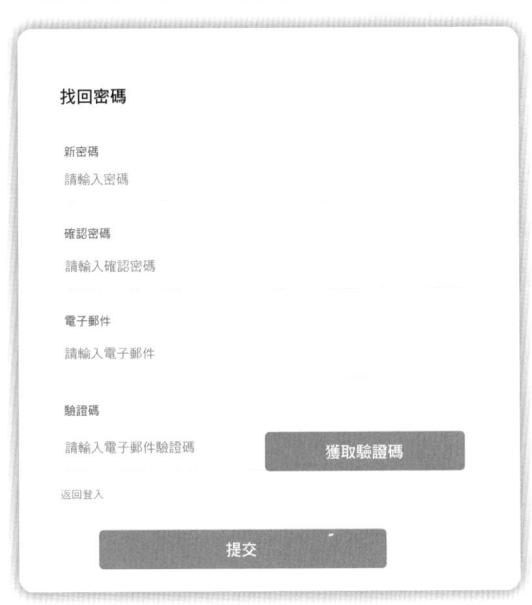

▲ 圖 10-3 找回密碼功能

在註冊頁面需要二次確認密碼並填寫電子郵件，接收到電子郵件驗證碼後，允許註冊並登入，如圖 10-4 所示。

▲ 圖 10-4 註冊功能

2. 商品管理

在商品管理模組，商品清單按照建立時間排序，列表中展示的欄位包括商品編號、商品名稱、商品圖片、狀態、價格、建立時間和操作，使用者可以對商品進行上下架、編輯、刪除操作，如圖 10-5 所示。

▲ 圖 10-5 商品列表

點擊「新建」按鈕，可以新增門店的商品資訊，如圖 10-6 所示。

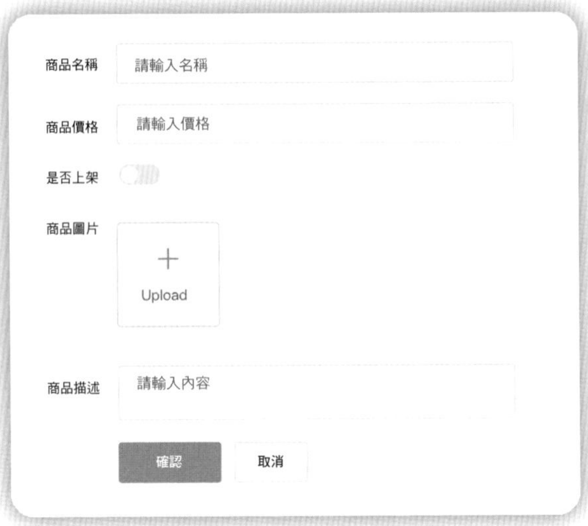

▲ 圖 10-6 新增商品

在頁面左側點擊「排行榜」選項，可以查看當前熱銷的商品情況，如圖 10-7 所示。

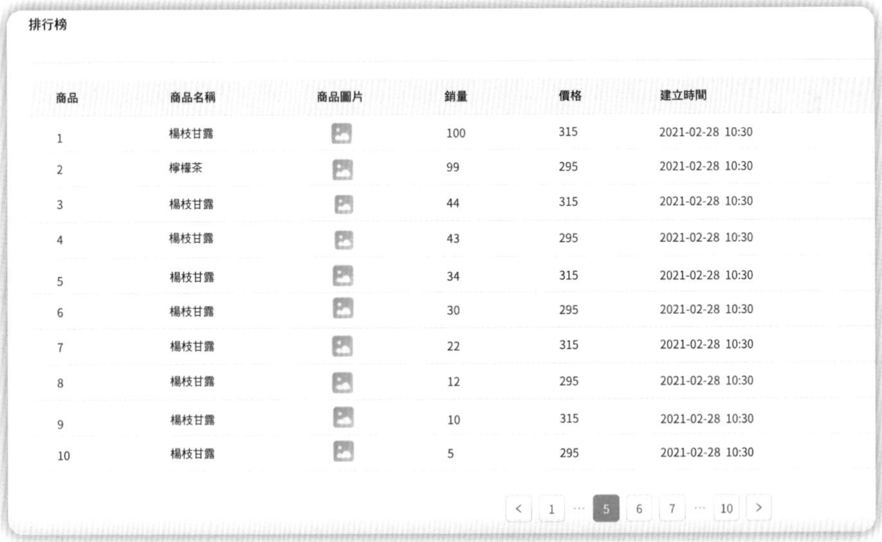

排行榜

商品	商品名稱	商品圖片	銷量	價格	建立時間
1	楊枝甘露		100	315	2021-02-28 10:30
2	檸檬茶		99	295	2021-02-28 10:30
3	楊枝甘露		44	315	2021-02-28 10:30
4	楊枝甘露		43	295	2021-02-28 10:30
5	楊枝甘露		34	315	2021-02-28 10:30
6	楊枝甘露		30	295	2021-02-28 10:30
7	楊枝甘露		22	315	2021-02-28 10:30
8	楊枝甘露		12	295	2021-02-28 10:30
9	楊枝甘露		10	315	2021-02-28 10:30
10	楊枝甘露		5	295	2021-02-28 10:30

< 1 ··· 5 6 7 ··· 10 >

▲ 圖 10-7 商品排行榜

3. 訂單管理

訂單資料通常由收銀員建立，根據狀態可分為未付款、已付款和已取消狀態，清單中展示的欄位包括訂單編號、商品名稱、連結員工、狀態、價格、建立時間和操作，如圖 10-8 所示。

▲ 圖 10-8 訂單清單

點擊「開單收銀」按鈕，建立訂單，表單欄位如圖 10-9 所示。

點擊首列「訂單編號」，可以查看當前訂單的詳細資訊，如圖 10-10 所示。

▲ 圖 10-9 建立訂單　　　　　▲ 圖 10-10 訂單詳情

4. 活動管理

店鋪活動通常由店長建立，可以設置開始時間和結束時間，根據狀態活動可分為未開始、進行中和已結束，店長可以設置活動提前結束。列表中展示的欄位包括活動編號、活動名稱、商品名稱、狀態、開始時間、結束時間和操作，如圖 10-11 所示。

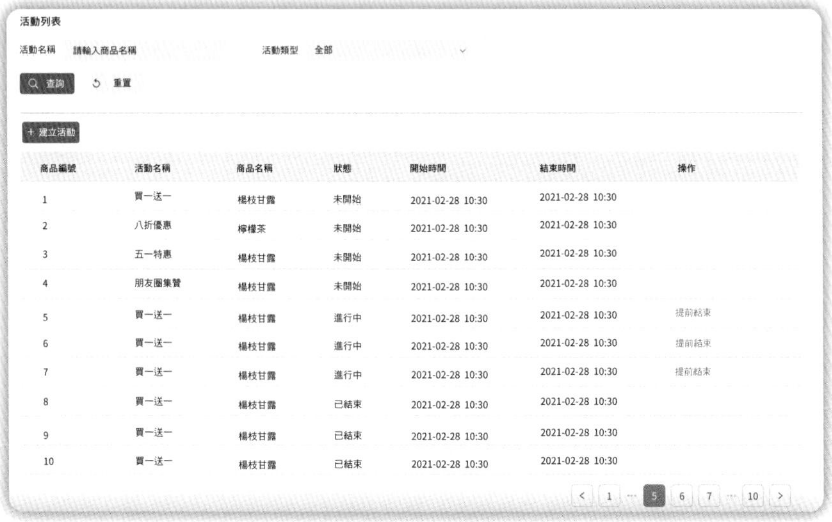

商品編號	活動名稱	商品名稱	狀態	開始時間	結束時間	操作
1	買一送一	楊枝甘露	未開始	2021-02-28 10:30	2021-02-28 10:30	
2	八折優惠	檸檬茶	未開始	2021-02-28 10:30	2021-02-28 10:30	
3	五一特惠	楊枝甘露	未開始	2021-02-28 10:30	2021-02-28 10:30	
4	朋友圈集贊	楊枝甘露	未開始	2021-02-28 10:30	2021-02-28 10:30	
5	買一送一	楊枝甘露	進行中	2021-02-28 10:30	2021-02-28 10:30	提前結束
6	買一送一	楊枝甘露	進行中	2021-02-28 10:30	2021-02-28 10:30	提前結束
7	買一送一	楊枝甘露	進行中	2021-02-28 10:30	2021-02-28 10:30	提前結束
8	買一送一	楊枝甘露	已結束	2021-02-28 10:30	2021-02-28 10:30	
9	買一送一	楊枝甘露	已結束	2021-02-28 10:30	2021-02-28 10:30	
10	買一送一	楊枝甘露	已結束	2021-02-28 10:30	2021-02-28 10:30	

▲ 圖 10-11 活動列表

點擊「建立活動」按鈕，可以新增店鋪活動，並且指定一個商品參加活動，表單欄位如圖 10-12 所示。

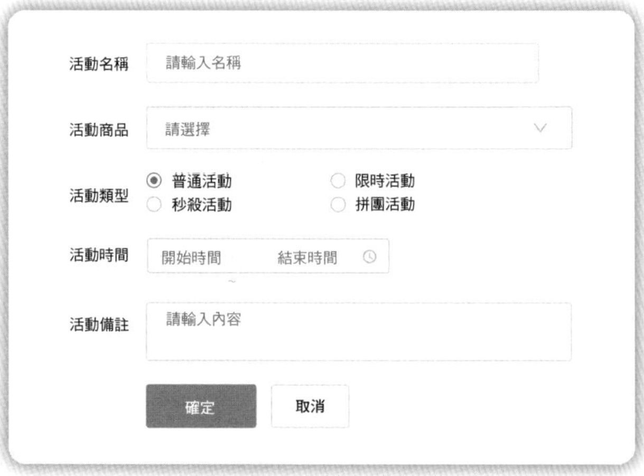

▲ 圖 10-12 建立活動

5. 使用者管理

登入使用者可以在「使用者管理」中查看個人資訊，列表欄位如圖 10-13 所示。

▲ 圖 10-13 使用者列表

點擊「編輯」按鈕，使用者可以修改個人資訊，而管理員可以修改任意使用者的資訊，如圖 10-14 所示。

▲ 圖 10-14 編輯使用者資訊

6. 角色管理

角色通常由管理員建立，每個角色對應一份許可權集，列表欄位如圖 10-15 所示。

▲ 圖 10-15 角色列表

管理員可以重新編輯角色來分配許可權，如圖 10-16 所示。

▲ 圖 10-16 編輯角色

除模組的基本互動外，我們還需要定義一些全域互動共識，例如點擊所有操作欄中的凍結、刪除等敏感操作按鈕，統一需要使用者二次確認後才能操作，如圖 10-17 所示。

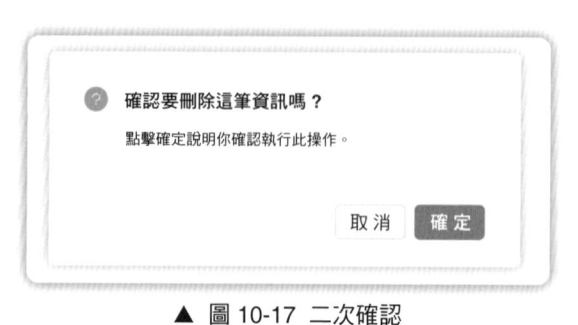

▲ 圖 10-17 二次確認

10.2 技術選型與專案準備

在 10.1 節中，我們從產品經理的角度分析了實際業務需求。本節將帶領讀者轉換角色，從技術的角度出發，分別對前端和後端的技術方案進行選型，引導讀者從 0 到 1 學會架設全端專案。

10.2.1 前端技術選型

為了與各大公司的前端技術堆疊契合，我們選擇主流的 React 作為開發框架，並利用 Ant Design + ProComponents 快速撰寫 UI 頁面。我們採用 TypeScript 作為類型檢測工具，選擇 Vite 作為建構工具，以獲得極致的建構體驗，並使用 Axios 作為 HTTP 請求函數庫。在開發過程中，為了方便偵錯介面，我們採用 VS Code 外掛程式中的 REST Client 作為介面請求工具。有關更多輔助相依的詳細資訊，請查看專案中的 package.json 檔案。

10.2.2 初始化前端專案

首先用 Vite 初始化 React 專案，執行「pnpm create vite store-web-frontend --template react」命令，結果如圖 10-18 所示。

```
$ pnpm create vite store-web-frontend --template react
.../Library/pnpm/store/v3/tmp/dlx-36975    |  Progress: resolved 1, reused 0, downl.../Library/pnpm
/store/v3/tmp/dlx-36975    |  +1 +
.../Library/pnpm/store/v3/tmp/dlx-36975    |  Progress: resolved 1, reused 0, downl.../Library/pnpm
/store/v3/tmp/dlx-36975    |  Progress: resolved 1, reused 0, downlPackages are hard linked from th
e content-addressable store to the virtual store.
  Content-addressable store is at: /Users/jmin/Library/pnpm/store/v3
  Virtual store is at:              ../../../Library/pnpm/store/v3/tmp/dlx-36975/node_modules/.pn
pm
.../Library/pnpm/store/v3/tmp/dlx-36975    |  Progress: resolved 1, reused 0, downl.../Library/pnpm
/store/v3/tmp/dlx-36975    |  Progress: resolved 1, reused 0, downloaded 1, added 1, done

Scaffolding project in /Users/jmin/development/nest-book/第十章/store-web-frontend...

Done. Now run:

  cd store-web-frontend
  pnpm install
  pnpm run dev
```

▲ 圖 10-18 建立 React 專案

　　接下來，安裝「antd @ant-design/pro-components @ant-design/icons axios」相依。其中，antd 是基礎 UI 框架版本，@ant-design/pro-components 是對 antd 的更高等級抽象，配置更高效；@ant-design/icons 是對應的字型圖示庫，axios 則是用於前端的請求函數庫。執行「pnpm add antd @ant-design/pro-components @ant-design/icons axios -S」命令進行安裝。安裝成功後，在專案根目錄下執行「pnpm dev」命令啟動專案，效果如圖 10-19 所示。

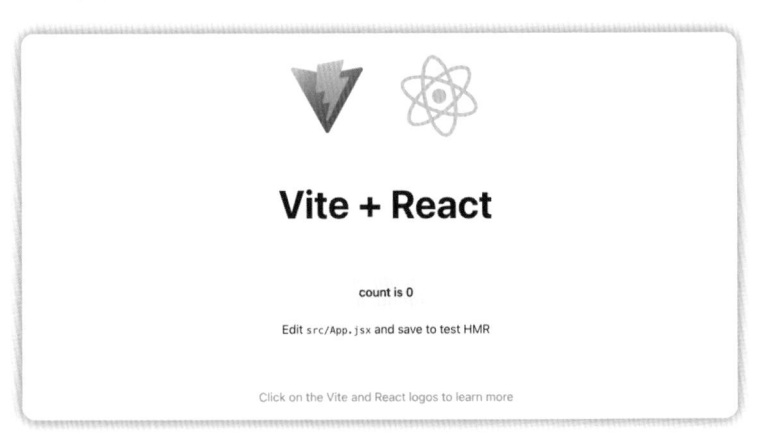

▲ 圖 10-19 預覽效果

10.2.3 前端架構設計

　　前端專案基於 React 框架和 Vite 建構，我們設計了以下的目錄結構：

```
├── README.md
├── config
│   ├── defaultSettings.ts
```

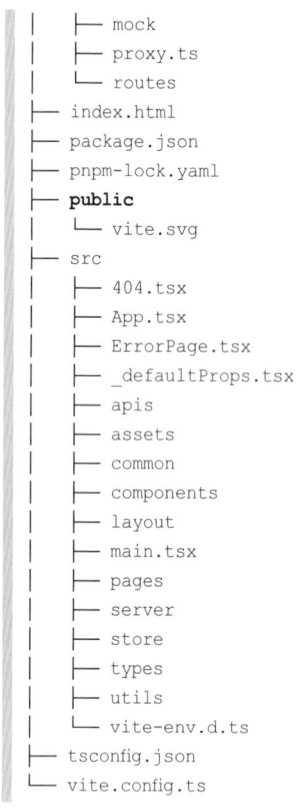

```
│   ├── mock
│   ├── proxy.ts
│   └── routes
├── index.html
├── package.json
├── pnpm-lock.yaml
├── public
│   └── vite.svg
├── src
│   ├── 404.tsx
│   ├── App.tsx
│   ├── ErrorPage.tsx
│   ├── _defaultProps.tsx
│   ├── apis
│   ├── assets
│   ├── common
│   ├── components
│   ├── layout
│   ├── main.tsx
│   ├── pages
│   ├── server
│   ├── store
│   ├── types
│   ├── utils
│   └── vite-env.d.ts
├── tsconfig.json
└── vite.config.ts
```

其中，config 目錄用於配置頁面路由資訊，管理不同開發環境下的代理配置，同時支援以 mock 方式模擬介面資料。在 src 目錄中，根據模組職能設計了不同目錄進行管理：apis 負責統一定義介面，common 用於管理全域通用屬性，layout 用於架設系統骨架，pages 用於管理每個頁面元件。頁面元件需要借助 components 中的積木元件完成架設過程。為了更方便地管理請求工具，在 server 中封裝了 Axios 邏輯，並使用 store 作為資料倉儲進行公共資料管理。最後，結合 utils 和 types 提供開發所需的工具和類型支援。

10.2.4 後端技術選型

在本專案的後端開發中，我們精心挑選了以下技術堆疊來建構穩定、高效、易於維護的服務。

- Nest：選擇 Nest 作為後端服務的開發框架，它提供了模組化和高性能的特點，非常適合建構企業級應用程式。

- MySQL + TypeORM：使用 MySQL 作為資料庫儲存解決方案，並結合 TypeORM 作為物件關係映射（ORM）工具，以簡化資料庫的增刪改查（CRUD）操作。
- RBAC 模型：為了有效管理使用者和許可權，採用基於角色的存取控制（RBAC）模型，以實現靈活的許可權分配和管理。
- Redis：考慮到性能最佳化和回應速度，引入了 Redis 作為資料快取層，這不僅提升了性能，同時也減輕了資料庫的壓力。
- Winston：為了增強系統的穩定性並便於問題排除，採用 Winston 作為日誌記錄工具，它支援豐富的日誌統計功能。
- Docker Compose：在開發完成後，使用 Docker Compose 來部署專案，它簡化了容器化應用程式的部署和管理。
- PM2：採用 PM2 作為處理程序守護工具，確保後端服務的高可用性和穩定性。
- Nginx：使用 Nginx 作為閘道層，它不僅託管系統的靜態資源檔，還提供了負載平衡、請求轉發等功能。

透過這些技術選型，可以確保後端服務的高性能、高穩定性以及良好的可維護性，為建構全端專案打下堅實的基礎。

10.2.5 初始化後端專案

首先用 CLI 初始化 Nest 專案，執行「nest n store-web-backend -p pnpm」命令，結果如圖 10-20 所示。

```
$ nest n store-web-backend -p pnpm
⚡  We will scaffold your app in a few seconds..

CREATE store-web-backend/.eslintrc.js (663 bytes)
CREATE store-web-backend/.prettierrc (51 bytes)
CREATE store-web-backend/README.md (3347 bytes)
CREATE store-web-backend/nest-cli.json (171 bytes)
CREATE store-web-backend/package.json (1958 bytes)
CREATE store-web-backend/tsconfig.build.json (97 bytes)
CREATE store-web-backend/tsconfig.json (546 bytes)
CREATE store-web-backend/src/app.controller.spec.ts (617 bytes)
CREATE store-web-backend/src/app.controller.ts (274 bytes)
CREATE store-web-backend/src/app.module.ts (249 bytes)
CREATE store-web-backend/src/app.service.ts (142 bytes)
CREATE store-web-backend/src/main.ts (208 bytes)
CREATE store-web-backend/test/app.e2e-spec.ts (630 bytes)
CREATE store-web-backend/test/jest-e2e.json (183 bytes)

✔ Installation in progress... ☕

🚀  Successfully created project store-web-backend
👉  Get started with the following commands:

$ cd store-web-backend
$ pnpm run start
```

▲ 圖 10-20 建立 Nest 專案

然後安裝相依，操作資料庫需要用到 mysql2、typeorm、@nestjs/typeorm，實現介面快取需要用到 Redis，實現請求資料驗證需要用到 class-validator、class-transformer，實現本地策略驗證和 JWT 認證需要用到 @nestjs/passport、passport、passport-local、@types/passport-local、@nestjs/jwt passport-jwt、types/passport-jwt，實現檔案上傳需要用到 @types/multer，實現密碼雜湊加密需要用到 bcryptjs @types/bcryptjs，實現路徑比對需要用到 path-to-regexp。

執行下面兩筆命令：

```
pnpm add mysql2 typeorm @nestjs/typeorm redis class-validator class-transformer passport
@nestjs/passport passport-local @nestjs/jwt passport-jwt bcryptjs path-to-regexp -S
pnpm add @types/passport-local @types/multer @types/bcryptjs -D
```

安裝成功後，效果如圖 10-21 所示。

```
dependencies:
+ @nestjs/jwt 10.2.0
+ @nestjs/passport 10.0.3
+ bcryptjs 2.4.3
+ class-transformer 0.5.1
+ class-validator 0.14.1
+ mysql2 3.9.2
+ passport 0.7.0
+ passport-jwt 4.0.1
+ passport-local 1.0.0
+ path-to-regexp 6.2.1
+ redis 4.6.13
+ typeorm 0.3.20

The integrity of 1386 files was checked. This might have caused installation to take longer.
Done in 3.2s
Packages: +5
+++++
Progress: resolved 738, reused 738, downloaded 0, added 5, done

devDependencies:
+ @types/bcryptjs 2.4.6
+ @types/multer 1.4.11
+ @types/passport-local 1.0.38

Done in 1.6s
```

▲ 圖 10-21 安裝後端相依

在專案根目錄下執行「pnpm start:dev」命令啟動服務，並確保測試能夠正常執行，如圖 10-22 所示。

```
$ pnpm start:dev

> store-web-backend@0.0.1 start:dev /Users/jmin/development/nest-book/第十章/store-web-backend
> nest start --watch
[18:24:42] Starting compilation in watch mode...

[18:24:43] Found 0 errors. Watching for file changes.

[Nest] 37860  - 2024/03/14 18:24:43     LOG [NestFactory] Starting Nest application...
[Nest] 37860  - 2024/03/14 18:24:43     LOG [InstanceLoader] AppModule dependencies initialized +5ms
[Nest] 37860  - 2024/03/14 18:24:43     LOG [RoutesResolver] AppController {/}: +6ms
[Nest] 37860  - 2024/03/14 18:24:43     LOG [RouterExplorer] Mapped {/, GET} route +0ms
[Nest] 37860  - 2024/03/14 18:24:43     LOG [NestApplication] Nest application successfully started +1ms
```

▲ 圖 10-22 執行專案

10.2.6 後端架構設計

服務端基於 Nest 及多種中介軟體開發，我們設計了以下的目錄結構：

```
├── README.md
├── nest-cli.json
├── package.json
├── pnpm-lock.yaml
├── src
│   ├── activity
│   │   ├── activity.controller.ts
│   │   ├── activity.module.ts
│   │   ├── activity.service.ts
│   │   ├── dto
│   │   ├── entities
│   │   └── start-activity.service.ts
│   ├── app.controller.spec.ts
│   ├── app.controller.ts
│   ├── app.module.ts
│   ├── app.service.ts
│   ├── auth
│   │   ├── auth.module.ts
│   │   ├── auth.service.ts
│   │   ├── jwt-auth.guard.ts
│   │   ├── jwt.strategy.ts
│   │   └── role-auth.guard.ts
│   ├── common
│   │   ├── decorators
│   │   ├── enums
│   │   ├── http-exceptions.filter.ts
│   │   ├── logger
│   │   ├── logger.middleware.ts
│   │   ├── mail
│   │   ├── redis
│   │   ├── response.interceptor.ts
│   │   └── utils
│   ├── logs
│   │   ├── application-2024-04-05.error.log.gz
│   │   ├── application-2024-04-05.info.log.gz
│   │   ├── application-2024-04-06.error.log
│   │   ├── application-2024-04-06.info.log.gz
│   │   ├── application-2024-04-07.error.log
│   │   └── application-2024-04-07.info.log
│   ├── main.ts
│   ├── order
│   │   ├── dto
│   │   ├── entities
│   │   ├── order.controller.ts
```

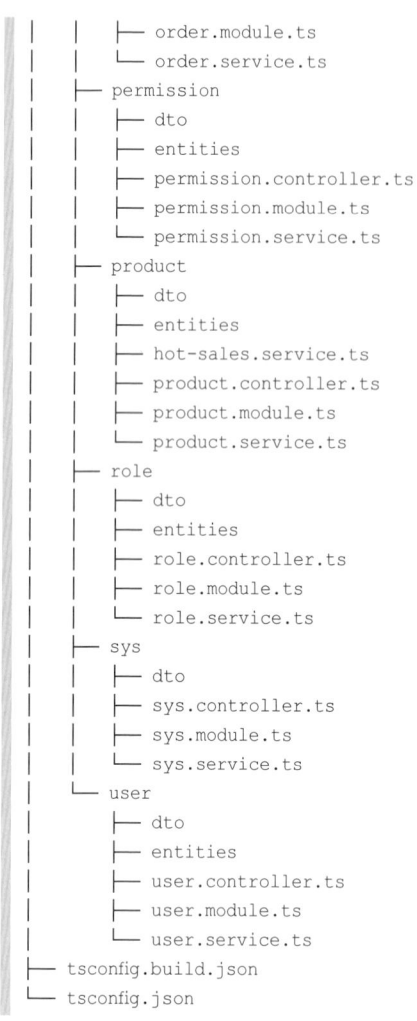

```
│   │   ├── order.module.ts
│   │   └── order.service.ts
│   ├── permission
│   │   ├── dto
│   │   ├── entities
│   │   ├── permission.controller.ts
│   │   ├── permission.module.ts
│   │   └── permission.service.ts
│   ├── product
│   │   ├── dto
│   │   ├── entities
│   │   ├── hot-sales.service.ts
│   │   ├── product.controller.ts
│   │   ├── product.module.ts
│   │   └── product.service.ts
│   ├── role
│   │   ├── dto
│   │   ├── entities
│   │   ├── role.controller.ts
│   │   ├── role.module.ts
│   │   └── role.service.ts
│   ├── sys
│   │   ├── dto
│   │   ├── sys.controller.ts
│   │   ├── sys.module.ts
│   │   └── sys.service.ts
│   └── user
│       ├── dto
│       ├── entities
│       ├── user.controller.ts
│       ├── user.module.ts
│       └── user.service.ts
├── tsconfig.build.json
└── tsconfig.json
```

　　由於 Nest 具有優秀的架構設計理念，因此我們可以很輕鬆在此基礎上架設服務端專案。在 src 目錄下，每個功能模組用單獨的目錄來管理自身的 DTO（驗證規則）、entities（實體）、controller（控制器）、service（服務）和 module（模組）；common 目錄用於管理全域公共攔截器、篩檢程式、守衛，它們會被註冊到全域中，此外還有統一的裝飾器、列舉類別、工具方法、日誌記錄器、Redis 服務和郵件服務等，可以很方便地維護它們。

　　為了後續能夠快速進行專案開發，基於前面學習的內容，我們完成了專案的基礎模組建設，包含 RBAC 許可權控制邏輯、自訂日誌器、MySQL 配置和 Redis 配置等，詳情可在 GitHub 中查看。

至此，我們完成了前後端技術方案的選型及專案初始化，下一節將設計 API 介面和資料庫表結構。

10.3 API 介面及資料庫表設計

技術方案確定之後，接下來設計資料表結構，整理清楚表與表之間的關係，並按照功能劃分 API 介面。

10.3.1 API 介面功能劃分

按照功能劃分，首先實現登入註冊模群組，包含使用者登入、註冊使用者和找回密碼功能，涉及的介面如表 10-1 所示。

▼ 表 10-1 登入註冊模組

請求路徑	請求方式	描述
/sys/registry	POST	註冊使用者
/sys/login	POST	使用者登入
/sys/forgot	POST	找回密碼
/sys/sendMail	GET	發送電子郵件驗證碼

接著實現使用者管理模群組，包含使用者清單、刪除使用者和編輯使用者資訊等功能，涉及的介面如表 10-2 所示。

▼ 表 10-2 使用者管理模組

請求路徑	請求方式	描述
/user/currentUser	GET	獲取當前登入使用者資訊
/user/delete	GET	刪除使用者
/user/list	GET	使用者列表
/user/edit	POST	編輯使用者資訊
/user/freeze	GET	凍結使用者

餐飲行業會銷售商品，對應商品管理模組，包含商品清單、新增 / 刪除商品和編輯商品資訊等功能，涉及的介面如表 10-3 所示。

▼ 表 10-3 商品管理模組

請求路徑	請求方式	描述
/product/create	POST	新增商品
/product/list	GET	商品列表
/product/edit	POST	編輯商品資訊
/product/delete	GET	刪除商品

商品銷售出去會產生訂單，所以有訂單管理模組，包含建立訂單、訂單清單和訂單詳情等功能，涉及的介面如表 10-4 所示。

▼ 表 10-4 訂單管理模組

請求路徑	請求方式	描述
/order/list	GET	訂單清單
/order/detail	GET	訂單詳情
/order/create	POST	建立訂單
/order/delete	DELETE	刪除訂單

店鋪通常涉及行銷，推出拓客活動來吸引顧客，對應活動管理模組，包含活動清單、建立 / 刪除活動和編輯活動資訊等功能，涉及的介面如表 10-5 所示。

▼ 表 10-5 活動管理模組

請求路徑	請求方式	描述
/activity/list	GET	活動列表
/activity/create	POST	建立活動
/activity/edit	POST	編輯活動
/activity/delete	GET	刪除活動

基於 RBAC 模型，還要實現角色管理模組，包含角色清單、新增 / 刪除角色和編輯角色資訊等功能，涉及的介面如表 10-6 所示。

▼ 表 10-6 角色管理模組

請求路徑	請求方式	描述
/role/list	GET	活動列表
/role/create	POST	建立活動
/role/edit	POST	編輯活動
/role/delete	GET	活動詳情

10.3.2 資料庫設計

完成功能 API 介面的定義後，接下來設計資料庫，理清表與表之間的關係。

基於 RBAC 模型，使用者與角色是多對多關係，角色與許可權也是多對多關係。在實際業務中，我們將不使用物理外鍵約束來處理表與表之間的關係，而是使用業務邏輯來處理。首先建立使用者表，其欄位如表 10-7 所示。

▼ 表 10-7 使用者表的欄位

欄　位	類　型	描　述
id	int	使用者 ID
username	varchar(30)	使用者帳號
password	varchar(50)	使用者密碼
salt	varchar(50)	雜湊加密的鹽
userType	int	使用者類型
email	varchat(50)	使用者電子郵件
freezed	boolean	是否凍結
createTime	Date	建立時間
updateTime	Date	更新時間

角色表的欄位如表 10-8 所示。

▼ 表 10-8 角色表的欄位

欄　位	類　型	描　述
id	int	角色 ID
name	varchar(50)	角色名稱
desc	varchar(255)	角色描述
createTime	Date	建立時間
updateTime	Date	更新時間

使用者與角色的關係需要一個中間表來維護，欄位如表 10-9 所示。

▼ 表 10-9 中間表的欄位

欄　位	類　型	描　述
id	int	中間表 ID
userId	int	使用者 ID
roleId	int	角色 ID

還要建立許可權表，用於維護每個模組的許可權資訊，欄位如表 10-10 所示。

▼ 表 10-10 許可權表的欄位

欄位	類型	描述
id	int	許可權 ID
title	varchar	前端功能表列的名稱
code	varchar	選單程式
type	int	選單類型（選單 / 頁面 / 元件 / 按鈕）

角色與許可權集的關係用中間表來維護，欄位如表 10-11 所示。

▼ 表 10-11 中間表的欄位

欄位	類型	描述
id	int	中間表 ID
permissionId	int	許可權集 ID
roleId	int	角色 ID

許可權表中的每一筆記錄對應頁面上的功能操作，等於一個後端介面，我們用專門的表來維護這種關係，欄位如表 10-12 所示。

▼ 表 10-12 介面記錄表的欄位

欄位	類型	描述
id	int	介面記錄 ID
apiUrl	varchar	介面 URL
apiMethod	varchar	介面請求方式
permissionId	int	功能許可權 ID

接下來建立商品表，用於存放店鋪建立的商品資訊，欄位如表 10-13 所示。

▼ 表 10-13 商品表的欄位

欄位	類型	描述
id	int	表自動增加 ID
name	varchar(255)	商品名稱
price	bigint	商品價格
desc	varchar(255)	商品描述
images	ProductImage[]	商品圖片
createTime	Date	建立時間
updateTime	Date	更新時間
status	int	商品狀態（未上架、已上架、已下架）
avtivityId	int	商品對應的活動 ID

商品的圖片會透過單獨表來維護，屬於一對多關聯性，欄位如表 10-14 所示。

▼ 表 10-14 商品的圖片

欄 位	類 型	描 述
id	int	圖片 ID
imageUrl	varchar	圖片路徑
productId	int	圖片對應的商品 ID

前臺收銀後會生成訂單，訂單表資訊的欄位如表 10-15 所示。

▼ 表 10-15 訂單表資訊的欄位

欄 位	類 型	描 述
id	bigint	訂單 ID
name	varchar(50)	商品名稱
count	int	商品數量
status	int	訂單狀態
price	bigint	訂單金額
discount	int	商品折扣
operator	varchar	操作員
createTime	Date	訂單建立時間

商品與訂單之間的關係透過中間表來維護，欄位如表 10-16 所示。

▼ 表 10-16 中間表的欄位

欄 位	類 型	描 述
id	init	活動 ID
productId	int	商品 ID
orderId	bigint	訂單 ID

最後，店鋪經常會以做活動的形式對商品進行促銷，同一時間段一個活動綁定一種商品，是一對一關聯性，透過在商品表中綁定 activity_id 可以找到活動資訊，活動表的欄位如表 10-17 所示。

▼ 表 10-17 活動表的欄位

欄 位	類 型	描 述
id	init	活動 ID
name	varchar（255）	活動名稱
status	int	活動狀態
type	int	活動類型

欄　位	類　型	描　述
desc	text	活動描述
startTime	Date	活動開始時間
endTime	Date	活動結束時間
createTime	Date	建立時間
updateTime	Date	更新時間
productId	int	商品 ID

　　資料表結構設計完成了。可能有讀者對表與表之間的關係並不清楚，沒關係，我們透過圖來展示各種表之間的關係，讓讀者更容易理解，如圖 10-23 所示。

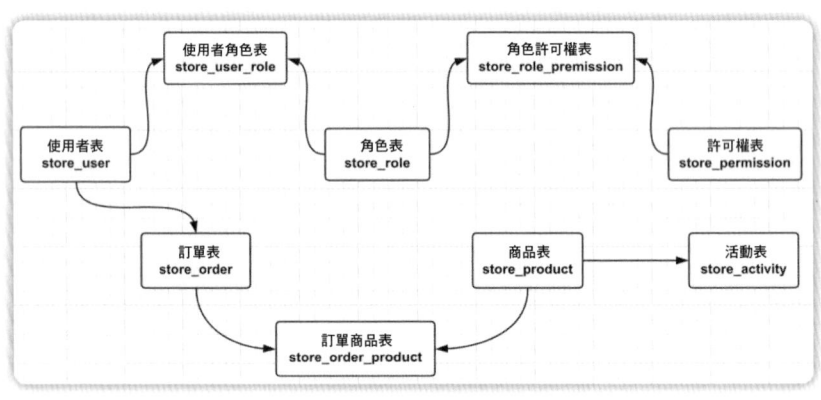

▲ 圖 10-23 資料表關係

　　我們清楚了各表之間的關係和作用，從下一節開始，將進入搭積木階段，逐一撰寫各模組的程式。

10.4　實現註冊登入

　　在前文中，我們已經完成了專案需求分析和原型設計，同時設計了 API 介面和資料庫表。在本節中，我們將進入實際開發階段。前端將實現登入、註冊和找回密碼頁面，而服務端則會實現相應頁面的介面，並完成資料對接工作。

10.4.1　頁面效果展示

　　首先來實現前端頁面，使用者透過使用者名稱、密碼、電子郵件和電子郵件驗證碼進行註冊，電子郵件驗證碼以 6 位數字的形式發送到使用者電子郵件中。

該電子郵件還將用於密碼找回、發送通知等功能。使用者註冊頁面如圖 10-24 所示。

▲ 圖 10-24 使用者註冊頁面

註冊頁面路由元件的具體實現程式如下：

```
import { getCaptchaCode, login, registry } from '@/apis/login'
import { ComponTypeEnum } from '@/layout/BasicLayout'
import { storeGlobalUser } from '@/store/globalUser'
import { storage } from '@/utils/Storage'
import {
  LockOutlined,
  MailOutlined,
  UserOutlined
} from '@ant-design/icons'
import {
  LoginForm,
  ProFormCaptcha,
  ProFormInstance,
  ProFormText
} from '@ant-design/pro-components'
import { RouteType } from '@config/routes'
import { routers } from '@config/routes/routers'
import { message, Tabs } from 'antd'
import { useRef } from 'react'
import { useNavigate } from 'react-router-dom'
import logo from '@/assets/mouse.jpg'

const Registry = () => {
  const navigate = useNavigate()
  const formRef = useRef<ProFormInstance>()
  const backToLogin = () => {
    navigate('/login')
```

```tsx
  }
  const handleLogin = async (val: Login.LoginEntity) => {
    const { data = {} } = await login(val)
    storage.set('token', data?.access_token)
    console.log('data', data)
    /** 跳躍有許可權的第一個選單 */
    await storeGlobalUser.getUserDetail()
    const flattenRoutes: (routes: RouteType[]) => RouteType[] = (routes: RouteType[]) => {
      const flattenedRoutes: RouteType[] = []
      function traverse(routes: RouteType[]) {
        routes.forEach(route => {
          flattenedRoutes.push(route)
          if (route.children) {
            traverse(route.children)
          }
        })
      }

      traverse(routes)

      return flattenedRoutes
    }
    const resRoutes = flattenRoutes(routers)
    const findPath =
      resRoutes?.[
        resRoutes?.findIndex(
          item =>
            item?.name ===
            storeGlobalUser?.userInfo?.menus?.filter(
              citem => citem?.type === ComponTypeEnum.MENU
            )?.[0]?.title
        )
      ]?.path
    navigate(findPath || '/')
  }
  const handleRegistry = async (val: Login.LoginEntity) => {
    const { success } = await registry(val)
    if (success) {
      handleLogin(val)
    }
  }
  return (
    <div style={{ backgroundColor: 'white', height: '100vh' }}>
      <LoginForm
        logo={logo}
        title=" 數字門店管理平臺 "
        subTitle=" 基於 NestJS + React 的全端專案 "
        formRef={formRef}
```

```
      onFinish={async (val: Login.LoginEntity) => {
        await handleRegistry(val)
      }}
      submitter={{
        searchConfig: {
          submitText: '註冊'
        }
      }}
    >
      <>
        <Tabs
          centered
          activeKey={'registry'}
        >
          <Tabs.TabPane key={'registry'} tab={'使用者註冊'} />
        </Tabs>
        <>
          <ProFormText
            name="username"
            fieldProps={{
              size: 'large',
              prefix: <UserOutlined className={'prefixIcon'} />
            }}
            placeholder={'請輸入帳號/使用者名稱'}
            rules={[
              {
                required: true,
                message: '請輸入使用者名稱!'
              }
            ]}
          />
          <ProFormText.Password
            name="password"
            fieldProps={{
              size: 'large',
              prefix: <LockOutlined className={'prefixIcon'} />
            }}
            placeholder={'請輸入註冊密碼'}
            rules={[
              {
                required: true,
                message: '請輸入註冊密碼!'
              }
            ]}
          />
          <ProFormText.Password
            name="confirmPassword"
            fieldProps={{
```

```jsx
              size: 'large',
              prefix: <LockOutlined className={'prefixIcon'} />
            }}
            placeholder={'請輸入確認密碼'}
            rules={[
              {
                required: true,
                message: '請輸入確認密碼！',
              }
            ]}
          />
          <ProFormText
            name="email"
            fieldProps={{
              size: 'large',
              prefix: <MailOutlined />,
            }}
            placeholder={'請輸入電子郵件'}
            rules={[
              {
                required: true,
                message: '請輸入使用者名稱！',
              }
            ]}
          />
          <ProFormCaptcha
            name="code"
            rules={[
              {
                required: true,
                message: '請輸入電子郵件驗證碼',
              },
            ]}
            placeholder=" 請輸入電子郵件驗證碼 "
            onGetCaptcha={async () => {
              // 獲取驗證碼
              const { success } = await getCaptchaCode({ email: formRef.current?.getFieldValue('email') })
              if (success) {
                message.success('驗證碼已發送至電子郵件')
              }
            }}
          />
        </>
      </>
      <div
        style={{
          marginBlockEnd: 24,
```

```
          }}
        >
          <a onClick={backToLogin}>
            傳回登入
          </a>
        </div>
      </LoginForm>
    </div>
  )
}

export default Registry
```

如果已經註冊過了，則點擊「傳回登入」按鈕進入登入頁面，輸入帳號和密碼進行登入，如圖 10-25 所示。

▲ 圖 10-25 使用者登入頁面

為了方便讀者預覽效果，可以使用預設的管理員帳號和密碼（admin/admin）來登入以體驗效果。在實際業務中，應嚴格保護好使用者的帳號和密碼（隱私保護）。具體登入組件程式實現如下：

```
import { login } from '@/apis/login'
import { ComponTypeEnum } from '@/layout/BasicLayout'
import { storeGlobalUser } from '@/store/globalUser'
import { storage } from '@/utils/Storage'
import { LockOutlined, UserOutlined } from '@ant-design/icons'
import { LoginForm, ProFormText } from '@ant-design/pro-components'
import { RouteType } from '@config/routes'
import { routers } from '@config/routes/routers'
import { Tabs } from 'antd'
import { useState } from 'react'
```

```tsx
import { useNavigate } from 'react-router-dom'
import logo from '@/assets/mouse.jpg'

type LoginType = <username> | <email>

const Login = () => {
  const [loginType, setLoginType] = useState<LoginType>('username')
  const navigate = useNavigate()

  const handleForgot = () => {
    navigate('/forgetPassword')
  }
  const handleLogin = async (val: Login.LoginEntity) => {
    const { data = {} } = await login(val)
    storage.set('token', data?.access_token)
    console.log('data', data)
    /** 跳躍有許可權的第一個選單 */
    await storeGlobalUser.getUserDetail()
    const flattenRoutes: (routes: RouteType[]) => RouteType[] = (routes: RouteType[]) => {
      const flattenedRoutes: RouteType[] = []
      function traverse(routes: RouteType[]) {
        routes.forEach(route => {
          flattenedRoutes.push(route)
          if (route.children) {
            traverse(route.children)
          }
        })
      }

      traverse(routes)

      return flattenedRoutes
    }
    const resRoutes = flattenRoutes(routers)
    const findPath =
      resRoutes?.[
        resRoutes?.findIndex(
          item =>
            item?.name ===
            storeGlobalUser?.userInfo?.menus?.filter(
              citem => citem?.type === ComponTypeEnum.MENU
            )?.[0]?.title
        )
      ]?.path
    navigate(findPath || '/')
  }
  const handleRegistry = () => {
    navigate('/registry')
```

```jsx
  }
  return (
    <div style={{ backgroundColor: 'white', height: '100vh' }}>
      <LoginForm
        logo={logo}
        title=" 數字門店管理平臺 "
        subTitle=" 基於 NestJS + React 的全端專案 "
        onFinish={async (val: Login.LoginEntity) => {
          await handleLogin(val)
        }}
      >
        <>
          <Tabs
            centered
            activeKey={loginType}
            onChange={(activeKey: LoginType) => setLoginType(activeKey as LoginType)}
          >
            <Tabs.TabPane key={'username'} tab={' 帳號密碼登入 '} />
          </Tabs>
          {loginType === 'username' && (
            <>
              <ProFormText
                name="username"
                fieldProps={{
                  size: 'large',
                  prefix: <UserOutlined className={'prefixIcon'} />
                }}
                placeholder={' 使用者名稱：admin'}
                rules={[
                  {
                    required: true,
                    message: ' 請輸入使用者名稱！'
                  }
                ]}
              />
              <ProFormText.Password
                name="password"
                fieldProps={{
                  size: 'large',
                  prefix: <LockOutlined className={'prefixIcon'} />
                }}
                placeholder={' 密碼：admin'}
                rules={[
                  {
                    required: true,
                    message: ' 請輸入密碼！'
                  }
                ]}
```

```
            />
          </>
        ))}
      </>
      <div
        style={{
          marginBlockEnd: 24
        }}
      >
        <a onClick={handleRegistry}>立即註冊</a>
        <a style={{ float: 'right' }} onClick={handleForgot}>
          忘記密碼
        </a>
      </div>
    </LoginForm>
  </div>
  )
}

export default Login
```

在登入過程中，如果使用者忘記密碼，可以透過點擊「忘記密碼」按鈕來重新設置密碼。使用者需要使用註冊時填寫的電子郵件來接收驗證碼，如果帳號與電子郵件不匹配，系統將拒絕該提交請求。找回密碼的頁面如圖 10-26 所示。

▲ 圖 10-26 使用者找回密碼的頁面

最後，在路由管理檔案 routers.tsx 中註冊這 3 個頁面元件，程式如下：

```
{
  path: '/login',
```

```
    name: '登入',
    element: <Login />
  },
  {
    path: '/registry',
    name: '註冊',
    element: <Registry />
  },
  {
    path: '/forgetPassword',
    name: '找回密碼',
    element: <ForgetPassword />
  }
```

前端頁面元件撰寫完畢後，接下來實現服務端的介面邏輯。

10.4.2 介面實現

在實現介面之前，需要定義使用者資料實體 userEntity，程式如下：

```
import { Column, CreateDateColumn, Entity, PrimaryGeneratedColumn } from "typeorm";

@Entity('store_user')
export class UserEntity {
  @PrimaryGeneratedColumn({type: 'int'})
  id: number;

  @Column({ type: 'varchar', length: 32, comment: '使用者登入帳號' })
  username: string;

  @Column({ type: 'varchar', length: 200, nullable: false, comment: '使用者登入密碼' })
  password: string;

  @Column({ type: 'varchar', length: 50, nullable: false, comment: '雜湊加密的鹽' })
  salt: string;

  @Column({ type: 'int', comment: '使用者類型 0 管理員 1 普通使用者', default: 1 })
  userType: number;

  @Column({ type: 'varchar', comment: '使用者電子郵件', default: '' })
  email: string;

  @Column({ type: 'int', comment: '是否凍結使用者 0 不凍結 1 凍結', default: 0 })
  freezed: number;

  @Column({ type: 'varchar', comment: '使用者圖示', default: '' })
  avatar: string;
```

```
  @Column({ type: 'varchar', comment: '使用者備註', default: '' })
  desc: string;

  @CreateDateColumn({ type: 'timestamp', comment: '建立時間' })
  createTime: Date
}
```

建立 sys 模組,用於管理使用者註冊、登入、找回密碼和發送驗證碼介面等公共路由方法,程式如下:

```
import { Controller, Get, Post, Body, Inject, Query } from '@nestjs/common';
import { SysService } from './sys.service';
import { LoginUserDto } from './dto/login-user.dto';
import { CreateUserDto } from 'src/user/dto/create-user.dto';
import { UserService } from 'src/user/user.service';
import { ForgotUserDto } from './dto/forgot-user.dto';
import { AllowNoToken } from 'src/common/decorators/token.decorator';
@Controller('sys')
export class SysController {
  @Inject(UserService)
  private userService: UserService;

  constructor(private readonly sysService: SysService) {}

  // 使用者註冊
  @Post('registry')
  @AllowNoToken()
  registry(@Body() createUserDto: CreateUserDto) {
    return this.userService.registry(createUserDto);
  }

  // 使用者登入
  @Post('login')
  @AllowNoToken()
  login(@Body() loginUserDto: LoginUserDto) {
    return this.userService.login(loginUserDto);
  }

  // 找回密碼
  @Post('forgot')
  @AllowNoToken()
  forgot(@Body() forgotUserDto: ForgotUserDto) {
    return this.userService.updatePassword(forgotUserDto);
  }
  // 發送找回密碼電子郵件驗證碼
  @Get('sendEmailForGorgot')
  @AllowNoToken()
  sendEmailForGorgot(@Query() dto: { email: string }) {
    return this.sysService.sendEmailForGorgot(dto.email);
```

```typescript
  }

  // 發送註冊電子郵件驗證碼
  @Get('sendEmailForRegistry')
  @AllowNoToken()
  sendEmailForRegistry(@Query() dto: { email: string }) {
    return this.sysService.sendMail(dto.email,'註冊驗證碼');
  }
}
```

接下來,在生成的 user 模組中,實現具體的 registry 方法,程式如下:

```typescript
/**
 * 註冊新使用者
 *
 * @param createUserDto 建立使用者 DTO
 * @returns 傳回使用者資訊
 * @throws 如果使用者名稱或註冊電子郵件已存在,則傳回衝突錯誤
 * @throws 如果兩次輸入的密碼不一致,則傳回預期失敗錯誤
 * @throws 如果驗證碼有誤或已過期,則傳回預期失敗錯誤
 */
async registry(createUserDto: CreateUserDto) {
  const { username, email } = createUserDto;
  // 1. 判斷使用者是否存在,參數為電子郵件或使用者名稱,使用 createQueryBuilder 一次性查詢兩個欄位
  const user = await this.userRepository
    .createQueryBuilder('su')
    .where('su.username = :username OR su.email = :email', { username, email })
    .getOne()
  // 2. 若存在則傳回錯誤資訊
  if (user) {
    throw new HttpException('使用者名稱或註冊電子郵件已存在,請重新輸入', HttpStatus.CONFLICT);
  }
  if (createUserDto.confirmPassword !== createUserDto.password) {
    throw new HttpException('兩次輸入的密碼不一致,請重新輸入',
HttpStatus.EXPECTATION_FAILED);
  }
  // 3. 驗證註冊驗證碼
  const codeRedisKey = getRedisKey(RedisKeyPrefix.REGISTRY_CODE, createUserDto.email)
  const code = await this.redisService.get(codeRedisKey)
  if (!code) {
    throw new HttpException('驗證碼有誤或已過期', HttpStatus.EXPECTATION_FAILED);
  }
  // 4. 雜湊加密
  const salt = await genSalt()
  createUserDto.password = await hash(createUserDto.password, salt);

  const newUser = plainToClass(UserEntity, { salt, ...createUserDto },
{ignoreDecorators: true});
  // 5. 若不存在則建立使用者
  const { password, salt: salter, ...rest } = await this.userRepository.save(newUser);
```

```
// 6. 快取使用者資訊
const redisKey = getRedisKey(RedisKeyPrefix.USER_INFO, rest.id)
await this.redisService.hSet(
  redisKey,
  rest
)
return rest;
}
```

在註冊介面中,我們對使用者密碼進行加密操作,同時把加密的鹽持久化到資料庫中。當需要修改使用者密碼時,可以使用這個鹽對新密碼進行加密。儲存到資料庫後,將使用者資訊快取到 Redis 中,以便下一次存取時直接從 Redis 中獲取,最後,傳回除密碼和鹽外的非敏感性資料給用戶端。

在登入介面邏輯中,首先根據使用者名稱查詢使用者是否存在。若使用者存在,則使用參數中的密碼和加密後的雜湊密碼進行比較,判斷是否相同。如果使用者已經被凍結,則禁止登入。所有判斷透過後,將使用使用者資訊生成 access_token,JWT 策略會把使用者資訊儲存在 Request 物件中,以便後續使用。最後,將 access_token 傳回到用戶端。具體程式如下:

```
/**
 * 登入方法
 *
 * @param loginUserDto 登入使用者資訊
 * @returns 傳回生成的 Token
 * @throws 當帳號或密碼錯誤時,拋出 HttpException 例外
 * @throws 當帳號被凍結時,拋出 HttpException 例外
 */
async login(loginUserDto: LoginUserDto) {
  const user = await this.userRepository.findOne({
    where: {
      username: loginUserDto.username,
    },
  });
  // 1. 判斷使用者是否存在
  if (!user) {
    throw new HttpException('帳號或密碼錯誤', HttpStatus.EXPECTATION_FAILED);
  }
  // 2. 判斷密碼是否正確
  const checkPassword = await compare(loginUserDto.password, user.password);
  if (!checkPassword) {
    throw new HttpException('帳號或密碼錯誤', HttpStatus.EXPECTATION_FAILED);
  }
  // 3. 判斷使用者是否被凍結
  if (user.freezed) {
    throw new HttpException('帳號已被凍結,請聯繫管理員', HttpStatus.EXPECTATION_FAILED);
```

```
  }
  // 4. 生成 Token
  const { password, salt, ...rest } = user
  const access_token = this.generateAccessToken(rest)
  return {
    access_token
  };
}
```

　　使用者忘記密碼並希望重新設置密碼以便登入時，同樣地，需要判斷使用者是否存在、是否已經凍結。同時需要驗證用戶端提供的驗證碼 code 與快取在 Redis 中快取的電子郵件驗證碼是否匹配，並檢查驗證碼是否在 5 分鐘的有效期內。更新密碼成功後，最後刪除 Redis 中舊的使用者資訊及使用過的驗證碼。具體程式如下：

```
/**
 * 更新使用者密碼
 *
 * @param dtoForgotUserDto 使用者資訊物件
 * @returns 修改成功資訊
 * @throws 驗證碼錯誤或已過期，使用者不存在或已刪除，使用者被凍結，兩次輸入的密碼不一致，修改失敗
 */
async updatePassword(dto: ForgotUserDto) {
  const { password, confirmPassword, code, username } = dto
  if (password !== confirmPassword) {
    throw new HttpException('兩次輸入的密碼不一致', HttpStatus.EXPECTATION_FAILED);
  }
  // 1. 判斷使用者是否存在
  const exists = await this.userRepository.findOneBy({ username })
  if (!exists) {
    throw new HttpException('使用者不存在或已刪除', HttpStatus.EXPECTATION_FAILED);
  }
  const { id, freezed } = exists
  if (freezed) {
    throw new HttpException('使用者已被凍結，請解凍後再修改', HttpStatus.EXPECTATION_FAILED);
  }
  // 2. 驗證 Redis 中的驗證碼與使用者輸入的驗證碼是否一致
  const cacheCode = await this.redisService.get(
    getRedisKey(RedisKeyPrefix.PASSWORD_RESET, id)
  );
  if (cacheCode !== code) {
    throw new HttpException('驗證碼錯誤或已過期', HttpStatus.EXPECTATION_FAILED);
  }
  // 3. 更新密碼
  const newPassword = await hash(password, exists.salt)
  const { affected } = await this.userRepository.update({ id }, { password: newPassword });
  if (!affected) {
    throw new HttpException('修改失敗，請稍後重試', HttpStatus.EXPECTATION_FAILED);
```

```typescript
}
// 4. 刪除 Redis 快取的驗證碼和使用者資訊
const codeRedisKey = getRedisKey(RedisKeyPrefix.PASSWORD_RESET, id)
this.redisService.del(codeRedisKey)

const userRedisKey = getRedisKey(RedisKeyPrefix.USER_INFO, id)
this.redisService.del(userRedisKey)

return '修改成功，請重新登入'
}
```

我們在多個場景中需要使用電子郵件驗證碼功能，可以透過 nodemailer 套件來實現郵件收發，並將 MailService 抽象為 Provider 注入需要的模組中。MailService 的具體實現程式如下：

```typescript
import { HttpException, HttpStatus, Injectable } from "@nestjs/common";
import { ConfigService } from "@nestjs/config";
import * as nodemail from "nodemailer";

@Injectable()
export class MailService {
  private transporter: nodemail.Transporter;
  constructor(private config: ConfigService) {
    this.transporter = nodemail.createTransport({
      host: config.get<string>('EMAIL_HOST'),
      port: config.get<Number>('EMAIL_PORT'),
      secure: config.get<string>('EMAIL_SECURE'),
      auth: {
        user: config.get<string>('EMAIL_USER'),
        pass: config.get<string>('EMAIL_PASS'),
      },
    });
  }

  /**
   * 發送郵件
   *
   * @param email 收件人電子郵件
   * @param subject 郵件主題
   * @param html 郵件正文，可選
   * @returns 傳回一個 Promise 物件，解析後得到一個包含驗證碼和郵件信封資訊的物件
   * @throws HttpException 當郵件發送失敗時，會拋出一個 HttpException 例外
   */
  sendMail(email: string, subject: string, html?: string): Promise<Record<string, string>> {
    const code = Math.random().toString().slice(-6);
    const mailOptions = {
      from: this.config.get<string>('EMAIL_USER'),
      to: email,
```

```
      text: '使用者驗證碼為:${code},有效期為 5 分鐘,請及時使用!',
      subject,
      html,
    };
    return new Promise((resolve, reject) => {
      this.transporter.sendMail(mailOptions, (error: Error, info:
{ envelope: Record<string, string[]> }) => {
        if (error) {
          reject(new HttpException('發送郵件失敗:${error}',
HttpStatus.INTERNAL_SERVER_ERROR));
        } else {
          resolve({
            code,
            ...info.envelope
          })
        }
      });
    })
  }
}
```

郵件發送相關配置使用 .env 設定檔進行管理,配置及解釋如下:

```
# 電子郵件配置
EMAIL_PASS=EBBOFYAQDWBLOTZY          # 電子郵件授權碼,需要從電子郵件平臺中申請
EMAIL_HOST=smtp.163.com              # 郵件伺服器地址
EMAIL_PORT=465                       # 郵件伺服器通訊埠,預設為 465
EMAIL_SECURE=true                    # 是否使用預設的電子郵件通訊埠
EMAIL_USER=jmin95@163.com            # 發送給使用者郵件的電子郵件
EMAIL_ALIAS=store_web_project        # 電子郵件別名
```

在註冊和找回密碼的過程中,我們需要實現發送電子郵件驗證碼的邏輯。具體過程包括生成驗證碼,將其儲存在 Redis 中,並使用不同的鍵(key)來區分不同的使用者或場景。為了確保安全性,這些驗證碼設置了一個 5 分鐘的有效時間。以下是實現這一功能的範例程式:

```
import { HttpException, HttpStatus, Inject, Injectable } from '@nestjs/common';
import { InjectRepository } from '@nestjs/typeorm';
import { RedisKeyPrefix } from 'src/common/enums/redis-key.enum';
import { MailService } from 'src/common/mail/mail.service';
import { RedisService } from 'src/common/redis/redis.service';
import { getRedisKey } from 'src/common/utils';
import { UserEntity } from 'src/user/entities/user.entity';
import { Repository } from 'typeorm';

@Injectable()
export class SysService {
  @Inject(MailService)
  private mailService: MailService;
```

```typescript
  @Inject(RedisService)
  private redisService: RedisService;

  @InjectRepository(UserEntity)
  private userRepository: Repository<UserEntity>;

  /**
   * 發送找回密碼郵件
   *
   * @param email 使用者電子郵件
   * @returns 傳回驗證碼已發送至電子郵件的提示訊息
   * @throws 當電子郵件為空時,拋出 HttpException 例外
   * @throws 當前使用者未綁定該電子郵件時,拋出 HttpException 例外
   */
  async sendEmailForGorgot(email: string) {
    if (!email) {
      throw new HttpException(' 電子郵件不能為空 ', HttpStatus.EXPECTATION_FAILED);
    }
    const exists = await this.userRepository.findOneBy({email})
    // 1. 判斷當前的電子郵件與使用者註冊電子郵件是否一致
    if (!exists) {
      throw new HttpException(' 當前使用者未綁定該電子郵件,請檢查後重試 ',
HttpStatus.EXPECTATION_FAILED);
    }
    // 2. 發送電子郵件驗證碼
    const { code } = await this.mailService.sendMail(email, ' 找回密碼驗證碼 ')

    // 快取 Redis
    const redisKey = getRedisKey(RedisKeyPrefix.PASSWORD_RESET, exists.id);
    await this.redisService.set(redisKey, code, 60*5); // 5 分鐘有效
    return ' 驗證碼已發送至電子郵件,請注意查收 '
  }


  /**
   * 發送註冊郵件驗證碼
   *
   * @param email 電子郵件位址
   * @param text 郵件內容
   * @returns 傳回發送成功資訊
   */
  async sendMailForRegistry(email: string, text: string) {
    const { code } = await this.mailService.sendMail(email, text)
    // 快取 Redis
    const redisKey = getRedisKey(RedisKeyPrefix.REGISTRY_CODE, email);
    await this.redisService.set(redisKey, code, 60*5);
    return ' 發送成功 ';
  }
}
```

完成開發後，接下來測試一遍流程，可以使用 REST Client 模擬使用者請求來獲取註冊驗證碼，如圖 10-27 所示。

▲ 圖 10-27 獲取註冊驗證碼

使用這個驗證碼發送註冊請求，成功註冊之後，傳回使用者資訊，如圖 10-28 所示。

▲ 圖 10-28 使用者註冊帳號

註冊完畢後，接下來使用帳號和密碼進行登入操作，成功換取了 access_token，如圖 10-29 所示。

▲ 圖 10-29　使用者登入

同理，找回密碼之前，需要獲取電子郵件驗證碼，如圖 10-30 所示。

▲ 圖 10-30　獲取驗證碼

用獲取的驗證碼重置密碼，將密碼從 test 改為 test2，介面傳回「修改成功，請重新登入」，如圖 10-31 所示。

```
30
39    # 找回密碼
      Send Request
40    POST http://localhost:3333/sys/forgot
41    Content-Type: application/json
42
43    {
44      "username": "test",
45      "password": "test2",
46      "confirmPassword": "test2",
47      "email": "yc_micro_front@163.com",
48      "code": 143495
49    }
50            2 周前 · 門店管理系統

終端  偵錯主控台  輸出

[NEST] 2024-04-07 22:58:58 info [Response ResponseInterceptor]
##################################################################
Request original url: /sys/forgot
Method: POST
IP: ::ffff:127.0.0.1
Response data: "修改成功,請重新登入"
##################################################################
```

▲ 圖 10-31 找回密碼

至此，我們完成了登入註冊模組中前端頁面元件的撰寫和服務端介面的實現。關於本節更詳細的介面程式實現，請存取提供的 GitHub 原始程式。下一節將繼續實現使用者模組的功能。

10.5 實現使用者與角色模組

使用者與角色是 RABC 許可權設計中必不可少的部分，本節將實現使用者與角色管理模組。使用者管理功能包含獲取當前登入使用者資訊、獲取使用者列表、編輯使用者資訊、凍結使用者和刪除使用者。角色管理包含獲取角色列表、編輯角色資訊和刪除角色功能。在操作許可權上，普通使用者不能越權操作，例如不能修改管理員資訊，凍結管理員帳號或刪除管理員帳號。此外，普通使用者也不能凍結或刪除自己的帳號。系統內建的角色不允許刪除，並且所有列表支持模糊查詢。

10.5.1 頁面效果展示

在實現使用者與角色管理模組之前，我們需要先實現左側的多級功能表列功能，如圖 10-32 所示。

▲ 圖 10-32 系統功能表列

使用者登入成功後，系統將獲取當前使用者的詳細資訊，包括使用者的基本資訊、角色以及許可權選單資料。前端應用將根據使用者路由表中預設的配置資訊與服務端傳回的許可權資料進行對比。透過計算這兩個資料集的交集，前端將篩選出使用者具有相應許可權的選單項，並據此著色功能表選項。核心實現程式如下：

```
/** 處理選單許可權隱藏選單 */
const reduceRouter = (routers: RouteType[]): RouteType[] => {
  // 過濾出屬於選單和頁面類型的資料
  const authMenus = storeGlobalUser?.userInfo?.menus
    ?.filter(item => item?.type === ComponTypeEnum.MENU || item?.type === ComponTypeEnum.PAGE)
    ?.map(item => item?.title)

  return routers?.map(item => {
    if (item?.children) {
      const { children, ...extra } = item
      return {
        ...extra,
        routes: reduceRouter(item?.children),
        hideInMenu: item?.hideInMenu || !item?.children?.find(citem =>
authMenus?.includes(citem?.name))
      }
    }
    return {
      ...item,
      hideInMenu: item?.hideInMenu || !authMenus?.includes(item?.name)
    }
  }) as any
}
```

接下來，在選單元件中綁定路由資料，範例程式如下：

```
<ProLayout
   ...
   route={reduceRouter(router?.routes)?.[1]}
   ...
>
  ...
</ProLayout>
```

ProLayout 元件接收 route 等屬性並自動著色出頁面框架。除了圖 10-32 中展示的使用者列表外，點擊「編輯」按鈕可以對使用者資訊進行修改，如圖 10-33 所示。

▲ 圖 10-33 編輯使用者資訊

在列表上方，使用者可以透過輸入使用者名稱來執行模糊搜尋。此外，清單還提供了操作按鈕，如點擊「刪除」按鈕可以刪除指定使用者，點擊「凍結」按鈕可以凍結使用者。被凍結的使用者將無法登入系統，只有在執行「解凍」操作後才能恢復登入許可權。

角色模組中的清單欄位版面配置和功能與此類似，如圖 10-34 所示。

▲ 圖 10-34 角色管理

在專案中，清單頁面的互動基於 ProComponents 的 ProTable 來實現。我們在此基礎上封裝了更加簡單、高效且可重複使用的業務元件 ExcelTable，使得可以用更少的程式實現表格互動效果。以下是封裝的核心範例程式：

```
<ExcelTable
  columns={[
    {
      title: ' 角色名稱 ',
      dataIndex: 'name',
      hideInTable: true
    },
    /** search */
    {
      title: ' 角色名稱 ',
      dataIndex: 'name',
      hideInSearch: true
    },
    {
      title: ' 描述 ',
      dataIndex: 'desc',
      hideInSearch: true
    },
    {
      title: ' 是否系統內建 ',
      dataIndex: 'isSystem',
      hideInSearch: true,
      render(_, entity) {
        return <Tag color={entity.isSystem ? 'green' : 'red'}>{entity.isSystem ? ' 是 ' : ' 否 '}</Tag>
      }
    },
    {
```

```
        title: ' 建立時間 ',
        dataIndex: 'createTime',
        hideInSearch: true,
        valueType: 'dateTime'
      },
      {
        title: ' 操作 ',
        key: 'option',
        valueType: 'option',
        render: (_, record) => (!record.isSystem && [
          <Button key="edit" type="link" onClick={() => showModal(record)}>
            編輯
          </Button>,
          <Popconfirm
            key="delete"
            placement="topRight"
            title=" 確定要刪除嗎 ?"
            onConfirm={async () => {
              const res = await delRole({ id: record?.id })
              if (res?.code === 200) {
                message.success(' 刪除成功 ')
                actionRef?.current?.reloadAndRest?.()
                return Promise.resolve()
              }
              return Promise.reject()
            }}
            okText=" 確定 "
            okType="danger"
            cancelText=" 取消 "
          >
            <Button type="link" danger key="delete">
              刪除
            </Button>
          </Popconfirm>
        ])
      }
    ]}
    requestFn={async (params) => {
      const data = await getRoleList(params)
      return data
    }}
    actionRef={actionRef}
    rowSelection={false}
    toolBarRenderFn={() => [
      <Button key="add" type='primary' onClick={() => showModal()}>
        新增角色
      </Button>
    ]}
  />
```

關於自定義元件更詳細的實現過程，請查看 GitHub 專案中的原始程式部分。

點擊「新增角色」按鈕，使用者可以自訂角色並為其分配初始許可權，如圖 10-35 所示。

▲ 圖 10-35 新增角色

點擊「編輯」按鈕，使用者可以對角色資訊進行二次修改和分配許可權，如圖 10-36 所示。

▲ 圖 10-36 編輯角色

角色許可權支援樹形互動選擇，勾選後可以為角色分配不同的許可權，如圖 10-37 所示，使用者擁有相關許可權才能進行相應的操作。

▲ 圖 10-37 分配許可權的互動過程範例

我們可以使用 Modal 元件輕鬆實現以上彈窗的互動效果，範例程式如下：

```
Modal.confirm({
  title: record ? '編輯角色' : '新增角色',
  onOk: async () => onSubmit(record),
  okText: '確定',
  cancelText: '取消',
  maskClosable: true,
  width: 800,
  content: (
    <ProForm
      labelCol={{ span: 6 }}
      wrapperCol={{ span: 12 }}
      submitter={false}
      layout="horizontal"
      initialValues={{
        status: 1,
        ...record,
        permissions: record?.permissions?.map(item => item?.id)
      }}
      formRef={modalFormRef}
    >
      <ProFormText label="角色名稱" name="name" rules={[{ required: true }]} />
      <ProFormTreeSelect
        label="角色許可權"
        name="permissions"
        fieldProps={{
          multiple: true,
          showCheckedStrategy: TreeSelect.SHOW_ALL,
          fieldNames: {
            label: 'title',
```

```
            value: 'id'
          },
          treeCheckable: true
        }}
        allowClear
        rules={[{ required: true, message: '請選擇' }]}
        request={async () => {
          const { data } = await getPermissionList()
          return data.list
        }}
      />
      <ProFormTextArea label=" 描述 " name="desc" />
    </ProForm>
  )
})
```

撰寫完前端互動頁面後,接下來實現服務端的介面邏輯。

10.5.2 表關係設計

獲取使用者資訊通常涉及多表聯集查詢。在 RBAC 許可權系統中,除了使用者表,還包括角色表和許可權表,這些表透過中間表來維護它們之間的多對多關係。角色表實體如下:

```
import { Column, CreateDateColumn, Entity, PrimaryGeneratedColumn,
UpdateDateColumn } from "typeorm";

@Entity('store_role')
export class RoleEntity {
  @PrimaryGeneratedColumn()
  id: number

  @Column({ type: 'varchar', length: 50, comment: '角色名稱' })
  name: string

  @Column({ type: 'varchar', length: 255, comment: '角色描述' })
  desc: string

  @CreateDateColumn({ type: 'timestamp', comment: '建立時間' })
  createTime: Date

  @UpdateDateColumn({ type: 'timestamp', comment: '更新時間' })
  updateTime: Date

  @Column({ type: 'int', comment: '是否為系統內建 0 否 1 是 ', default: 0 })
  isSystem: number
}
```

系統通常會預設一些預設角色，例如管理員、店長、收銀員等，但這些角色的設置並非固定不變，而是根據具體的應用場景來確定。在 RBAC 許可權系統中，使用者與角色之間的關係透過一個中間表 store_user_role 來維護，實體欄位定義如下：

```typescript
import { Column, Entity, PrimaryGeneratedColumn } from "typeorm";

@Entity('store_user_role')
export class UserRoleEntity {
  @PrimaryGeneratedColumn()
  id: number

  @Column({ type: 'int', comment: '使用者 id' })
  userId: number

  @Column({ type: 'int', comment: '角色 id' })
  roleId: number
}
```

　　許可權表用於儲存系統中包含哪些選單、頁面、按鈕或元件的許可權資訊，通常也被稱為「選單表」，其實體欄位如下：

```typescript
import { PermissionType } from "src/common/enums/common.enum";
import { Column, Entity, PrimaryGeneratedColumn } from "typeorm";

@Entity('store_permission')
export class PermissionEntity {
  @PrimaryGeneratedColumn()
  id: number

  @Column({ type: 'varchar', length: 10, comment: '許可權名稱（選單名稱）' })
  title: string

  @Column({ type: 'varchar', length: 50, comment: '許可權碼' })
  code: string

  @Column({ type: 'int', comment: '許可權類型 0 選單 1 頁面 2 元件 3 按鈕' })
  type: PermissionType

  @Column({ type: 'int', comment: '父級 id', default: 0 })
  parentId: number
}
```

　　在設計 RBAC 許可權系統時，需要注意許可權資料具有多種類型，這樣的設計有助實現更細粒度的許可權控制。元件或按鈕等級的細粒度許可權通常不會直接展示在功能表列中，而是透過許可權碼在頁面中進行單獨控制。舉例來說，許可權碼可以傳遞給高階元件，該元件將自動判斷當前使用者是否有許可權著色對應的元件或按鈕。

對 Vue 開發者來說，一種常見的做法是使用自訂指令 v-permission 來接收許可權碼，並據此實現許可權控制。

許可權表定義好後，角色與許可權之間還需要一個中間表 store_role_permission 來連結，實體欄位如下：

```
import { Column, Entity, PrimaryGeneratedColumn } from "typeorm";

@Entity('store_role_permission')
export class RolePermissionEntity {
  @PrimaryGeneratedColumn()
  id: number

  @Column({ type: 'int', comment: '角色 id' })
  roleId: number

  @Column({ type: 'int', comment: '許可權集 id' })
  permissionId: number
}
```

至此，我們大致完成了 RBAC 基本的表關係設計。選單許可權將分配給不同的角色，而不同的角色又將分配給不同的使用者，從而實現了不同使用者只能操作特定的許可權功能。這一設計看起來沒問題，但實際上在服務端還會有一個隱憂。

儘管在使用者頁面上看不到具體的功能頁面或按鈕，無法直接操作，但某些功能對應的介面卻沒有進行許可權控制，使得一些使用者可以透過手動發送 HTTP 請求的方式存取需要許可權的介面因此，我們還需要一個表（store_permission_api），用來維護服務端所有的路由介面。該表的主要欄位如下：

```
import { Column, Entity, PrimaryGeneratedColumn } from "typeorm";
@Entity('store_permission_api')
export class PermissionApiEntity {
  @PrimaryGeneratedColumn()
  id: number

  @Column({ type: 'varchar', length: 50, comment: '許可權名稱' })
  apiUrl: string

  @Column({ type: 'varchar', length: 100, comment: '許可權碼' })
  apiMethod: string
```

```
  @Column({ type: 'int', comment: '功能 id' })
  PermissionId: number
}
```

這樣設計的目的在於,當使用者每一次請求介面時,我們需要判斷當前使用者發送的請求 URL 是否在他的角色控制範圍內,從而確定是否放行該用戶端的請求。我們定義了一個全域守衛(RoleAuthGuard)來處理這個邏輯,程式如下:

```
import { CanActivate, Inject, ExecutionContext, Injectable,
ForbiddenException } from '@nestjs/common'
import { Reflector } from '@nestjs/core'
import { pathToRegexp } from 'path-to-regexp'
import { ALLOW_NO_PERMISSION } from 'src/common/decorators/permission.decorator'
import { PermissionService } from 'src/permission/permission.service'
import { ALLOW_NO_TOKEN } from 'src/common/decorators/token.decorator'
import { UserType } from 'src/common/enums/common.enum'

@Injectable()
export class RoleAuthGuard implements CanActivate {
  constructor(
    private readonly reflector: Reflector,
    @Inject(PermissionService)
    private readonly permissionService: PermissionService,
  ) {}

  async canActivate(ctx: ExecutionContext): Promise<boolean> {
    // 若函數請求標頭配置了 @AllowNoToken() 裝飾器,則無須驗證 Token 許可權
    const allowNoToken = this.reflector.getAllAndOverride<boolean>(ALLOW_NO_TOKEN,
[ctx.getHandler(), ctx.getClass()])
    if (allowNoToken) return true

    // 若函數請求標頭配置了 @AllowNoPermission() 裝飾器,則無須驗證許可權
    const allowNoPerm = this.reflector.getAllAndOverride<boolean>
(ALLOW_NO_PERMISSION, [ctx.getHandler(), ctx.getClass()])
    if (allowNoPerm) return true

    const req = ctx.switchToHttp().getRequest()
    const user = req.user
    // 沒有攜帶 Token 直接傳回 false
    if (!user) return false
    // 管理員擁有所有介面許可權,不需要判斷
    if (user.userType === UserType.ADMIN_USER) return true

    // 獲取該使用者所擁有的介面許可權
    const userApis = await this.permissionService.getPermApiList(user)
    console.log('當前使用者擁有的介面許可權集:',userApis);

    const index = userApis.findIndex((route) => {
```

```
      // 請求方法類型相同
      if (req.method.toUpperCase() === route.method.toUpperCase()) {
        // 比較當前請求 URL 是否在使用者介面許可權集中
        const reqUrl = req.url.split('?')[0]
        console.log(' 當前請求 URL：', reqUrl);

        return !!pathToRegexp(route.url).exec(reqUrl)
      }
      return false
    })
    if (index === -1) throw new ForbiddenException(' 您無許可權存取該介面 ')
    return true
  }
}
```

在程式實現中，透過註釋明確了幾種特殊情況：管理員帳戶、無需進行 token 驗證的帳戶以及無需進行許可權驗證的帳戶，這些帳戶將被直接允許存取。對於其他使用者的請求，則必須透過 API 許可權驗證。如果驗證失敗，守衛將阻止存取。在整理清楚許可權關係後，接下來實現具體的介面邏輯。

10.5.3 介面實現

使用者管理與角色管理的介面實現極其相似，本節以使用者模組為例，當使用者登入成功後，首先獲取使用者資訊，包含使用者基本資訊以及對應的角色和許可權資料。具體實現程式如下：

```
/**
 * 獲取當前使用者資訊
 *
 * @param currentUser 當前使用者實體
 * @returns 傳回當前使用者資訊、使用者角色以及許可權
 */
async getCurrentUser(currentUser: UserEntity) {
  // 同時查詢使用者角色和許可權
  const queryBuilder = this.dataSource
    .createQueryBuilder()
    .select([
      'user.id AS userId',
      'user.username AS userName',
      'role.id AS roleId',
      'role.name AS roleName',
      'p.id AS permissionId',
      'p.title AS permissionTitle',
      'p.type AS permissionType',
      'p.code AS permissionCode'
```

```
          ])
          .from('store_user', 'user')
          .leftJoin('store_user_role', 'userRole', 'user.id = userRole.userId')
          .leftJoin('store_role', 'role', 'userRole.roleId = role.id')
          .leftJoin('store_role_permission', 'rolePerm', 'role.id = rolePerm.roleId')
          .leftJoin('store_permission', 'p', 'rolePerm.permissionId = p.id')
          .where('user.id = :userId', { userId: currentUser.id })
          .orderBy('p.id', 'ASC');

        const enrichedData = await queryBuilder.getRawMany();
        // 這裡，我們需要對 enrichedData 進行一些後處理，以將其轉為所需的格式
        const roles = enrichedData.reduce((acc, row) => {
          if (!acc.includes(row.roleId)) {
            acc.push(row.roleId);
          }
          return acc;
        }, [] as number[]);

        const permissions = enrichedData.map(row => ({
          id: row.permissionId,
          title: row.permissionTitle,
          type: row.permissionType,
          code: row.permissionCode
        }));

        return {
          ...currentUser,
          menus: permissions,
          roles: roles.map((roleId: number) => ({ id: roleId }))
        };
      }
```

在上述程式中，我們首先根據當前登入使用者的 ID，使用 createQueryBuilder 方法查詢並獲取使用者的許可權和角色資訊，然後將這些資訊傳回給用戶端。用戶端在請求成功後，會根據傳回的使用者許可權著色相應的功能表列，並將使用者資訊儲存到全域狀態管理中，以便於後續的重複使用。

為了增強使用者體驗和系統性能，在查詢使用者列表時，我們實現了模糊查詢和分頁功能。此外，還需要查詢並傳回每個使用者對應的角色集合。接下來詳細查看具體的程式實現。

```
/**
 * 獲取使用者列表
 *
 * @returns 傳回使用者列表、使用者總數、當前頁碼
 */
async getUserList(dto: UserListDto) {
```

```
  const { username, page, pageSize } = dto
  const skipCount = pageSize * (page - 1)
  let queryBuilder = this.dataSource
    .createQueryBuilder('store_user', 'u')
    .leftJoin('store_user_role', "ur", "ur.userId = u.id")
    .leftJoin('store_role', "r", "r.id = ur.roleId")
    .select(['u.*', "JSON_ARRAYAGG(JSON_OBJECT('id', r.id, 'name', r.name)) as roles"])
    .groupBy('u.id')

    // 如果有模糊檢索條件
    if (username) {
      queryBuilder = queryBuilder.where('u.username Like :username').
setParameter('username', `%${username}%`)
    }
    const list = await queryBuilder.skip(skipCount).take(pageSize).getRawMany()

  return {
    list: list.map(user => {
      Reflect.deleteProperty(user, 'password')
      Reflect.deleteProperty(user, 'salt')
      return user
    }),
    total: list.length,
    page
  }
}
```

在查詢使用者及其連結角色的資料時，程式使用了 JSON_ARRAYAGG 和 JSON_OBJECT 敘述來生成一個 JSON 格式的 roles 陣列。此外，根據提供的查詢準則，透過使用 LIKE 敘述實現模糊搜尋。為了最佳化性能和使用者體驗，程式中還結合了 skip 和 take 敘述來實現分頁功能。在傳回查詢結果之前，確保對敏感性資料進行了適當的清理。

對於使用者資訊的二次編輯，除了更新資料庫中使用者的基本資訊之外，還需要考慮同步更新使用者角色的中間表，並及時刷新 Redis 中快取的使用者資訊，以確保資料的一致性。範例程式如下：

```
/**
 * 更新使用者資訊
 *
 * @param updateUserDto 更新使用者 DTO
 * @param currentUser 當前使用者
 * @returns 更新成功提示
 * @throws 使用者不存在異常
 * @throws 沒有許可權修改管理員資訊異常
 */
```

```
async update(updateUserDto: UpdateUserDto, currentUser: UserEntity) {
  // 1. 判斷使用者是否存在
  const user = await this.userRepository.findOne({where: { id: updateUserDto.id }});
  if (!user) {
    throw new HttpException(' 使用者不存在 ', HttpStatus.EXPECTATION_FAILED);
  }
  // 2. 普通使用者不能修改管理員資訊
  if (user.userType === UserType.ADMIN_USER && currentUser.userType ===
UserType.NORMAL_USER) {
    throw new HttpException(' 你沒有許可權修改管理員資訊 ', HttpStatus.FORBIDDEN);
  }
  // 3. 更新資料
  const { password, ...rest } = await this.userRepository.save(
    plainToClass(UserEntity, {
      ...user,
      ...updateUserDto,
    }, { ignoreDecorators: true })
  );
  // 4. 更新使用者角色表
  if (updateUserDto.roleIds) {
    // 先刪除使用者角色表
    await this.userRoleRepository.delete({ userId: user.id })
    const roles = updateUserDto.roleIds.map(item => ({ userId: user.id, roleId: item }))
    const result = await this.userRoleRepository.save(roles)
    if (!result) {
      throw new HttpException(' 更新失敗，請稍後重試 ', HttpStatus.EXPECTATION_FAILED);
    }
  }
  // 5. 更新 Redis 快取
  const redisKey = getRedisKey(RedisKeyPrefix.USER_INFO, updateUserDto.id)
  await this.redisService.hSet(
    redisKey,
    classToPlain(rest)
  );
  return ' 更新成功 ';
}
```

有許可權的使用者可以對其他使用者的帳號執行凍結操作。一旦使用者帳號被凍結，該使用者將無法登入系統，直到有許可權的使用者執行解凍操作。重要的是，在執行凍結操作後，必須及時更新 Redis 快取，以確保系統的回應與資料狀態保持一致。範例程式如下：

```
/**
 * 更新使用者凍結狀態
 *
 * @param id 使用者 ID
 * @param freezed 是否凍結
```

```
 * @param currUserId 當前使用者 ID
 * @returns 操作成功
 * @throws 使用者不能凍結自己
 * @throws 使用者不存在
 * @throws 你沒有許可權修改管理員資訊
 * @throws 凍結或解凍失敗,請稍後重試
 */
async updateFreezedStatus(id: number, freezed: number, currUserId: number) {
  // 1. 使用者不能凍結自己
  if (id === currUserId) {
    throw new HttpException('你不能凍結自己', HttpStatus.EXPECTATION_FAILED);
  }
  // 2. 判斷使用者是否存在
  const user = await this.userRepository.findOne({where: {id}});
  if (!user) {
    throw new HttpException('使用者不存在', HttpStatus.EXPECTATION_FAILED);
  }
  // 3. 管理員是有最高許可權的,不能被凍結
  if (user.userType === UserType.ADMIN_USER) {
    throw new HttpException('你沒有許可權修改管理員資訊', HttpStatus.FORBIDDEN);
  }
  // 4. 更新資料
  const { affected } = await this.userRepository.update({ id }, { freezed });
  if (!affected) {
    throw new HttpException(`${freezed ? '凍結' : '解凍'}失敗,請稍後重試`,
HttpStatus.EXPECTATION_FAILED);
  }
  // 5. 更新 Redis 快取
  const redisKey = getRedisKey(RedisKeyPrefix.USER_INFO, id);
  const { password, ...rest } = user
  await this.redisService.hSet(
    redisKey,
    classToPlain({ ...rest, freezed })
  );

  return '操作成功';
}
```

管理員可以刪除指定使用者,使用者被刪除之後,需要及時更新使用者角色表及 Redis 快取,程式如下:

```
/**
 * 刪除指定 ID 的使用者
 *
 * @param id 使用者 ID
 * @returns 傳回刪除結果,成功傳回 '刪除成功',失敗拋出例外
 * @throws 當使用者不存在時,拋出 '使用者不存在' 的 HTTP 例外
 * @throws 當刪除失敗時,拋出 '刪除失敗,請稍後重試' 的 HTTP 例外
```

```
   */
  async delete(id: number) {
    // 1. 判斷使用者是否存在
    const user = await this.userRepository.findOne({where: {id}});
    if (!user) {
      throw new HttpException('使用者不存在', HttpStatus.EXPECTATION_FAILED);
    }
    if (user.userType === UserType.ADMIN_USER) {
      throw new HttpException('你沒有許可權刪除管理員', HttpStatus.FORBIDDEN);
    }
    const { affected } = await this.userRepository.delete({id});
    if (!affected) {
      throw new HttpException('刪除失敗,請稍後重試', HttpStatus.EXPECTATION_FAILED);
    }
    // 2. 刪除角色連結表
    const result = await this.userRoleRepository.delete({userId: id});
    if (!result) {
      throw new HttpException('刪除失敗,請稍後重試', HttpStatus.EXPECTATION_FAILED);
    }
    // 3. 刪除 Redis 快取
    const redisKey = getRedisKey(RedisKeyPrefix.USER_INFO, id)
    await this.redisService.del(redisKey)

    return '刪除成功'
  }
```

至此,已完成了使用者模組介面邏輯的實現。相信讀者已經學會了如何編輯角色模組的介面。下一節將繼續開發門店系統中核心的商品與訂單模組。

10.6 實現商品與訂單模組

商品模組是門店經營系統中的核心模組。商家在經營過程中會銷售各種各樣的商品,收銀員為指定商品開單後會產生一筆銷售訂單。門店系統可以根據銷售訂單資料進行統計分析,以便進行銷售提成計算、門店財務統計以及時調整經營策略。

10.6.1 頁面效果展示

商品管理選單衍生出商品列表和熱銷排行兩個子功能表。在商品清單頁面實現對商品表的操作,如查詢、新增、編輯、刪除商品,以及對商品進行上下架操作,如圖 10-38 所示。

▲ 圖 10-38 商品列表

點擊「新增商品」按鈕，使用者可以新增門店對外銷售的商品資訊，如圖 10-39 所示。

▲ 圖 10-39 新增商品

需要注意的是，如果在建立商品時啟用了「是否上架」按鈕，這表示可以對商品進行開單操作，未上架的商品不會出現在開單商品列表中。

點擊「編輯」按鈕，使用者可以對商品資訊進行二次修改後儲存，如圖 10-40 所示。

▲ 圖 10-40 編輯商品

另外，使用者可以直接在列表中更新商品的上下架狀態和刪除指定商品資訊。

在完成商品清單模組的開發之後，接下來實現訂單管理模組，效果如圖 10-41 所示。

▲ 圖 10-41 訂單清單

使用者可以根據訂單編號查詢指定的訂單資訊，同時支援透過訂單狀態篩選資料。收銀員根據客戶付款情況選擇操作對應的訂單狀態，包括未付款、已付款和已取消。

點擊「開單收銀」按鈕可以為客戶購買某個商品進行開單，如圖 10-42 所示。

▲ 圖 10-42 建立訂單

在開單時，系統要確保所選商品不包括任何已下架的商品。使用者選定商品、數量和折扣後，系統將自動計算出商品的總價。此外，選擇一個指定的連結員工是必要的，將會使當前訂單的銷售額與該員工的業績掛鉤，為可能需要的員工業績統計或提成計算提供依據。

訂單建立完成後，使用者可以透過點擊訂單清單中的「訂單編號」按鈕來查看訂單的詳細資訊，如圖 10-43 所示。

▲ 圖 10-43 訂單詳情

完成前端頁面的撰寫後，接下來實現服務端的介面邏輯。

10.6.2 表關係設計

在本專案中，商品與訂單之間存在一對多關聯性，這種連結關係在不同的應用場景中存在差異，比如在許多的電子商務系統中，商品與訂單之間存在多對多關係。首先定義商品實體表欄位，程式如下：

```typescript
import { Column, CreateDateColumn, Entity, PrimaryGeneratedColumn,
UpdateDateColumn } from "typeorm";

@Entity('store_product')
export class ProductEntity {
  @PrimaryGeneratedColumn()
  id: number;

  @Column({ type: 'varchar', length: 50, comment: '商品名稱' })
  name: string;

  @Column({ type: "decimal", precision: 10, scale: 2, comment: '商品價格' })
  price: number

  @Column({ type: 'simple-array', comment: '商品圖片', nullable: true })
  images: string[]

  @Column({ type: 'text', comment: '商品描述' })
  desc: string

  @Column({ type: 'int', comment: '商品狀態 0 未上架  1 已上架 2 已下架' })
  status: number

  @CreateDateColumn({ type: 'datetime', comment: '建立時間' })
  createTime: Date

  @UpdateDateColumn({ type: 'datetime', comment: '更新時間' })
  updateTime: Date
}
```

10.6.3 介面實現

首先完成商品模組的介面開發，在 ProductController 中定義了增刪改查相關的路由方法，程式如下：

```typescript
import {
  Controller,
  Get,
  Post,
  Body,
```

```typescript
  Patch,
  Param,
  Query,
} from '@nestjs/common';
import { ProductService } from './product.service';
import { CreateProductDto } from './dto/create-product.dto';
import { UpdateProductDto } from './dto/update-product.dto';
import { ProductListDto } from './dto/product-list.dto';
import { HotSalesService } from './hot-sales.service';

@Controller('product')
export class ProductController {
  constructor(
    private readonly productService: ProductService,
    private readonly hotSalesService: HotSalesService,
  ) {}

  @Post('create')
  create(@Body() createProductDto: CreateProductDto) {
    return this.productService.create(createProductDto);
  }

  @Get('list')
  getProductList(@Query() productListDto: ProductListDto) {
    return this.productService.getProductList(productListDto);
  }

  @Patch('edit')
  update(@Body() updateProductDto: UpdateProductDto) {
    return this.productService.update(updateProductDto);
  }

  @Get('delete/:id')
  delete(@Param('id') id: number) {
    return this.productService.delete(id);
  }

  @Patch('updateStatus')
  updateStatus(@Body() updateProductDto: UpdateProductDto) {
    return this.productService.updateStatus(
      updateProductDto.id,
      updateProductDto.status,
    );
  }
}
```

新增商品比較簡單，只需將用戶端發送的資訊儲存到資料庫中即可。實現 create 方法的程式如下：

```
/**
 * 建立商品
 *
 * @param createProductDto 建立商品的 DTO
 * @returns 傳回建立成功的訊息
 * @throws 當建立失敗時，拋出 HttpException 例外
 */
async create(createProductDto: CreateProductDto) {
  const product = await this.productRepository.save(createProductDto)
  if (!product) {
    throw new HttpException('建立失敗，請稍後重試', HttpStatus.EXPECTATION_FAILED)
  }
  return '建立成功'
}
```

需要注意的是，在併發請求大的場景下，可以考慮將商品資料儲存到 Redis 中。在需要獲取商品詳情時，首先從快取中獲取，以減少對資料庫的存取次數。

接下來，撰寫獲取商品清單資料的介面。該介面需要支援使用者輸入商品名稱進行模糊查詢，並根據指定的商品狀態篩選資料，最後傳回經過分頁和排序處理的資料，實現程式如下：

```
/**
 * 獲取產品清單
 *
 * @param productListDto 產品清單 DTO
 * @returns 傳回產品清單和總數
 */
async getProductList(productListDto: ProductListDto) {
  const { name, status, page, pageSize } = productListDto
  const where = {
    ...(name ? { name: Like(`%${name}%`) } : null),
    ...(status ? { status } : null),
  }
  const [list, total] = await this.productRepository.findAndCount({
    where,
    order: { id: 'DESC' },
    skip: pageSize * (page - 1),
    take: pageSize,
  });
  return {
    list,
    total
  }
}
```

在上述程式中，根據商品名稱和商品狀態組合查詢準則（where），呼叫實體方法 findAndCount 查詢出清單（list）並統計資料筆數（total）。

接下來，對商品進行二次修改。根據用戶端傳遞的商品 id 判斷是否存在被編輯的記錄，如果存在，則將資料庫資訊更新為新傳遞的商品資訊。範例程式如下：

```
/**
 * 更新商品資訊
 *
 * @param updateProductDto 更新商品資訊所需的資料傳輸物件
 * @returns 傳回更新成功的資訊
 * @throws 如果商品不存在或已刪除，則拋出 HttpException 例外
 * @throws 如果更新失敗，則拋出 HttpException 例外
 */
async update(updateProductDto: UpdateProductDto) {
  const { id } = updateProductDto
  const exists = await this.productRepository.findOneBy({ id });
  if (!exists) {
    throw new HttpException('商品不存在或已刪除', HttpStatus.EXPECTATION_FAILED)
  }
  const { affected } = await this.productRepository.update({ id }, updateProductDto)
  if (!affected) {
    throw new HttpException('更新失敗，請稍後重試', HttpStatus.EXPECTATION_FAILED)
  }
  return '更新成功';
}
```

除了編輯商品資訊外，系統還允許對商品執行快速上下架操作。為了更精細地控制這一操作的許可權，我們將其設計為一個獨立的介面進行管理，程式如下：

```
/**
 * 更新商品狀態
 *
 * @param id 商品 ID
 * @param status 商品狀態，1 表示上架，2 表示下架
 * @returns 傳回一個字串，表示上架或下架是否成功
 * @throws 如果商品不存在或已刪除，則拋出 HttpException 例外
 * @throws 如果更新狀態失敗，則拋出 HttpException 例外
 */
async updateStatus(id: number, status: 1 | 2) {
  const exists = await this.productRepository.findOneBy({ id });
  if (!exists) {
    throw new HttpException('商品不存在或已刪除', HttpStatus.EXPECTATION_FAILED)
  }
  const { affected } = await this.productRepository.update({ id }, { status })
  const text = status === 1 ? '上架' : '下架'
  if (!affected) {
    throw new HttpException(`${text}失敗，請稍後重試`, HttpStatus.EXPECTATION_FAILED)
  }
  return `${text}成功`;
}
```

然後在 store_web.sql 中執行這行 SQL 敘述，向 store_permission 表中插入一筆

資料：

```
INSERT INTO 'store_permission' VALUES (17, '上下架商品', 'updateStatus:product', 3, 2);
```

此時許可權表中多了一筆資料，可以在編輯角色的彈窗中看到「上下架商品」選項，為指定的角色分配許可權，如圖 10-44 所示。

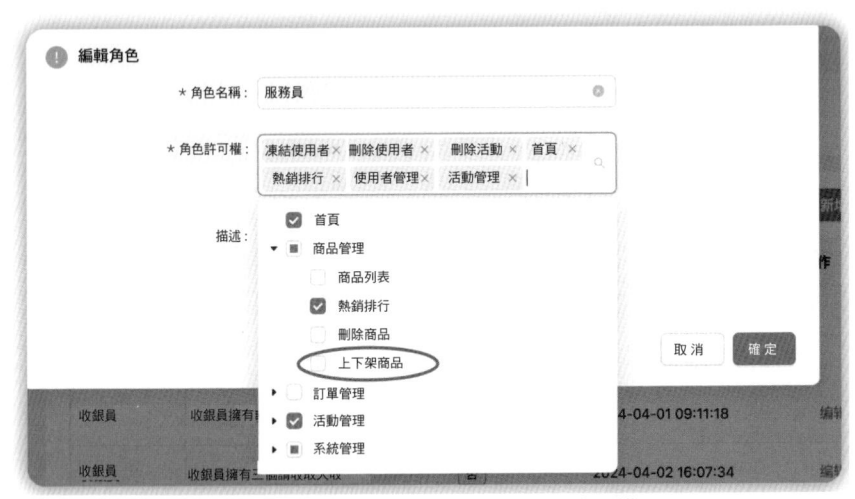

▲ 圖 10-44 分配「上下架商品」許可權

除此之外，我們還需要在 store_permission_api 表中增加 updateStatus 介面資料：

```
INSERT INTO `store_permission_api` VALUES (24, '/product/updateStatus', 'PATCH', 17);
```

這筆資料連結 store_permission 表中 id=17 的記錄，此時守衛會在每次請求時，使用當前使用者角色對應的許可權集驗證是否有該介面的請求許可權，以此決定是否放行，實現介面層的許可權控制。

同理，商品模組中的刪除介面的許可權控制也是如此，這裡不再贅述，介面的具體實現程式如下：

```
/**
 * 刪除商品
 *
 * @param id 商品 ID
 * @returns 傳回刪除結果，成功傳回'刪除成功'，失敗拋出例外
 * @throws 當商品不存在或已刪除時，拋出 HttpException 例外，狀態碼為
HttpStatus.EXPECTATION_FAILED
 * @throws 當刪除失敗時，拋出 HttpException 例外，狀態碼為 HttpStatus.EXPECTATION_FAILED
 */
async delete(id: number) {
```

```
  const exists = await this.productRepository.findOneBy({ id });
  if (!exists) {
    throw new HttpException(' 商品不存在或已刪除 ', HttpStatus.EXPECTATION_FAILED)
  }
  const { affected } = await this.productRepository.delete({ id })
  if (!affected) {
    throw new HttpException(' 刪除失敗，請稍後重試 ', HttpStatus.EXPECTATION_FAILED)
  }
  return ' 刪除成功 ';
}
```

刪除商品與編輯商品類似，首先根據 id 判斷商品是否存在，如果存在，則按照正常刪除邏輯刪除商品即可。

前面介紹了商品模組的介面實現，接下來看訂單模組的介面實現過程。前面提到了商品與訂單是一對多關聯性，在設計上，訂單表中會記錄當前訂單對應商品的 id，透過 id 就能查詢商品資訊，具體實體表欄位如下：

```
import { Column, CreateDateColumn, Entity, PrimaryGeneratedColumn } from "typeorm";

@Entity('store_order')
export class OrderEntity {
  @PrimaryGeneratedColumn()
  id: number;

  @Column({ type: 'varchar', length: 50, comment: ' 商品名稱 ' })
  name: string;

  @Column({ type: 'int', default: 1, comment: ' 商品數量 ' })
  count: number;

  @Column({ type: "decimal", precision: 5, scale: 2, default: 1, comment: ' 訂單折扣 ' })
  discount: number;

  @Column({ type: "decimal", precision: 10, scale: 2, comment: ' 訂單價格 ' })
  price: number

  @Column({ type: "decimal", precision: 10, scale: 2, comment: ' 訂單折扣價 ' })
  discountPrice: number

  @Column({ type: 'int', comment: ' 訂單狀態 0 未付款 1 已付款 2 已取消 ' })
  status: number;

  @Column({ type: 'varchar', comment: ' 操作員 ' })
  operator: string;

  @CreateDateColumn({ type: 'timestamp', comment: ' 建立時間 ' })
  createTime: Date;
```

```ts
  @Column({ type: 'text', comment: '訂單備註', nullable: true })
  desc: string;

  @Column({ type: 'int', comment: '商品id' })
  productId: number;
}
```

其中,訂單價格表示在未參與折扣時的訂單總價,而訂單折扣價表示在參與折扣後客戶應該實付的價格。定義完實體表後,在 OrderController 中定義需要實現的路由介面,程式如下:

```ts
import { Controller, Get, Post, Body, Patch, Param, Req, Query } from '@nestjs/common';
import { OrderService } from './order.service';
import { CreateOrderDto } from './dto/create-order.dto';
import { UpdateOrderDto } from './dto/update-order.dto';
import { OrderListDto } from './dto/order-list.dto';

@Controller('order')
export class OrderController {
  constructor(private readonly orderService: OrderService) {}

  @Post('create')
  create(@Body() createOrderDto: CreateOrderDto) {
    return this.orderService.create(createOrderDto);
  }

  @Get('list')
  getOrderList(@Query() orderListDto: OrderListDto) {
    return this.orderService.getOrderList(orderListDto);
  }

  @Get('detail/:id')
  getOrderDetail(@Param('id') id: string) {
    return this.orderService.getOrderDetail(+id);
  }

  @Patch('updateOrder')
  updateOrder(@Body() updateOrderDto: UpdateOrderDto, @Req() req) {
    return this.orderService.updateOrder(updateOrderDto, req.user);
  }

  @Get('delete/:id')
  delete(@Param('id') id: string, @Req() req) {
    return this.orderService.delete(+id, req.user);
  }
}
```

接下來,根據上面的方法分別實現服務介面邏輯。在實際經營場景中,一般收銀員、店長或管理員才擁有開單許可權，在本專案中，我們將此許可權分配給收銀員這個角色。開單介面的具體實現程式如下：

```
/**
 * 建立訂單
 *
 * @param createOrderDto 建立訂單所需的參數
 * @returns 傳回成功資訊
 * @throws HttpException 拋出 HTTP 例外
 */
async create(createOrderDto: CreateOrderDto) {
  const { productId, discount = 1, status = 0, count } = createOrderDto
  // 獲取商品資訊
  const product = await this.productRepository.findOneBy({ id: productId })
  if (!product) {
    throw new HttpException('商品不存在', HttpStatus.NOT_FOUND)
  }
  // 獲取開單商品計算價格
  let totalPrice = product.price * count
  let discountPrice = totalPrice * discount
  let orderItem = {
    ...createOrderDto,
    name: product.name,
    price: totalPrice,
    status,
    count,
    productId,
    discountPrice,
    discount
  }
  const order = await this.orderRepository.save(plainToClass(OrderEntity, orderItem))
  if (!order) {
    throw new HttpException('開單失敗', HttpStatus.INTERNAL_SERVER_ERROR)
  }
  return '開單成功';
}
```

在上述程式中，首先驗證所選商品是否存在。如果商品存在，系統將基於商品的單價、數量以及訂單折扣重新計算訂單的總價。這一步驟至關重要，因為它確保了資料的準確性，而非簡單地儲存用戶端發送的價格。

完成訂單總價的計算後，我們使用 plainToClass 方法將包含訂單詳情的普通物件轉換成 OrderEntity 類別的實例。這一轉換確保了訂單物件擁有 OrderEntity 類別定義的所有屬性，從而可以被正確地儲存到資料庫中。

訂單建立成功後，獲取訂單清單的操作就變得相對簡單。這一過程與之前獲取商品列表的操作類似。訂單清單獲取介面的程式如下：

```
/**
 * 獲取訂單清單
 *
 * @param orderListDto 訂單清單 DTO
 * @returns 傳回一個包含訂單清單和總筆數的物件
 */
async getOrderList(orderListDto: OrderListDto) {
  const { page, pageSize, id, status } = orderListDto
  const where = {
    ...(id ? { id } : null),
    ...(status ? { status } : null)
  }
  const [list, total] = await this.orderRepository.findAndCount({
    where,
    order: { id: 'DESC' },
    skip: pageSize * (page - 1),
    take: pageSize
  });
  return {
    list,
    total
  }
}
```

在通常情況下，清單中不能夠完全展示訂單的所有欄位，為此可以透過查看訂單詳情看到更詳細的訂單資訊，介面程式如下：

```
/**
 * 獲取訂單詳情
 *
 * @param id 訂單 ID
 * @returns 傳回訂單詳情物件，包含訂單資訊和商品資訊
 * @throws 當訂單不存在時，拋出 HttpException 例外，狀態碼為 NOT_FOUND
 * @throws 當查詢訂單商品失敗時，拋出 HttpException 例外，狀態碼為 NOT_FOUND
 */
async getOrderDetail(id: number) {
  const order = await this.orderRepository.findOneBy({ id })
  if (!order) {
    throw new HttpException(' 訂單不存在 ', HttpStatus.NOT_FOUND)
  }
  const product = await this.productRepository.findOneBy({ id: order.productId })
  if (!product) {
    throw new HttpException(' 查詢訂單商品失敗 ', HttpStatus.NOT_FOUND)
  }
  return {
    ...order,
    product
```

```
    };
}
```

在上述程式中，先根據訂單編號獲取對應的訂單資訊，由於訂單中綁定著商品 id，再根據商品 id 查詢出連結的商品資訊，重新組裝後再傳回給用戶端。

收銀員可以為客戶生成未付款訂單，隨後對使用者訂單進行收款或取消訂單操作，介面的實現程式如下：

```
/**
 * 更新訂單狀態及描述
 *
 * @param updateOrderDto 更新訂單資訊
 * @param currentUser 當前使用者資訊
 * @returns 傳回更新結果
 * @throws 當訂單不存在或已刪除時，拋出 HttpException 例外，狀態碼為 NOT_FOUND
 * @throws 當修改失敗時，拋出 HttpException 例外，狀態碼為 INTERNAL_SERVER_ERROR
 */
async updateOrder(updateOrderDto: UpdateOrderDto, currentUser: UserEntity) {
  const { id, status, desc = '' } = updateOrderDto
  const exists = this.orderRepository.findOneBy({ id })
  if (!exists) {
    throw new HttpException('訂單不存在或已刪除', HttpStatus.NOT_FOUND);
  }
  const { affected } = await this.orderRepository.update({ id }, { status, desc })
  if (!affected) {
    throw new HttpException('修改失敗', HttpStatus.INTERNAL_SERVER_ERROR);
  }
  return '修改成功';
}
```

在上述程式中，允許對訂單的狀態和備註進行修改。然而，根據業務規則，一旦訂單狀態被更改，該變更通常是不可逆的。舉例來說，一旦訂單被取消，就無法將其狀態重新修改為「待付款」。這種設計旨在確保交易過程的完整性，並防止可能的不當商家行為，如濫用訂單狀態更改進行詐騙或其他不當操作。

此外，收銀員有權根據訂單 id 刪除已建立的訂單記錄。這通常在訂單因錯誤或其他原因需要撤銷時進行。訂單狀態修改和訂單記錄刪除的範例程式如下：

```
/**
 * 刪除訂單
 *
 * @param id 訂單 ID
 * @param currentUser 當前使用者
 * @returns 刪除成功或失敗的資訊
 * @throws 如果訂單不存在或已刪除，則拋出 404 錯誤；如果刪除失敗，則拋出 500 錯誤
 */
```

```
async delete(id: number) {
  const exists = await this.orderRepository.findOneBy({ id })
  if (!exists) {
    throw new HttpException('訂單不存在或已刪除', HttpStatus.NOT_FOUND);
  }
  const { affected } = await this.orderRepository.delete({ id })
  if (!affected) {
    throw new HttpException('刪除失敗', HttpStatus.INTERNAL_SERVER_ERROR);
  }
  return '刪除成功';
}
```

至此，我們完成了商品與訂單模組的開發。在此基礎上，我們可以新增幾行程式邏輯，實現用 Redis 來統計熱銷的商品排行榜，詳細的實現過程將在下一節介紹。

10.7 基於 Redis 實現商品熱銷榜

排行榜是日常生活中普遍存在的一種列表形式，例如在新聞平臺如今日頭條、遊戲排行榜、熱門文章榜單以及商品銷量排行榜中都能找到它們的身影。這類排行榜在很大程度上展現了使用者對不同內容或產品的關注熱度。透過對這些資料進行細緻的統計分析，企業可以及時調整其行銷和營運策略，從而有效提高行銷的轉化效率。

學習本節內容後，讀者將能夠理解並掌握在多種場景下實現排行榜的原理和方法。

10.7.1 頁面效果展示

商品熱銷排行榜的前端頁面比較簡單，只需要一個列表即可展示，如圖 10-45 所示。

▲ 圖 10-45 商品熱銷排行榜

清單頁面的實現程式如下：

```
import { getHotProductList } from '@/apis/product'
import ExcelTable from '@/components/exportExcel'
import { Tag, Image } from 'antd'
import { observer } from 'mobx-react'
import { ProductStatus } from '@/common/enums'

const HotProductList: React.FC = () => {
  return (
    <>
      <ExcelTable
        hideSearch
        columns={[
          {
            title: '排名',
            key: 'index',
            render(_, __, index) {
              return index + 1
            }
          },
          {
            title: '商品名稱',
            dataIndex: 'name',
            hideInSearch: true
          },
          {
            title: '商品圖片',
            dataIndex: 'images',
            hideInSearch: true,
            render(_, record) {
              return record.images?.length ? record.images?.map((item: string) => (
```

```jsx
                <Image width={60} src={item} alt={item} key={item} />
              )) : '-'
            },
          },
          {
            title: '狀態',
            dataIndex: 'status',
            hideInSearch: true,
            render(_, record) {
              const green = record.status === ProductStatus.ON_SALE
              const text = record.status === ProductStatus.NOT_ON_SALE ? '未上架' :
                record.status === ProductStatus.ON_SALE ? '已上架' : '已下架'
              return <Tag color={green ? 'green' : 'red'}>{ text }</Tag>
            },
          },
          {
            title: '價格',
            dataIndex: 'price',
            hideInSearch: true
          },
          {
            title: '銷量',
            dataIndex: 'score',
            hideInSearch: true
          },
          {
            title: '建立時間',
            dataIndex: 'createTime',
            hideInSearch: true,
            valueType: 'dateTime'
          }
        ]}
        requestFn={async () => {
          const data = await getHotProductList({ topN: 10 })
          return data
        }}
        rowSelection={false}
      />
    </>
  )
}

export default observer(HotProductList)
```

上述程式定義了排行榜相關的列欄位，並在 requestFn 方法中請求了 top10 的商品銷量資料，展示銷量從高到低的排行資料。

10.7.2 介面實現

服務端想要實現這種排名效果，通常會使用 Redis 中的 Sorted Set，也就是有序集合。有序集合的每個成員都與一個分數相連結，並根據這個分數進行排序，因此非常適合這種場景。

首先建立一個單獨的熱銷服務類別 HotSalesService，並實現一個增加統計商品銷量的介面 addProductSales，當每次對商品進行開單時，將商品的數量增加到 Redis 中，程式如下：

```
/**
 * 增加產品銷量
 *
 * @param productId 產品 ID
 * @param saleCount 銷量
 * @returns Promise<void>
 */
async addProductSales(productId: string, saleCount: number) {
  const client = this.redisService.getClient()
  const redisKey = getRedisKey(RedisKeyPrefix.HOT_SALES)
  await client.zIncrBy(redisKey, saleCount, productId)
}
```

在上述程式中，介面接收商品 id（productId）與商品銷量（saleCount）兩個參數，並定義了名為 HOT_SALES 的 Redis 鍵，呼叫 zIncrBy 方法新增或更新指定成員（商品）的分數（銷量），最終 Redis 中快取的銷量資料如圖 10-46 所示。

▲ 圖 10-46 銷量資料

在圖 10-46 中，左側的 Member 代表商品 id，右側的 Score 代表商品對應的銷量數。

接下來，在開單介面中新增一行程式，呼叫 addProductSales 方法來統計資料，程式如下：

```
/**
 * 建立訂單
 *
 * @param createOrderDto 建立訂單所需的參數
 * @returns 傳回成功資訊
 * @throws HttpException 拋出 HTTP 例外
 */
async create(createOrderDto: CreateOrderDto) {
  const { productId, discount = 1, status = 0, count } = createOrderDto
  // 獲取商品資訊
  const product = await this.productRepository.findOneBy({ id: productId })
  if (!product) {
    throw new HttpException('商品不存在', HttpStatus.NOT_FOUND)
  }
  // 獲取開單商品計算價格
  let totalPrice = product.price * count
  let discountPrice = totalPrice * discount
  let orderItem = {
    ...createOrderDto,
    name: product.name,
    price: totalPrice,
    status,
    count,
    productId,
    discountPrice,
    discount
  }
  const order = await this.orderRepository.save(plainToClass(OrderEntity, orderItem))
  if (!order) {
    throw new HttpException('開單失敗', HttpStatus.INTERNAL_SERVER_ERROR)
  }
+ // 快取銷量到 Redis 做排行榜
+ await this.hotSalesService.addProductSales(String(productId), count)
  return '開單成功';
}
```

有了 Redis 銷量排行資料後，接下來實現介面獲取 topN 的熱銷商品資料，程式如下：

```
/**
 * 獲取熱銷產品清單前 n 個
 *
```

```
 * @param n 獲取數量
 * @returns 傳回包含熱銷產品清單的物件,每個產品包含 id、name、price、score 等屬性,按照 score 從高
到低排序
 */
async getTopNProducts(n: number) {
  const client = this.redisService.getClient()
  const redisKey = getRedisKey(RedisKeyPrefix.HOT_SALES)
  const scoreList = await client.zRangeWithScores(redisKey, 0, -1)
  const hotIdList = scoreList.reverse().slice(0, n).map(item => item.value)
  const hotList = await this.productRepository.find({where: { id: In(hotIdList) }})
  return {
    list: hotList.map((item, index) => ({ ...item, score: scoreList[index].score }))
.sort((a, b) => b.score - a.score)
  }
}
```

由於我們使用 node-redis 套件來操作 Redis 用戶端,為了獲取 topN 排行資料,程式中用 zRangeWithScores 方法來獲取商品 id 及銷量資料(其他 Redis 套件可能有現成的 API 直接呼叫)。在預設情況下,這些資料按銷量昇冪排列。

查詢出 topN 的商品 id 清單後,最後從資料庫中查詢指定 id 的商品資料,根據分數由高到低排序並傳回用戶端。

至此,我們使用 Redis 中一個高效的集合完成了商品熱銷排行的功能。除此之外,按日排行或按月排行無非是將 redisKey 加入一個時間的維度來實現儲存不同的排行資料,如 daily_ranking:${year}-${month}-${day} 或 monthly_ranking:${year}-${month} 等。

10.8 實現活動模組與定時任務

在本專案的最終模組中,我們將聚焦於活動管理模組的開發。商家為了提升銷售額,經常推出多種行銷策略,比如團購、限時優惠或買一贈一等促銷活動。在建立這些活動時,可以指定參與的商品並設置活動的有效期,包括具體的開始和結束時間。

為了自動化管理活動的時間限制,我們計畫引入定時任務功能,以便在預設時間自動啟動或終止活動。此外,我們還將開發一個展示當前活動列表的功能,並允許使用者編輯活動、更改活動狀態或刪除活動。

10.8.1 頁面效果展示

活動清單頁面的對話模式與商品清單類似，區別在於對列表資料執行的操作不同，如圖 10-47 所示。

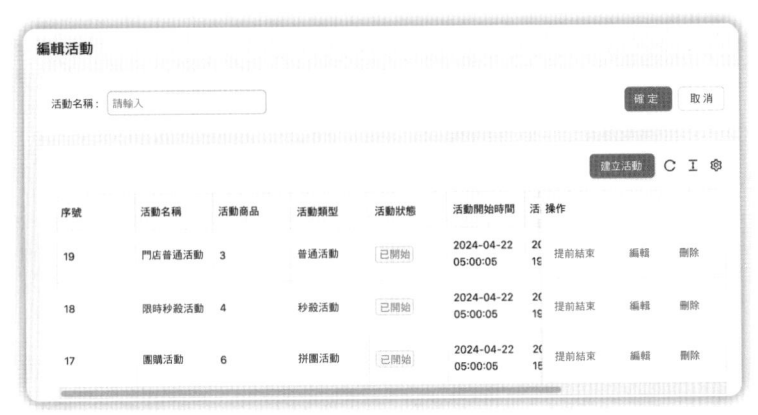

▲ 圖 10-47 活動列表

使用者可以透過輸入活動名稱的部分或全部內容來對活動清單進行模糊搜尋。系統也允許使用者提前結束正在進行中的活動，但請注意，活動一旦結束，其狀態將無法更改。另外，使用者還可以透過點擊「建立活動」按鈕來增加新的門店活動，如圖 10-48 所示。

▲ 圖 10-48 建立活動

同樣地，使用者可以對未結束的活動詳情進行二次編輯操作，如圖 10-49 所示。

▲ 圖 10-49 編輯活動

在圖 10-48 的建立表單與圖 10-49 的編輯表單中都可以選擇參與活動的商品，未上架的商品不會在此顯示，同時活動時間需要設置在合理的範圍內，例如活動結束時間不能夠早於當前時間，否則服務端會拒絕請求，詳細的實現過程將在後文實現介面時介紹。

最後，為了預防使用者誤操作導致資料遺失，系統將在使用者點擊「刪除」按鈕時出現一個二次確認的互動提示。這樣可以確保使用者在刪除前再次確認，如圖 10-50 所示。

▲ 圖 10-50 刪除操作的互動過程

10.8.2 表關係設計

在行銷活動中需要綁定一個參與活動優惠的商品，在活動表中記錄一個商品 id，實體欄位定義如下：

```
import { ActivityStatus } from "src/common/enums/common.enum";
import { Column, CreateDateColumn, Entity, PrimaryGeneratedColumn,
```

```typescript
UpdateDateColumn } from "typeorm";

@Entity('store_activity')
export class ActivityEntity {
  @PrimaryGeneratedColumn()
  id: number;

  @Column({ type: 'varchar', length: 30, comment: '活動名稱' })
  name: string;

  @Column({ type: 'int', comment: '活動狀態 0 未開始 1 進行中 2 已結束', default: 0 })
  status: ActivityStatus

  @Column({ type: 'int', comment: '活動類型 0 普通活動 1 拼團活動' })
  type: number

  @Column({ type: 'text', comment: '活動描述' })
  desc: string

  @Column({ type: 'timestamp', comment: '活動開始時間' })
  startTime: Date

  @Column({ type: 'timestamp', comment: '活動結束時間' })
  endTime: Date

  @CreateDateColumn({ type: 'timestamp', comment: '建立時間' })
  createTime: Date

  @UpdateDateColumn({ type: 'timestamp', comment: '更新時間' })
  updateTime: Date

  @Column({ type: 'int', comment: '參與活動的商品 id' })
  productId: number

}
```

10.8.3 介面實現

　　門店活動的建立通常由店長或管理員負責。在建立活動的過程中，需要判斷活動時間的有效性。結束時間小於當前時間或開始時間大於結束時間，活動將無法成功建立。此外，如果開始時間已經大於當前時間，則應該把活動狀態設置為已開始。具體的實現程式如下：

```
/**
 * 建立活動
 *
```

```
 * @param createActivityDto 建立活動的 DTO
 * @param currentUser 當前使用者
 * @returns 傳回建立成功提示
 * @throws 當使用者類型不是管理員時,拋出許可權例外
 * @throws 當結束時間小於當前時間時,拋出期望失敗例外
 * @throws 當開始時間大於結束時間時,拋出期望失敗例外
 * @throws 當建立活動失敗時,拋出內部伺服器錯誤例外
 */
async create(createActivityDto: CreateActivityDto, currentUser: UserEntity) {
  if (currentUser.userType !== UserType.ADMIN_USER) {
    throw new HttpException('您沒有許可權建立活動,請聯繫管理員', HttpStatus.FORBIDDEN)
  }
  const { startTime, endTime } = createActivityDto
  let status = ActivityStatus.NOT_START
  if (new Date(endTime).getTime() < Date.now()) {
    throw new HttpException('結束時間不能小於當前時間', HttpStatus.EXPECTATION_FAILED)
  }
  if (startTime > endTime) {
    throw new HttpException('開始時間不能大於結束時間', HttpStatus.EXPECTATION_FAILED)
  }
  if (new Date(startTime).getTime() < Date.now()) {
    status = ActivityStatus.IN_PROGRESS
  }
  const activity = await this.activityRepository.save({ ...createActivityDto, status })
  if (!activity) {
    throw new HttpException('建立失敗', HttpStatus.INTERNAL_SERVER_ERROR)
  }
  return '建立成功';
}
```

前文在設計活動表時記錄了商品 id。在獲取活動清單時,需要同時查詢出 id 對應的商品資訊。由於我們沒有使用 @OneToMany 或 @ManyToOne 宣告活動表與商品表的連結關係,因此在查詢時需要手動使用 QueryBuilder 進行連接。具體的實現程式如下:

```
/**
 * 獲取活動列表
 *
 * @param dto 活動列表查詢參數
 * @returns 傳回活動列表及總數
 */
async getActList(dto: ActivityListDto) {
  const { page, pageSize, name, type, status } = dto
  const where = {
    ...(name ? { name: Like(`%${name}%`) } : null),
    ...(type ? { type } : null),
    ...(status ? { status } : null),
```

```
  }
  const queryBuilder = this.dataSource
  .createQueryBuilder('store_activity', 'act')
  .leftJoinAndSelect('store_product', 'p', 'p.id = act.productId')
  .select(['act.*', "JSON_OBJECT('name', p.name, 'price', p.price) as product"])

  const total = await queryBuilder.getCount()

  const list = await queryBuilder
  .where(where)
  .skip(pageSize * (page - 1))
  .take(pageSize)
  .orderBy('act.id', 'DESC')
  .getRawMany()

  return {
    list,
    total
  }
}
```

在程式中，首先根據查詢準則定義了 where 敘述。接著建立了一個查詢活動表的 queryBuilder，並使用 leftJoinAndSelect 方法查詢出對應的 product 資訊，將其命名為物件 product，其中包含 name 和 price 兩個欄位。最後，透過分頁和排序後將結果傳回給用戶端。

當使用者對活動進行「提前結束」或「編輯」操作時，需要重新修改活動資訊。此時需要判斷活動時間是否在合理範圍內，並及時更新活動狀態。具體的實現程式如下：

```
/**
 * 更新活動資訊
 *
 * @param updateActivityDto 更新活動資訊的 DTO
 * @returns 傳回更新成功的提示訊息
 * @throws 當活動不存在或已刪除時，拋出 HTTP 例外，狀態碼為 404
 * @throws 當活動已結束時，拋出 HTTP 例外，狀態碼為 417
 * @throws 當結束時間小於當前時間時，拋出 HTTP 例外，狀態碼為 417
 * @throws 當更新失敗時，拋出 HTTP 例外，狀態碼為 417
 */
async update(updateActivityDto: UpdateActivityDto) {
  const { id, startTime, endTime } = updateActivityDto
  const exists = await this.activityRepository.findOneBy({id})
  if (!exists) {
    throw new HttpException(' 活動不存在或已刪除 ', HttpStatus.NOT_FOUND)
  }
  if (exists.status === ActivityStatus.END) {
```

```typescript
    throw new HttpException(' 活動已結束，不能修改 ', HttpStatus.EXPECTATION_FAILED)
  }
  if (new Date(endTime).getTime() < Date.now()) {
    throw new HttpException(' 結束時間不能小於當前時間 ', HttpStatus.EXPECTATION_FAILED)
  }
  let newStatus = updateActivityDto.status || exists.status
  if (new Date(startTime).getTime() < Date.now()) {
    newStatus = ActivityStatus.IN_PROGRESS
  }
  const activity = await this.activityRepository.save(plainToClass(ActivityEntity,
{ ...exists, ...updateActivityDto, newStatus }))
  if (!activity) {
    throw new HttpException(' 更新失敗 ', HttpStatus.EXPECTATION_FAILED)
  }

  return ' 更新成功 '
}
```

最後，具備刪除許可權的成員根據活動 id 對指定活動進行刪除。實現過程比較簡單，程式如下：

```typescript
/**
 * 刪除活動
 *
 * @param id 活動 ID
 * @returns 傳回刪除成功的字串
 * @throws HttpException 當使用者類型不是管理員時，拋出許可權不足例外
 * @throws HttpException 當活動不存在或已刪除時，拋出未找到例外
 * @throws HttpException 當刪除失敗時，拋出期望失敗例外
 */
async delete(id: number) {
  const exists = await this.activityRepository.findOneBy({id})
  if (!exists) {
    throw new HttpException(' 活動不存在或已刪除 ', HttpStatus.NOT_FOUND)
  }
  const { affected } = await this.activityRepository.delete({ id })
  if (!affected) {
    throw new HttpException(' 刪除失敗，請稍後重試 ', HttpStatus.EXPECTATION_FAILED)
  }
  return ' 刪除成功 ';
}
```

至此，我們完成了所有模組的前端頁面開發和介面程式邏輯的實現，相信讀者完全有能力撰寫屬於自己的模組。下一節將進入專案部署階段，將開發完畢的專案部署到伺服器上執行。

10.9 使用 Docker Compose 部署專案

在之前的章節中,我們探討了如何利用 Docker 將 Nest 服務部署到伺服器上。然而,在實際應用專案中,還涉及 MySQL 和 Redis 服務。如果透過執行單獨的「docker run」命令來啟動這些服務,不僅效率低下,還需要考慮服務之間啟動的先後順序。舉例來說,如果 Nest 服務嘗試啟動時,MySQL 或 Redis 服務尚未啟動,程式就會直接顯示出錯。Docker Compose 能有效地解決這類問題。本節將使用 Docker Compose 將數字門店專案部署到本地環境。

10.9.1 撰寫後端 Docker Compose 檔案

通常情況下,後端各種中介軟體會部署到測試、預發佈和線上環境中,這裡以 Docker 作為測試環境進行演示。在 Nest 中建構時,需要區分不同環境的通訊埠編號、MySQL 和 Redis 配置,使用 .env 和 .env.docker 設定檔來管理這些配置。.env 配置如下:

```
# JWT 配置
JWT_SECRET=store-web-secret
JWT_EXPIRE_TIME=7d

# MySQL 配置
MYSQL_HOST=localhost
MYSQL_PORT=3306
MYSQL_USER=root
MYSQL_PASSWORD=jminjmin
MYSQL_DATABASE=store_web_project

# Redis 配置
REDIS_HOST=localhost
REDIS_PORT=6379

# 電子郵件配置
EMAIL_PASS=EBBOFYAQDWBLOTZY
EMAIL_HOST=smtp.163.com
EMAIL_PORT=465
EMAIL_SECURE=true
EMAIL_USER=jmin95@163.com
EMAIL_ALIAS=store_web_project

API_PREFIX=/api
APP_PORT=3332
```

在上述程式中，在開發環境下，MySQL 與 Redis 的 HOST 指向的是 localhost，同時 Nest 服務啟動在 3332 通訊埠下。當專案部署在 Docker 容器中時，同樣會用 localhost 來存取。為了避免開發與測試環境通訊埠的衝突，需要進行適當區分。接下來，請看 .env.docker 的配置：

```
# JWT 配置
JWT_SECRET=store-web-secret
JWT_EXPIRE_TIME=7d

# MySQL 配置
MYSQL_HOST=mysql
MYSQL_PORT=3306
MYSQL_USER=root
MYSQL_PASSWORD=jminjmin
MYSQL_DATABASE=store_web_project

# Redis 配置
REDIS_HOST=redis
REDIS_PORT=6379

# 電子郵件配置
EMAIL_PASS=EBBOFYAQDWBLOTZY
EMAIL_HOST=smtp.163.com
EMAIL_PORT=465
EMAIL_SECURE=true
EMAIL_USER=jmin95@163.com
EMAIL_ALIAS=store_web_project

API_PREFIX=/api
APP_PORT=3333
```

與 .env 不同的是，Docker 中的 Nest 服務啟動在 3333 通訊埠，以確保 Docker 中的 Nest 服務與宿主機上的 Nest 服務同時執行時期不會發生衝突。此外，MySQL 的 HOST 指向名為 mysql 的容器，而 Redis 的 HOST 指向名為 redis 的容器。這兩個容器將在 docker-compose.yml 中定義並執行。docker-compose.yml 的配置如下：

```
version: "3"

services:

  servers:
    build:
      context: ./
      dockerfile: ./Dockerfile
```

```yaml
    restart: always
    ports:
      - 3333:3333
    networks:
      - nest-net
    volumes:
      - upload:/upload
    depends_on:
      - mysql
      - redis

  mysql:
    container_name: mysql
    image: mysql:8
    command: mysqld --character-set-server=utf8mb4 --collation-server=utf8mb4_unicode_ci
    restart: always
    environment:
      - MYSQL_ROOT_PASSWORD=jminjmin
      - MYSQL_DATABASE=store_web_project
    networks:
      - nest-net
    volumes:
      - mysql:/var/lib/mysql

  redis:
    container_name: redis
    image: redis:latest
    restart: always
    networks:
      - nest-net
    volumes:
      - redis:/data
networks:
  nest-net:
    driver: bridge
volumes:
  mysql:
  redis:
  upload:
```

在上述配置中可以看到，在 services 下定義了三個鏡像配置，分別為 Nest 服務鏡像、MySQL 鏡像和 Redis 鏡像。它們都被指定同處於 nest-net 網路下，以保證 Nest 服務可以正常存取 MySQL 和 Redis 服務。連接方式採用橋接網路。最後，還宣告了各鏡像對應的資料卷冊。

首先來看 Nest 鏡像配置，指定了使用 Dockerfile 為 Nest 的鏡像檔案，該鏡像向宿主機暴露 3333 通訊埠用於連接。在 networks 中定義了名為 nest-net 的網路，MySQL 和 Redis 透過這個網路名稱進行連接。最後，指定了資料卷冊的路徑，並宣告 Nest 鏡像相依於 mysql 和 redis 鏡像。只有在這兩者啟動後，Nest 服務才會啟動。Nest 的 Dockerfile 配置如下：

```
FROM node:18-alpine3.18
WORKDIR /servers
COPY . .
ENV TZ=Asia/Guangzhou
RUN npm set registry=https://registry.npmmirror.com
RUN npm i -g pnpm && pnpm i && pnpm run build
EXPOSE 3333
CMD pnpm start:docker
```

需要特別說明的是，容器啟動時會執行「pnpm start:docker」命令來啟動 Nest 服務。本質上，這個命令執行的是以下的 Node 命令，它在 package.json 的 scripts 中定義：

```
"start:docker": "cross-env NODE_ENV=docker node dist/main"
```

設置 Node 環境變數為 docker 後，啟動服務，此時 Nest 就會獲取 .env.docker 中的配置資訊。

MySQL 和 Redis 的鏡像配置過程大致相和，唯一的差別是這兩個服務不需要將通訊埠編號暴露給外部宿主機或其他服務進行連接。我們只需確保 Nest 服務能夠與它們連接即可。另外，請不要忘記配置 .dockerignore 檔案：

```
.vscode
.prettierrc
.eslintrc.js
node_modules
dist
package-lock.json
npm-debug.log
```

一切就緒後，執行「docker-compose up」命令啟動容器，效果如圖 10-51 所示。

```
$ docker-compose up
[+] Running 15/20
 ⠿ mysql 10 layers [▓▓▓▓▓ ⠂ ] 11.96MB/112.1MB Pulling                                27.6s
   ⠴ c6a0976a2dbe Downloading [======>          ] ...                                21.5s
   ✓ 8dd4f8e415ca Download complete                                                  10.6s
   ✓ 6e01a6ece3af Download complete                                                  13.0s
   ✓ 6cfdeffd9140 Download complete                                                  14.0s
   ✓ 73fed55ee93c Download complete                                                  15.0s
   ✓ 83404f4e4847 Download complete                                                  15.5s
   ⠴ aad53405df78 Downloading [====>            ] ...                                21.5s
   ✓ d9c5f6f4cc6e Download complete                                                  17.0s
   ⠴ e04d803ff9c7 Waiting                                                            21.5s
   ⠴ f06a309d43da Waiting                                                            21.5s
 ⠿ redis 8 layers [▓▓▓▓▓▓▓▓]      0B/0B      Pulled                                  19.7s
   ✓ 22d97f6a5d13 Already exists                                                      0.0s
   ✓ c7b117eba408 Pull complete                                                       1.5s
   ✓ 3549e2a23473 Pull complete                                                       1.5s
   ✓ ccadc22b76d5 Pull complete                                                       2.4s
   ✓ d4ed6d335745 Pull complete                                                      10.9s
   ✓ 00c30f88f8b5 Pull complete                                                       3.2s
```

▲ 圖 10-51 compose 啟動容器

在此過程中，會優先建立 MySQL 和 Redis 的鏡像並啟動相應的容器，而 Nest 服務將在最後啟動。一旦完成，我們將看到三個服務都已順利啟動，如圖 10-52 所示。

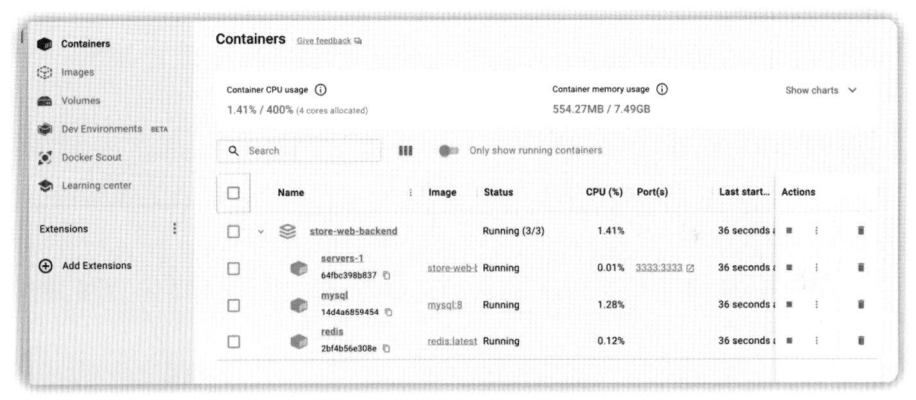

▲ 圖 10-52 容器啟動成功

在主控台下顯示 Nest 服務已正常啟動，如圖 10-53 所示。

▲ 圖 10-53 服務正常執行

我們撰寫的 Dockerfile 和 docker-compose.yml 等設定檔是可以重複使用的，預發佈和線上環境的部署類似。在部署完後端服務後，接下來部署前端靜態資源。

10.9.2 撰寫 Dockerfile 檔案

在前後端分離的開發模式中，前端採用單獨部署的方式，將打包後的靜態資源託管到 Nginx 服務下。首先來撰寫 Dockerfile 檔案：

```
FROM nginx:stable

COPY ./nginx.conf /etc/nginx/nginx.conf

COPY ./dist /usr/share/nginx/html

EXPOSE 4444

ENV TZ=Asia/Guangzhou
```

建構鏡像的過程相對簡單，首先指定一個穩定的 nginx 基礎鏡像，然後將 nginx.conf 檔案和打包後的 dist 資料夾複製到 nginx 的指定目錄下。在這裡，nginx.conf 檔案用於配置通訊埠和代理等重要資訊，而 /usr/share/nginx/html 資料夾則是 nginx 預設的靜態檔案儲存位置。最後，需要將 4444 通訊埠暴露給宿主機以便存取。

此外，還需要在專案的根目錄下建立 nginx.conf 檔案，並填寫以下配置資訊。特別需要關注的是 http 區塊下的 server 部分：

```
error_log /var/log/nginx/error.log notice;
pid /var/run/nginx.pid;

events {
  worker_connections 1024;
}

http {
  include /etc/nginx/mime.types;
  default_type application/octet-stream;

  map $time_iso8601 $logdate {
    '~^(?<ymd>\d{4}-\d{2}-\d{2})' $ymd;
    default      'date-not-found';
  }
```

```
    log_format main '$remote_addr [$time_local] "$request" '
    '$status $body_bytes_sent "$http_referer"';

    root /var/log/nginx;
    access_log /var/log/nginx/access-$logdate.log main;

    sendfile on;
    #tcp_nopush     on;

    keepalive_timeout 65;

    #gzip  on;

    server {
      listen 4444;
      server_name localhost;

      location / {
        root /usr/share/nginx/html;
        index index.html index.htm;
        try_files $uri $uri/ /index.html;
      }

      location /static {
        root /usr/share/nginx/html;
      }

      location /api/ {
        proxy_pass http://localhost:3333;
      }
    }
}
```

在上述程式中，server 中指定了需要監聽 4444 通訊埠，localhost 是本機的服務名稱，location 指定請求的處理邏輯，會先請求指定的路由資源，而 index.html 是作為兜底存取的檔案。/api/ 是反向代理，表示以 /api/ 開頭的請求會被轉發到 3333 通訊埠服務，用於跨域請求 Nest 服務。

為了保證在生產環境中能夠請求正確的介面路徑，在 .env.product 設定檔中還需要定義介面請求路徑為 VITE_APP_URL，以便 Axios 能夠存取，配置程式如下：

```
VITE_MODE='production'
VITE_APP_URL='http://localhost:3333'
VITE_APP_NAME='store-web-frontend'
```

然後在 Axios 的配置中引用這個變數，如圖 10-54 所示。

```
41  const Axios = new VAxios({
42    baseURL: import.meta.env.VITE_APP_URL,
43    timeout: 100 * 1000,
44    // 介面前綴
45    prefixUrl: urlPrefix,
```

▲ 圖 10-54 設置請求路徑

在建構鏡像之前，執行「pnpm build」命令進行打包，生成 dist 檔案後，再執行「docker build -t store-web-frontend」命令。完成之後，開啟 Docker Desktop，可以看到新增了前端專案鏡像，如圖 10-55 所示。

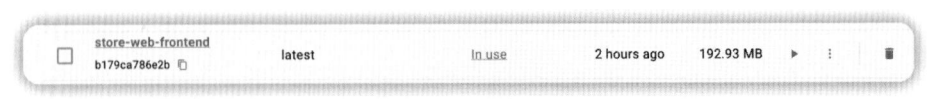

▲ 圖 10-55 建構前端鏡像

點擊 Run 按鈕填寫參數，把鏡像執行起來，然後設置通訊埠和資料卷冊資訊，如圖 10-56 所示。

▲ 圖 10-56 執行鏡像

執行成功後，在瀏覽器輸入 http://localhost:4444/activity 連結存取任意頁面，如圖 10-57 所示。

▲ 圖 10-57 成功存取專案

至此，我們已經成功完成了前端和後端專案的運行維護部署。透過學習本節內容，讀者們應能將專案部署至個人雲端服務器，並透過網際網路實現外網存取。

本節作為本書專案實戰章節的終章，展示了如何從基礎開始，利用 Nest 和 React 技術堆疊建構一個數字門店管理平臺。所有相關程式已經推送至筆者的 GitHub 倉庫，並將持續更新，包括 Nest 技術的最新進展。鼓勵並歡迎各位讀者 Fork 倉庫並給予 Star 支持。

完結語：

是終點，更是新的起點

首先，恭喜你堅持到了最後。請記住，堅持的過程可能會孤獨、枯燥，甚至痛苦，但堅持本身就是一種酷。我們都在做著很酷的事情。

一個小小的決定

當北京清華大學出版社的編輯詢問我是否願意嘗試撰寫技術圖書時，我內心既感到新奇又充滿疑問：我能否講好一門技術？我能否提供超越書籍價值的知識？這需要不斷的學習和實踐。經過深思熟慮，我決定立即行動。於是我與編輯一拍即合，決定撰寫這本書。

我在孩子四個月大時開始創作本書，現在他已經十一個月了。我用了七個月的時間完成這本書的創作，比預期多用了 1.5 個月。原因是我在中途決定重新打磨每個章節，豐富內容，最佳化表達，並在實戰專案中改進了程式結構。每一次小小的決定，每一塊時間的碎片，最終匯聚成了書中的每一頁。

時間的槓桿

時間是一種每個人都能利用的槓桿。在時間的作用下，每一個微小的投入或改變都將匯聚成可觀的收益。

與許多作者不同，這本書是時間槓桿的產物。超過一半的內容是在地鐵上構思和撰寫的，因為我每天需要花費 3 小時以上通勤。這並沒有給我帶來困擾，反而讓我享受這個過程，因為它為我提供了足夠的時間來構思和撰寫本書的內容。加上其餘時間的碎片投入，最終產生了累積效應，並呈現給每位讀者，這或許就是複利的力量。

結 語

　　技術學習不是短跑,而是一場馬拉松。本書的完成標誌著一個階段的圓滿,同時也預示著新的開始。願每位讀者帶著從書中汲取的知識與啟示,尋找屬於自己的那片璀璨。

Note

Note